GEOTECHNICAL AND ENVIRONMENTAL ASPECTS
OF WASTE DISPOSAL SITES

BALKEMA – Proceedings and Monographs
in Engineering, Water and Earth Sciences

PROCEEDINGS OF GREEN4 INTERNATIONAL SYMPOSIUM ON GEOTECHNICS
RELATED TO THE ENVIRONMENT, WOLVERHAMPTON, UK, 28 JUNE–1 JULY 2004

Geotcchnical and Environmental Aspects of Waste Disposal Sites

Editors

R.W. Sarsby
University of Wolverhampton, United Kingdom

A.J. Felton
University of Wolverhampton, United Kingdom

Taylor & Francis
Taylor & Francis Group

LONDON / LEIDEN / NEW YORK / PHILADELPHIA / SINGAPORE

Taylor & Francis is an imprint of the Taylor & Francis Group, an informa business

© 2007 Taylor & Francis Group London, UK

Typeset by Charon Tec Ltd (A Macmillan Company) Chennai, India
Printed and bound in Great Britain by Antony Rowe Ltd, Chippenham, Wiltshire

Published by: Taylor & Francis/Balkema
P.O. Box 447, 2300 AK Leiden, The Netherlands
e-mail: Pub.NL@tandf.co.uk
www.balkema.nl, www.taylorandfrancis.co.uk, www.crcpress.com

British Library Cataloguing in Publication Data
A catalogue record for this book is available from the British Library

Library of Congress Cataloging in Publication Data
A catalog record for this book has been requested

ISBN13 978-0-415-42595-7

Geotechnical and Environmental Aspects of Waste Disposal Sites – Sarsby & Felton (eds)
© 2007 Taylor & Francis Group, London, ISBN 978-0-415-42595-7

Table of Contents

GREEN4 foreword IX

Organisation XI

Sponsorship acknowledgements XIII

Landfill engineering

Stability and settlement behaviour of uncompacted soil for the recultivation of landfill
capping systems 3
A. Bieberstein, H. Reith & U. Saucke

Sanitary landfill sites in Northern Germany, developments in closure practice and after-care 17
W. Entenmann

Leachate leakage from landfill: Causes and mechanisms 29
H.-Y. Fang, A. Kaya & T.-H. Kim

Vegetable fibre degradation in polluted water 43
R.S. Karri, R.W. Sarsby & M.A. Fullen

Evaluation of geotechnical parameters for effective landfill design and risk assessment 49
T. Koliopoulos, V. Kollias, P. Kollias, G. Koliopoulou & S. Kollias

Hydraulic design of capillary covers to accommodate scatter in infiltration volumes 59
S.A. Reyes

Ground treatment by dynamic compaction in landfill materials 67
C.J. Serridge

The Mock-up-CZ experiment 81
J. Svoboda

Landfilling and re-use of a calcium sulphate waste 91
D.M. Tonks, A.P. Mott & D.A.C. Manning

Habitat creation on old landfill sites 101
I.C. Trueman, E.V.J. Cohn & P. Millett

Slope stabilisation measures at Hiriya waste site, Israel 107
A. Klein, D.M. Tonks & M.V. Nguyen

Landfill capping with microbiological treatment of landfill gases – materials
and performance in test cells 119
M.M. Leppänen, R.H. Kettunen & H.M. Martikkala

Contaminated land

Effect of rates of biological transformation of additives on the efficiency of injection
in soil under hydraulic and electric gradients 133
A.N. Alshawabkeh, X. Wu, D. Gent & J. Davis

Nickel attenuation dynamics through clayey soils 145
S. Ghosh, S.N. Mukherjee & R. Ray

Covering the mass graves at the Belzec Death Camp, Poland; geotechnical perspectives 149
A. Klein

The effects of persistent anthropogenic contamination on the geotechnical properties of soil 157
E. Korzeniowska-Rejmer

Improvements in long-term performance of permeable reactive barriers 163
K.E. Roehl, F.-G. Simon, T. Meggyes & K. Czurda

Geotechnical and environmental assessment of contaminated sites under migration
of polluting components 173
A.B. Shandyba

Experimental study of reactive solute transport in fine-grained soils under consolidation 179
G. Tang, A.N. Alshawabkeh & T.C. Sheahan

Groundwater protection by preventive extraction at an industrial waste deposit – history
and evaluation 187
C. Cammaer, L. De Ridder & T. Van Autenboer

Thermal properties of bentonite under different conditions – simulation of real states 199
R. Vasicek

Low leachability of As, Cd, Cr and Pb from soils at a contaminated site in Walsall,
West Midlands 207
R. Wesson, C. Roberts & C. Williams

Tailings

Self-weight consolidation behaviour of Golden Horn (Haliç) dredged material 217
S.A. Berılgen, P. Ipekoglu & M.M. Berılgen

Geotechnical investigation of tailings dams with non-destructive testing methods 225
A. Brink, M. Behrens, S. Kruschwitz & E. Niederleithinger

Tailings impoundments – A growing environmental concern 233
J. Engels & D. Dixon-Hardy

Some legal aspects of tailings dams safety – existing authorisation, management, monitoring
and inspection practices 247
R. Frilander, K. Kreft-Burman & J. Saarela

Shear strength of coalmining wastes used as coarse-grained building soils 261
A. Gruchot, P. Michalski & E. Zawisza

Novel solutions in tailings management 269
T. Meggyes & A. Debreczeni

Usability of sludges from coal mining industry for sealing Civil Engineering structures 285
P. Michalski, E. Zawisza & E. Kozielska-Sroka

Monitoring of spontaneous vegetation dynamics on post coal mining waste sites
in Upper Silesia, Poland 289
G. Woźniak & E.V.J. Cohn

Fly ash compaction 295
K. Zabielska-Adamska

Environmental management

Environmental insurance – addressing waste management liabilities 305
D.A. Brierley

Design for recyclability: Product function, failure, repairability, recyclability and disposability 311
A.J. Felton & E. Bird

Alternative ways of treating hazardous waste 319
N. Gaurina-Medimurec, G. Durn & H. Froschl

Construction applications of Foamed Waste Glass (FWG) – use in a rooftop vegetation system   331
Y. Hara, K. Onitsuka & M. Hara

Use of geosynthetics technology for river embankment protection using a cellular
confinement system   339
E. Korzeniowska-Rejmer, A. Kessler & Z. Szczepaniak

Leakage from the oil pipeline between Struzec pumping station and Sisak refinery   345
B. Muvrin & I.B. Kaličanin

Acid sulphate soil remediation techniques on the Broughton Creek Floodplain,
New South Wales, Australia   349
B. Indraratna, A. Golab, W. Glamore & B. Blunden

Basic study on development of permeable block pavement with purifying layer for
polluted rainwater on road   359
S. Iwai & X.Z. Xi

A multi-criterion approach for system and process design relating to waste management   369
A.A. Voronov

Author index   377

GREEN4 foreword

Worldwide there is a large range of forms of waste disposal and associated facilities and practices, e.g. engineered and non-engineered landfills for domestic, commercial and industrial refuse, tailings dams and lagoons for mineral wastes, spoil tips, infilling of voids and depressions within the ground, etc.

Geotechnical engineering may support the environment both directly (for example, through the design of remediation treatments for 'second-hand land') and indirectly (for instance, by providing a vehicle for technical qualification/assessment of different wastes as construction materials, or by enabling aesthetic amendments to be made to waste masses in a safe and sustainable way).

The complexity of the situations which have to be handled within the realm of environmental geotechnics, and the fact that the problems to be addressed are commonly ill-defined, must not be forgotten:

- Wastes used as construction 'materials' are not processed or manufactured according to specifications/codes as per conventional technical or construction materials.
- There is inherent variability within all geotechnical materials and re-used wastes and the latter may be susceptible to change of characteristics with time and ambient conditions.
- The history of sites will not be known/recorded in detail and there is a practical/economic limit to the extent of site investigation that can be conducted.

Nevertheless, realistic and feasible reliable solutions/designs have to be derived and it is fortuitous that for a number of practical situations there is a tendency for 'natural' improvement with time due to drainage, attenuation of pollution and contaminants, growth of vegetative matter, etc. However, there is also a negative side to the passage of time:

- Progressive climate change which leads to more intense precipitation, warming/drying/cracking of the ground and flooding. As a consequence there is increased mobility of contaminants, slope instability develops within tips, sliding within waste covers due to increased ground saturation leads to exposure of contaminated materials, etc.
- Pre-disposal processing (incineration, grinding, comminution, etc.) of refuse causes a change in physical characteristics of the material being disposed. This will affect the behaviour of the material with respect to settlement, biodegradation, structure of the deposits and hence the future use/rehabilitation of infilled areas.
- Re-mining of old wastes (particularly tailings deposits) and their re-processing to recover minerals leads to the creation of 'new' waste products for disposal.
- Developing Countries that become 'developed' have an inevitable rise in the quantities of waste materials that are produced and concentrated in location and they develop an industrial heritage with associated ground contamination and dereliction. Unfortunately these Developing Countries frequently fail to learn from the problems that have been (and continue to be) faced by developed countries.

The overall theme of the Fourth International Symposium on Geotechnics Related to the Environment (GREEN4) was 'Geotechnical and Environmental Aspects of Waste Disposal Sites'. It is hoped that these proceedings will act as a source of information about the latest developments within the foregoing topic area. It is also hoped that these proceedings may help individuals, companies, communities and countries to find solutions to problems related to waste disposal and to avoid repetition of former mistakes.

Bob Sarsby
Wolverhampton, 2006

Geotechnical and Environmental Aspects of Waste Disposal Sites – Sarsby & Felton (eds)
© 2007 Taylor & Francis Group, London, ISBN 978-0-415-42595-7

Organisation

Conference organising committee

Dr. A.J. Felton
Dr. R.S. Karri
M. Richardson
Prof. R.W. Sarsby (Chairman)
Dr. D.E. Searle

International scientific committee

Prof. H.-Y. Fang, University of North Carolina, USA
Prof. S. Iwai, Nihon University, Japan
A. Klein, Geotechnical and Environmental Consultant, Israel
Dr. T.A. Meggyes, German Federal Institute for Materials and Research, Germany
Prof. R.W. Sarsby, University of Wolverhampton, UK
Dr. T. Van Autenboer, SliM vzw, Belgium
Dr. E. Zawisza, University of Kraków, Poland

Sponsorship acknowledgements

The GREEN4 organising committee would like to acknowledge the assistance and sponsorship provided by the following organisations:

GeoAssist Limited
Specialist Geotechnical & Environmental Engineers

UNIVERSITY OF
WOLVERHAMPTON
School of Engineering and
the Built Environment

West Midlands
WMCCE
Centre for Constructing Excellence

Landfill engineering

Geotechnical and Environmental Aspects of Waste Disposal Sites – Sarsby & Felton (eds)
© 2007 Taylor & Francis Group, London, ISBN 978-0-415-42595-7

Stability and settlement behaviour of uncompacted soil for the recultivation of landfill capping systems

A. Bieberstein, H. Reith & U. Saucke

Institute of Soil Mechanics and Rock Mechanics, University of Karlsruhe, Germany

ABSTRACT: Increasing importance is being attached to the infiltration-reducing effect in optimised recultivation cappings (recultivation layer and vegetation) on landfills with regard to the long-term effectiveness of surface sealing systems. The prerequisite for the minimisation of infiltration is optimal vegetation, which requires the least possible density of the area available for roots to spread. This aim may stand in contradiction to the requirements for stability. A test field system has been built at a landfill site near Leonberg to compare loosely tipped or conventionally compacted recultivation covers with a thickness of 2 m (Anon, 2002b). Investigations carried out regarding the special aspects of stability and settlement behaviour of the recultivation substratum are described and the relevant results are given. In detail these involve large scale laboratory tests on the settlement and shear behaviour of loamy loess available in the region. In order to monitor the behaviour in the field, special devices were inserted in situ. This paper deals with the soil mechanics questions posed by putting the recultivation substrate in place, in particular without compaction, especially the settlement and shear behaviour of the loamy loess available in the region as used in the test fields at the regional landfill in Leonberg.

1 INTRODUCTION

In view of the current state of development with regard to the waste disposal sector and landfills in Germany, questions about landfill cappings are becoming increasingly important. In addition to barriers in the base area, insofar as they exist, capping systems contribute significantly to reducing the quantities of harmful substances discharged from layers of dumped waste into the environment. This contribution is not only achieved through infiltration-reducing elements (surface sealing) and through drainage which diverts the water but also through the recultivation cappings (recultivation layer and vegetation) which play an important role in infiltration with their influence on the presence of water. In view of the unanswered questions regarding the life span and long term effects of technical barrier systems, increasing importance is being attached to the lasting effect of recultivation cappings.

The decisive factor for a recultivation capping to achieve a lasting effect is constant vegetation which absorbs water and which uses up the highest possible proportion of any precipitation (and returns it to the atmosphere) and thus the formation of seepage water is reduced on a long term basis. Besides an extensive storage capacity for the infiltrating water, this requires the greatest possible ability of the roots to spread in the substrate forming the recultivation layer. In addition to an adequate thickness (of more than 1 m according to the stipulation of the TASI – Technical Instructions on the Recovery, Treatment and Other Management of Wastes from Human Settlements, Anon, 1993), the prerequisite here, on the other hand, is the least possible (stratification) density of the soil in question. With regard to plant ecology, the need for less density usually stands in contradiction to the requirement for adequate density to guarantee sufficient stability.

By using a test field system with differently prepared test fields, the behaviour of thick (at least 2 m) recultivation cappings was investigated in a comparative manner. Besides the mechanical

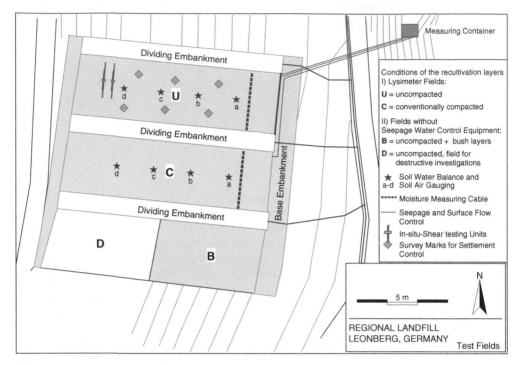

Figure 1.   Plan overall view of the test fields at the landfill in Leonberg.

Table 1.   Field sections for the examination of different recultivation systems.

Field	Area	Recultivation layer
U	$400\,m^2$	Uncompacted, i.e. put in place without any additional compacting
C	$400\,m^2$	Compacted three times according to conventional methods
D	$200\,m^2$	Loosely filled, field for more extensive destructive tests
B	$200\,m^2$	Loosely filled with a biological engineering construction (bush layers)

aspects, the investigations were concerned with questions regarding the water regime and the development of the vegetation (Wattendorf et al, 2003).

2   TEST FIELD SYSTEM AT THE LEONBERG LANDFILL

In order to investigate different recultivation cappings, a system of four sections was set up at the landfill in Leonberg (Figure 1). The different conditions in the sections of the testing area are listed in Table 1.

Two of the three 10 m wide sections were planned as a lysimeter by means of a sealing sheeting underneath and corresponding seepage water control equipment. This makes it possible to compare conventionally compacted (C) or uncompacted (U) conditions of the substrate. Fields U und C are lysimeter fields; seepage water discharges from the recultivation layers (drainage discharge) and the surface discharge (field U) can be measured.

Figure 2 gives an aerial view of the test field system immediately after putting the recultivation material into place: In section C the compacted surface has been roughened up afterwards. In section B the bush layers for the special plants investigated here have already been inserted (compare with Figure 1). The measuring container had not yet been installed when the aerial photo was taken,

4

Figure 2.   Aerial view of the test field system after installing the recultivation layers.

Figure 3.   Placing the recultivation substratum by means of a caterpillar.

Figure 3 shows the method used for placing the recultivation substrate by pushing it down the slope using mechanised plant (in one layer; thickness >2 m).

3   SOIL MECHANICS INVESTIGATIONS OF RECULTIVATION SUBSTRATE PLACED IN A LOOSE STATE

3.1   *Aims*

From a soil mechanics or soil statics view on the one hand and from an ecological perspective regarding plants on the other hand, the following problems were to be investigated:

- Determination of the achievable dry density with a loose substrate as well as clarification of the development of the density under static loads or water infiltration.
- Prognosis of the expected settlement to determine what thickness of material needs to be put in place on site.

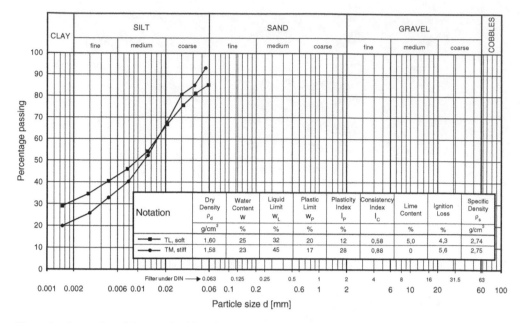

Figure 4.   Properties of the examined locally-available loamy loess.

- Investigation of the shear behaviour of the loosely-placed substrate using a suitable testing method in the laboratory.
- Proof of the stability for specific shear planes in the test fields in Leonberg.
- Verification of the settlement behaviour of the recultivation layer in the field.
- Development, construction and application of a measuring technique for checking the current shear strength on site.

3.2   Laboratory investigations

Figure 4 shows some of the properties of the material. In standard direct shear tests carried out for comparative purposes on material at the liquid limit, an internal friction angle of about 25°, with a cohesion of about $6 \, \text{kN/m}^2$, was determined.

The granular mass (the aggregates had a maximum size of 32 mm) initially had a dry density of around $\rho_d = 1.6 \, \text{g/cm}^3$ (corresponds to approx. $0.9 \rho_{\text{Proctor}}$) and a water content in the region of $w = 3$ to 25% (clearly above $w_{\text{Proctor}} \approx 17\%$). This was examined in as loose a state as possible in both a large oedometer (diameter 51 cm) and a large scale direct shear test apparatus (sample size w/l/t (width/length/thickness) = 1.2 m/1.2 m/0.4 m). The examination considered the settlement behaviour under load, the slump behaviour when water is added, and the shear behaviour under small loads. The results of the investigations formed the basis for the guidelines for placing the substrate in the field and for the evaluation of the deformations and stability on site.

3.3   Density of heaped granular material, settlement and slump behaviour

In various experiments, the dry density of the loosely heaped aggregates was determined to be $1.0 \leq \rho_d \leq 1.15 \, \text{g/cm}^3$ (a total porosity n of 60%). Correspondingly, it was seen that the heap had a large quantity of macropores, which are particularly important for the air regime and the ability of the roots to spread (porosity of the aggregates $n_A = 41\%$).

6

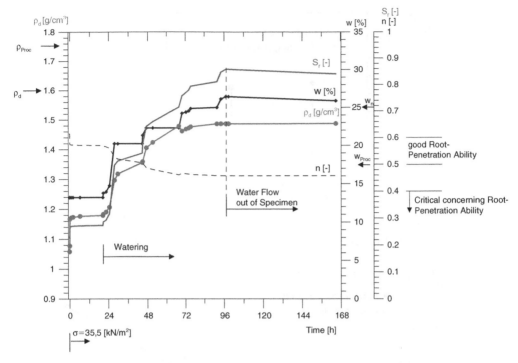

Figure 5. Loose substratum in a load test with subsequent watering ($\sigma = 35.5 \, kN/m^2$).

An example of the behaviour of the loosely-placed substrate under vertical load ($35 \, kN/m^2$) when water is added is described by the oedometer experiments (Figure 5). As can be seen from the given values entered at the start of the experiments:

– the initial dry density was $\rho_d = 1.05 \, g/cm^3$ (corresponding to $n \approx 60\%$), the initial water content was about $w = 13\%$ (thus during storage of the material until the time when it was used, from an initial quantity of 23 to 25% the water content had decreased noticeably through evaporation),
– the relative degree of saturation was $S_r = 26\%$.

Under the relatively low load, which was selected because of the low overburden stresses in the 2 to 3 m thick substrate layer in the field with $\sigma_{max} = 35 \, kN/m^2$, the dry density only increased to a limited extent, namely only under load up to around $\rho_d = 1.18 \, g/cm^3$ ($n \approx 56\%$). When the load settlement had stopped, the samples were gradually watered from above, in order to investigate the slump behaviour under the infiltration of precipitation. The watering took place in stages, until the capillary holding capacity had been exceeded and water ran out underneath the sample (hence the step-like course of the curves).

Changes in the parameters can be determined from Figure 5. The water content rose in stages and approached 26% (coincidentally about the same as the natural water content) and correspondingly the relative saturation rose towards 85%. The moistening brought about a slump of the sample, so that ρ_d rose further to about $1.48 \, g/cm^3$ (corresponding to a decrease of n to approx. 45%), which corresponds to approximately 0.85-times the value of the standard Proctor density. Nevertheless the total porosity fell – as shown in Figure 5 – it was only slightly below the range of 0.5 to 0.6, which is regarded as still being good for roots to spread. If n sinks below 0.4, then the possibility for the roots to spread no longer exists or it is at least regarded as being critical.

Comparable behaviour was seen in further oedometer tests with other vertical stresses and corresponding tests in a large scale direct shear apparatus. Altogether it was determined that the main

7

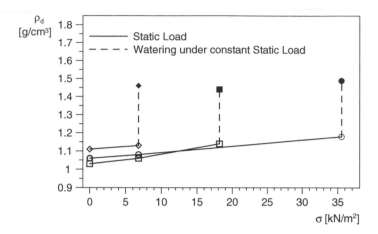

Figure 6.   Resulting dry densities depending on the load and the saturation condition.

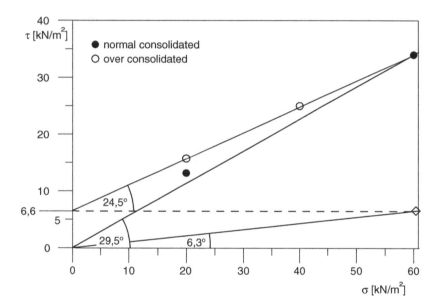

Figure 7.   Direct shear tests on prepared substrate samples (shear plane: 6 cm × 6 cm).

part of the deformations occur in connection with the addition of water, and that the densities, which then arise, depend only to an insignificant extent on the stresses selected (as indicated in Figure 6).

It was recorded that the loosely-placed substrates made of slightly cohesive material as recultivation layers maintain a high porosity, favourable for vegetation (root spreading), under loads which occur with granular mass heights of up to 3 m. This also applies under the effects of water infiltration. Apart from this, the behaviour under long term conditions in the field still remains to be investigated separately.

3.4   Shear behaviour

The shear behaviour of soils of the type in question is normally investigated in well-defined states of compaction and under saturated test conditions, e.g. in the direct shear tests with a "water bath" (Anon, 2002a). The result of a series of small scale direct shear tests (shear plane dimensions of 6 cm × 6 cm) on material prepared in a slurry like form and then consolidated is shown in Figure 7.

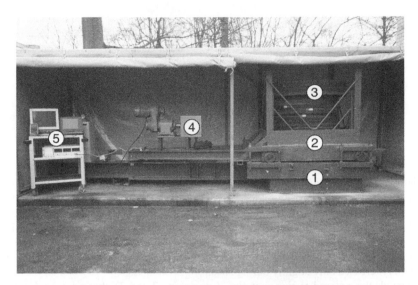

Figure 8. View of the testing apparatus for direct shear tests on a large scale. 1: lower non-shifting frame, 2: upper shiftable frame, 3: steel plate weights (load), 4: drive unit, 5: test control and data acquisition unit.

Figure 9. View into the lower part of the direct shear test apparatus (with an additional inserted grid frame for introducing shearing force over the whole plane).

The shearing resistance can be presented in the following ways:

– a global angle of total shear strength $\phi'_s \approx 30°$
– a angle of internal friction $\phi' \approx 25°$
– an angle of cohesion (depending on the prior load) $\phi'_c \approx 6°$

A question arises regarding the condition of the substrate of a loose filling used as a recultivation layer as to the available shear strength under "natural" conditions on site. In a large scale direct shear test an attempt was made to reconstruct the conditions of a loose filling in the laboratory. Without being compacted the material was placed in the large direct shear test apparatus (see Figures 8 and 9). Altogether three relevant specimens were investigated (Table 2).

9

Table 2. Testing conditions of direct shear tests on a large scale.

Specimen	Test type	Water content w [%]	Load σ [kN/m²]
1	Multi-stage test	19	6.9
		19	17.5
2	Single test	~25*	17.5
3	Single test	~25*	37.0

* By watering.

In order to better understand the shear behaviour, in Figure 10 the measured characteristic lines, as well as the settlement occurring under load or under shearing, have been entered. As may be expected, the shear resistance of the material interspersed with macropores is triggered via large shear paths. Thus the first stage of the two-stage test on non-watered material (specimen1, Table 2) was by no means carried out up to the limit. The material, which was dried from a natural moisture state to w ≈ 19% and in which condition it was put into place, was seen to settle when put under load, and this increased only slightly during shear.

As expected, the watered testing materials (specimens 2 and 3) showed considerably greater settlement when put under load. The available shear resistance was still not completely developed when shearing was stopped, even after a shear displacement of 160 mm each time.

The shear resistances reached at the end of each test have also been entered in the τ- σ-diagram of Figure 11 and are compared with the lines of the classical direct shear tests in Figure 7.

It would hardly be appropriate to connect the values of the two-stage test on specimen 1 to a shear line, in view of the characteristic lines (Figure 10). For the results of the individual tests (specimens 2 and 3), which were both run to a shearing displacement s of 160 mm, it may be more worthwhile. It is significant that the measured shear resistances, although they were not completely developed at the largest displacement, were all above the line marked ϕ'_s. Obviously the aggregates provide sufficient stability even in a watered state, so that the partially-saturated substrate structure, still interspersed with macropores, is overall able to develop a greater shear resistance than the soil in a (water-saturated) normal compacted state – a portion of "apparent cohesion" is naturally included here. Thus it can be concluded that it is not on the unsafe side to select inclinations of slopes for loosely-filled substrates using ϕ'_s determined from classical direct shear tests on disturbed material. This definitely applies to the tested material in this case so long as it is not anticipated that saturated soil conditions will arise.

On the basis of the investigations it was possible to prove the soil static requirements for the test fields on the landfill in Leonberg for all the shear planes to be observed and all observed loads. In view of the results obtained in the laboratory and bearing in mind the change to natural conditions, the question arises as to what will happen to moisture conditions, density and the development of the shear strength in the long term.

4   TESTING EQUIPMENT FOR FIELD INVESTIGATIONS

The section of the field testing area in which the non-compacted recultivation layer was placed on the landfill in Leonberg was equipped with measuring devices which made it possible to obtain relevant data. Initial results have become available.

4.1   *Measuring cables for moisture*

In addition to a number of the customary TDR-moisture-measuring-probes, which were positioned in the testing area in two places (indicated on the general plan, Figure 1) new types of moisture measuring cables were installed (Schlaeger, 2002; Schlaeger et al, 2001). These utilise a TDR-measuring-technique developed in the Institute of Meteorology and Climate Research at the

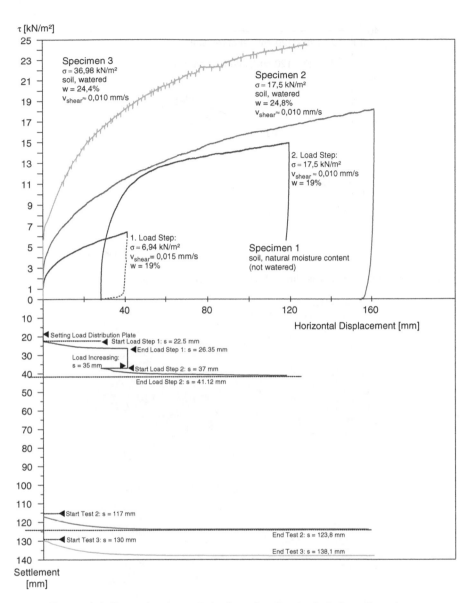

Figure 10. Characteristic lines and settlement behaviour when shearing in the large direct shear test apparatus.

University of Karlsruhe, which makes it possible to determine the moisture as a profile along a measuring cable many metres long (Schlaeger et al, 2001a; Schlaeger et al, 2001b). Figure 12 gives an impression of the installation of a cable of this kind in the lower area of the recultivation layer.

Several measurements have been obtained with these testing devices in Leonberg. They will have to be related to the climatic occurrences and to the development of the vegetation in the relevant section – in addition they are available for comparison with the measuring results from the conventional TDR-probes and with installed tensiometers.

4.2 Deformation behaviour

A total of six survey marks for settlement control, two each at three different levels of the surface of the test field with an incline of 1:2.7 (see Figure 1), were placed in the surface of the loosely

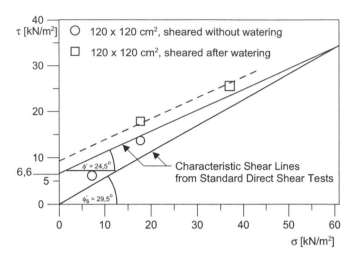

Figure 11.   Comparison of large scale direct shear tests and classical direct shear tests.

Figure 12.   Insertion of the moisture-measuring-cable into the recultivation layer.

filled recultivation layer. The intention was to follow the slump and translation movements of the surface of the recultivation capping over the course of time. The movements are followed in three directions in space. As expected, slump and movements were a significant number of centimetres (Figure 13).

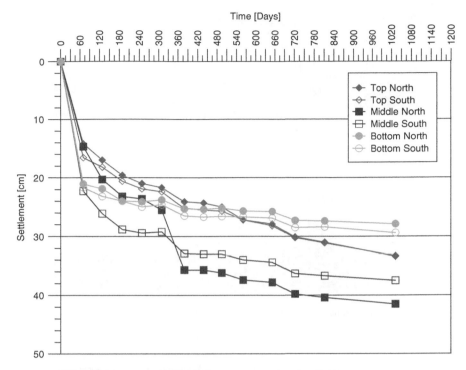

Figure 13.   Settlement of the survey marks in the uncompacted testing field.

The results obtained so far can be summarised as:

- The settlements are in the range of several centimetres.
- The main portion of the observed vertical deformations occurred during the first few month after placing the material (Figure 13).
- The settlements observed in the "long-term", e.g. one year after placing, correspond to a large extent with the settlements of the landfill structure.

4.3   *"Karlsruhe shear testing units"*

In order to investigate the actual shear parameter on site so-called "Karlsruhe shear testing units" (which were developed and constructed at the Institute for Rock Mechanics and Soil Mechanics) were installed at different points and at different depths (see Figure 1). These testing units consist of rods holding a number of concentric discs which are pulled through the soil by a wire cable (Figure 14) thus producing a cylindrical shear plane.

The reaction block for pulling the rods and causing shear off along a horizontal, cylindrical shear plane is formed by a completely-embedded, steel box. Within this box the hauling cable is turned through 90° in a vertical direction and is then passes through protective tubes to the surface of the site (Figure 14). When taking measurements (the time for which can be selected as desired according to the age of the material and also the moisture conditions) the shear test on site is carried out with the help of jacks which are also used for anchor tests (Figure 15).

In Figure 16 results are plotted in terms of the observed shear stress and the corresponding normal stress (due to the overburden). It is obvious that there is a scatter of values, as could be expected under the given conditions. Nevertheless, it can be observed that the results obtained by comparing the shear stress with the capping pressure conform well to the shearing behaviour according to the shear tests in the laboratory. Thus the shear behaviour in-situ confirms the knowledge derived from the laboratory.

13

Figure 14.   Installing "Karlsruhe shear testing units" in the recultivation layer.

Figure 15.   Performing a shear test on site using the "Karlsruhe shear testing unit".

The assumptions made for constructing the test fields have been checked by taking measurements on site. The results available up to now can be summarized as follows:

- The shear parameters obtained with the "Karlsruhe shear testing units" are in the expected range, as a whole they confirm the results of the laboratory tests.
- The settlement of the recultivation layer was largely finished after a few months.
- Technical equipment for measuring the moisture distribution has been installed. In order to reach conclusions regarding the water regime of the recultivation layer, longer measuring series are required.

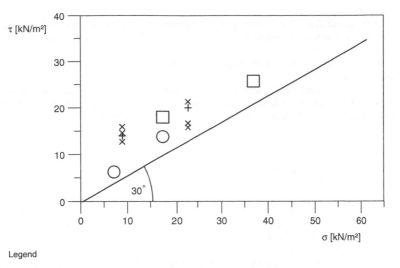

Legend

+   Values obtained with Karlsruhe shear testing units (date: 08.11.2001)

×   Values obtained with Karlsruhe shear testing units (date: 14.01.2002)

○   Shear plane 120 x 120 cm², sheared without watering ($\rho_d \sim 1{,}1$ g/cm³)

□   Shear plane 120 x 120 cm², sheared after watering ($\rho_d \sim 1{,}2$ g/cm³)   Laboratory tests

Figure 16. Comparative presentation of the results of large scale shear tests performed (a) in the laboratory and (b) tests with "Karlsruhe shear testing unit" on site.

5 CONCLUSIONS

From a soil mechanics and soil statics point of view the following conclusions can be drawn for loosely-placed recultivation layers:

- For loosely-placed cohesive substrates the decisive shear parameter can be determined by means of classical shear tests in the laboratory – the angle of the total shear strength (ϕ'_s) provides a good indicative value.
- In each individual case it is absolutely essential to check the shear parameters.
- The observed settlement behaviour in the laboratory and on site shows, that it is necessary to place layers with a thickness, which take the future vertical settlements into account, in order to achieve the required height of the recultivation substrate.
- The clay content (proportion of particles less than $2\,\mu$m) of recultivation substrate should not exceed 25% by weight.

ACKNOWLEDGEMENTS

The work reported in this paper was a joint project of; the Institute for Landscape Management, the Institute of Soil Science and Forest Nutrition (both at the University of Freiburg), the Institute of Soil Mechanics and Rock Mechanics (University of Karlsruhe), the Planungsbüro Umweltwirtschaft GmbH – Planning Office for Environmental Economics (Stuttgart). The project was supported by the state of Baden-Württemberg and the Rural District of Böblingen. The Authors are grateful for being allowed to publish the results of the project within this paper.

REFERENCES

Anon (1993) *Technische Anleitung zur Verwertung, Behandlung und sonstigen Entsorgung von Siedlungsabfällen*, TASI, Bundesanzeiger Verlags.Ges. mbH, Köln.

Anon (1997) *Forstwirtschaftliche Rekultivierung von Deponien mit TA-Siedlungsabfall-konformer Ober-flächenabdichtung,*Landesanstalt für umweltschutz baden-württemberg, handbuch Abfall, Band 13, Bezug über JVA Mannheim, Druckerei.

Anon (2002a) *Baugrund, Untersuchen von Bodenproben, Bestimmung der Scherfestigkeit,* DIN 18 137, Teil 3, Beuth Verlag, Berlin.

Anon (2002b) *Untersuchungen zur Gestaltung von Rekultivierungsschichten und Wurzelsperren,* Institut für landespflege, unpubl. Abschlussbericht des Forschungsvorhabens BWSD 99003, 187 S., Freiburg.

Meier J (1999) *Entwurf eines Deponietestfeldes,* Diplomarbeit am Institut für Bodenmechanik und Felsmechanik der Universität Karlsruhe (TH), unveröffentlicht.

Schlaeger S, Hübner C and Weber K (2001a) *Moisture profile determination with TDR – Development and application of Time-Domain-Reflectometry inversion algorithms for high resolution moisture profile determination,* Proceedings 2nd International Conference on Strategies and Techniques for On-Site Investigations and Monitoring of Contaminated Soil, Water, Air and Waste, Field Screening 2001, Karlsruhe, Germany, May, pp 341–346.

Schlaeger S, Hübner C, Scheuermann A and Gottlieb J (2001b) *Development and application of TDR inversion algorithms with high spatial resolution for moisture profile determination,* Proceedings 2nd International Symposium and Workshop on Time Domain Reflectometry for Innovative Geotechnical Applications, Evanston, September, pp 236–248.

Schlaeger S (2002) *Inversion von TDR-Messungen zur Rekonstruktion räumlich verteilter bodenphysikalischer,* Veröffentlichungen des Institutes für Bodenmechanik und Felsmechanik der Universität Karlsruhe, Heft 156.

Scheuermann A, Schlaeger S, Hübner C, Brandelik A and Brauns J (2001) *Monitoring of the spatial soil water distribution on a full-scale dike model,* Proceedings 4th International Conference on Electromagnetic Wave Interaction with Water and Moist Substances, Weimar, May, pp 343–350.

Wattendorf P, Konold W and Ehrmann O (2003) *Gestaltung von Rekultivierungsschichten und Wurzelsperren,* Culterra 32, 185 p, Freiburg.

Geotechnical and Environmental Aspects of Waste Disposal Sites – Sarsby & Felton (eds)
© 2007 Taylor & Francis Group, London, ISBN 978-0-415-42595-7

Sanitary landfill sites in Northern Germany, developments in closure practice and after-care

W. Entenmann
IGB Consulting Engineers, Hamburg, Germany

ABSTRACT: Most of the now existing sanitary landfills in Germany do not meet the requirements of the contemporary legislation and will be closed by 2005 at the latest. Therefore, closure and after-care plans have to be established. As in the last decades the safety standards of landfills have continuously been improved, there are landfills of very different kinds in type and composition that have to be reclaimed now. This task cannot be done schematically but must take into account the random conditions of each single site concerning environmental risk, derived from subsoil conditions, hydrogeological setting, contemporary condition of the waste, compaction and predicted alteration of chemical composure and especially its technical design. Investigations have to be carried out in the fields of risk assessment, site monitoring and prediction of chemical and emission behavior. These investigations are based on an evaluation of various sites that were reclaimed in the past and were provided with different sealing components. These revealed that the standard method of reclamation with a top sealing, consisting of two combined elements, regulated by ordinance, often is neither necessary nor efficient and sometimes even contradictory to the site's intended long-term safety. Therefore, each site has to be evaluated and assessed as an isolated case. This can lead to quite different solutions and can result in the proposal of semi-permeable covers, application of the principle of the "flushing bioreactor", vertical sealing elements or hydraulic measures to improve dewatering.

1 INTRODUCTION

In 2005 most of the German sanitary landfill sites will be closed as a result of the enormous requirements to the future running of sites, concerning safety aspects of the sites and pre-treatment technology, in order to obtain waste for deposition of minimum emission potential. Additionally, the amount of waste to be disposed of has continuously declined in the past few years. With improved combustion technology competing, this means that only few sites, most of them in rural areas, will remain. Figure 1 shows the future standard of a closed and already reclaimed landfill site according to our contemporary legislation. Figure 2 indicates the outer barriers with sealing liners and dewatering systems.

In view of many sites recently closed and due to the upcoming closure of various other sites in Northern Germany, master-plans have to be worked out for the phase of closure and aftercare. This is not possible without intensive working on the sites, usually carried out by Consulting Engineers and Hydrogeologists. This has not only to be done for planning purposes but also in order to close existing gaps, especially in knowledge of the emission behavior and hydrochemical and microbiological processes in the landfill itself.

Thus, in the last ten years, with 15 German sites, as necessary prerequisite, applied scientific research has been done on the topics of:

- risk assessment
- estimation of the consequences and performance of capping

17

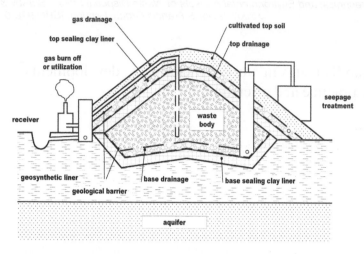

Figure 1. Closed and reclaimed landfill site according to contemporary German legislation.

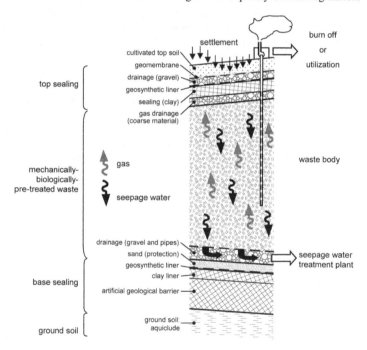

Figure 2. Base and top sealing and installations for dewatering and gas collection.

- performance monitoring during closure and after-care
- specific investigations to assess the soil mechanical and chemical behavior of the different waste types and their degradation to establish models of their further emission potential
- life cycle assessment.

2 CONTEMPORARY SITUATION OF THE SITES

One year before the total conversion of our waste disposal system, considering the site specifications and the emission situation, the contemporary situation with running of sanitary landfill sites has to be identified as manifold. All those landfills, yet to be reclaimed, have only two characteristics in common – they are filled in thin layers and compacted thoroughly with more or less success, depending on the condition of the waste. In addition it is necessary to deal with landfills of different types, structure and emission behavior.

2.1 Types of landfill

- *Landfill sites without artificial base sealing on natural subsoil of high permeability* (Figure 3a): As almost the complete amount of seepage is emitted into the subsoil, these landfills pose the greatest threat to the environment and are treated with special attention (Entenmann, 1993).
- *Landfill sites without artificial base sealing on natural subsoil of low permeability* (Figures 3b and 3c): This site specification, although easily discernible as a most significant feature, does not mean anything to its risk potential. Depending on other site aspects and hydraulic conditions some of those sites do not emit any leachate, others emit 100% of the seepage-water recharge. The most important element in this respect is whether or not a basal dewatering system is installed and how it works (Entenmann & Schwinn, 1997; Entenmann, 2001).
- *Landfill sites with base sealing* (Figure 3d): This landfill type, which has been installed since the early eighties, comprises two combined base sealing liners, usually clay and geomembranes, and a base drainage, consisting of a layer of gravel with drains and collectors (TASi, 1993). With usually lacking ground inclination in Northern Germany, active dewatering is obligatory. Landfills of this type are the only ones to remain in operation after 2005, but have also to be closed latest 2009 if they do not meet the requirements of suitable geological subsoil.

2.2 Types of waste

- *Separation*: Beginning in the outgoing eighties of the last century, the local authorities, in charge of running the landfills as well as private and semiprivate companies began with the separation of hazardous components from the waste delivered. Today this is obligatory, but, with different sites, very differing in efficiency. This means, even the same landfill often comprises waste of very different emission potential. At the same time recycling made great progress. Therefore, the content of potentially hazardous components, e.g. heavy metals has been declining. On the other hand, as a result of improving separation of inert material, such as demolition rubbish and waste soil, the organic content has been rising proportionately. Later, with an increasing proportion of separated green waste, the organic content has declined again.
- *Pre-treatment*: Pre-treatment of waste has been obligatory since 1993. But there is a transitional period that ends in 2005. This paper does not deal with landfills comprising slag from thermal pre-treatment, but only from mechanically-biologically pre-treated waste (MBP-waste). Facilities of various kinds and degree of efficiency have been installed on those sites where operation will be continued after 2005.

    Altogether, there is a vast variety of landfill types, depending on its installations, its geological setting and its type of waste-input. A precise classification and a differential assessment of the site must be the fundamental task for reclamation planning that has to deal with the technical measures in the course of closure.

3 LEGAL SITUATION

Compared with the European legislation (EU, 1999), the German legislation is even more complicated and demanding. What has to be done with the landfills in the phase of closure and after-care is scattered in different acts and ordinances, of which the most important are the ordinance on landfilling (DepV, 2002) and the ordinance on deposition (AbfAblV, 2001).

    In short, a combined top sealing system as well as a drainage of the runoff over the sealing has to be installed in order to minimize infiltration, and landfill gas has to be caught, see Figure 1. The

German legislation, although said to be based on the principle of multi-barriers, in fact depends strictly on the principle of entombment combined with pre-treatment.

After-care can be described as the further running of the leachate and the gas treatment plants and the collection of an enormous amount of groundwater and sewage water analyses data.

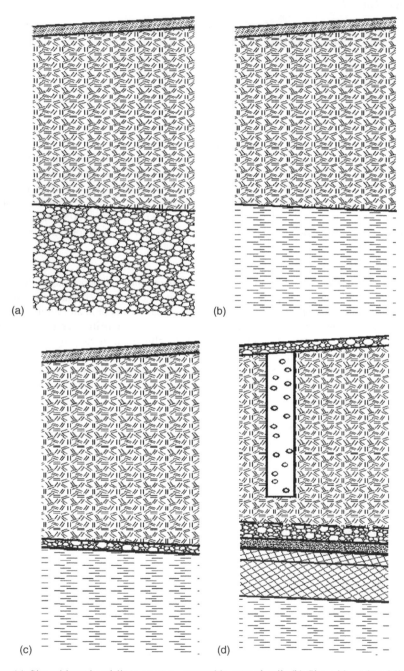

Figure 3. (a) Site without basal liner on very permeable natural soil. (b) Site without basal liner on low permeability natural soil. (c) No basal liner, low permeability sub-soil, basal drainage. (d) Site with basal liner.

4 DEVELOPMENT IN PRACTICAL EXPERIENCE AND RESEARCH

4.1 Risk assessment

Risk assessment is the first prerequisite of reclamation planning that has to be done with each site. The method and case-studies are presented elsewhere in detail (Entenmann, 1998) while the summarized most prominent characteristics of a selection of landfills are given here. As it is obvious that a gas collection system will be installed on each site, risk assessment can therefore be restricted to the emission path leachate – groundwater.

- *Vechta I sanitary landfill site*: The emission situation is given in the cross-section of Figure 4. The site without base sealing is characterized by a continuously spreading contamination plume in the groundwater. Even now, 20 years after closure some parameters of organic substances are rising in content. There are significant contents of toxic substances, at least periodically recurring, predominantly arsenic.
- *Wiefels sanitary landfill sites*: The emission situation is given in the cross-section of Figure 5. The site is characterized by deep-reaching sediments of low and very low permeability. With prevailing small hydraulic gradients and high retention potential of the sediments the emission rate even from the unsealed landfill Wiefels I into the groundwater is so small, that it will take centuries until compounds of the sewage water could be detected in lowest concentrations in the groundwater.
- *Varel I sanitary landfill site*: The emission situation of the unsealed site Varel I is given in the cross-section of Figure 6. The site is characterized by very thin underlying sediment strata of low permeability. However, these could not prevent or even diminish sewage water from discharge but only prolong its time. Nevertheless, the emission rate is greatly reduced by approximately 60%, as a result of good drainage to the ring drain ditch on this layer.
- *Hahn-Lehmden sanitary landfill site*: The emission situation is given in the cross-section of Figure 7. The unsealed landfill on sediments of very low permeability is characterized by a very simple, very poorly-dimensioned base drainage (Entenmann & Rappert, 1998). But this drainage has performed very well over a period of 30 years and shows no signs of incrustations. It is thought that this performance is the result of the poor dimensioning, as the pipes are constantly flushed with high discharge velocities. As a result of high hydraulic conductivities of the landfill, derived from a high proportion of demolition rubbish, the content of hazardous components in

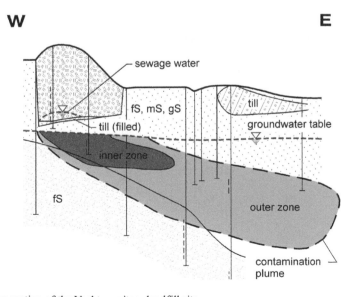

Figure 4.   Cross-section of the Vechta sanitary landfill site.

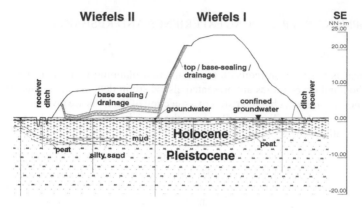

Figure 5. Cross-section of the Wiefels sanitary landfill site.

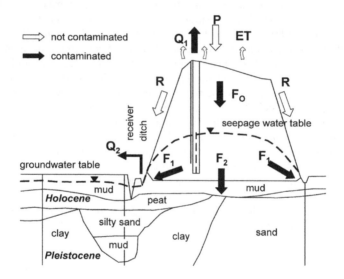

Figure 6. Cross-section of the Varel I sanitary landfill site.

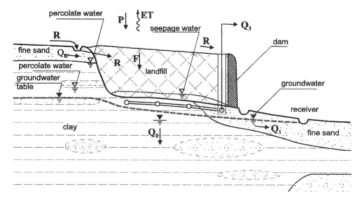

Figure 7. Cross-section of the Hahn-Lehmden sanitary landfill site.

the landfills is ultimately so small that no significant emission potential is left. To our knowledge this is the only central landfill site of considerable size in our region, which was released from the necessary top sealing liner with permission of the federal authorities as a result of intense arguing on the basis of risk assessment.

4.2 *Water balance and water storage*

The drawing-up of water balances has become a major tool in describing the emission situation and in predicting the future behavior of the landfill. With the Brake-Kaeseburg landfill a water balance was drawn up (Entenmann, 1998) prior to the reclamation. This revealed an infiltration rate of 43% of the total annual precipitation. The landfill was capped with a 0.5 m thick clay liner. Seven years later another water balance was drawn up. This revealed an infiltration rate of 34%. Thus, the capping without drainage has resulted in only 9% of infiltration reduction whilst the existing marginal dewatering by ditches regained 34% of the sewage water. This means that clay liners do not perform as sealing liners but only as coverings which could otherwise be established at much lower costs. This was also proved with other landfill sites.

With the Varel I landfill site investigations were carried out in order to establish subsequently drawn up water balances. These were not only done for the estimation of leachate discharge but also to predict the future behavior of the neighboring Varel II landfill that was begun 20 years after the beginning of first one. The infiltration rate into the open and provisionally covered landfill Varel I was in a range from 17.4% to 30.7% of the total annual precipitation (Entenmann, 1998). The just now finished capping with a geosynthetic liner will reduce this to almost 0%. Nevertheless, numerical simulations revealed a minimum 15 years period until the waste body will be totally drained as a result of a very high hydraulic head of sewage and high porosity. While the effective porosity is 39%, the total porosity must be assumed to be even much higher.

These results became part of reclamation planning of the Varel II landfill which will be confined in 2005. As closure is inevitable then, in the last years as much waste as possible has been deposited. This meant that the increase of height per year was very great. The waste deposited is not pre-treated but significant sorting of hazardous components has been done. It is compacted to a unit weight of 1.05 t/m^3 respectively 0.8 t/m^3 unit dry weight. As the water content ranges from 30% to 35% and the water content of saturated waste is approx. 55%, there is a deficit of at least 20%. With pre-treated waste the latter figure would be even significantly higher.

The low degree of saturation of the waste can easily be proven, as the base drainage produces much less leachate than expected from the water balance of Varel I and it can be reported from most landfills that the initial drainage in the first year of landfilling is almost zero. The results of continuously drawn up water balances between 1993 and 2001 are shown in Figure 8. In this time-span a water deficit of approx. 50,000 m^3 was built up. On the other hand, drainage discharge is much more than expected regarding the existing deficit. If the infiltration and saturation were homogeneous, the base drainage discharge would be much less. This is the proof of many observations from excavations in waste (Entenmann, 1996) that discrete flow-paths are prevailing and even completely dry zones can exist in positions significantly below the average seepage table. Only one half of the infiltration rate will result in compensation of the saturation deficit, while the other half will result in discharge. As Varel II is in the period of saturation which will last, as predicted by an extrapolation of the curve in Figure 8, at least until 2020, the retention in the waste body can be studied only with the Varel I landfill. It is in the range of only some weeks.

There is also hydrochemical proof of this result: The average concentration of leachate compounds, derived from samples from the ring drainage ditch, corrected for the influence of runoff and lateral infiltration from outside, is much less than that of those samples from leachate observation wells with permanent water columns.

4.3 *Investigation of the waste body*

Capping of landfills with an almost impermeable sealing means drying out. Many scientists have predicted mummification of the waste. As shown previously, the time elapsing until this state finally

23

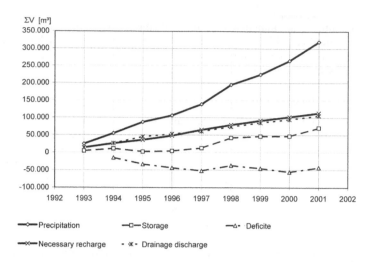

Figure 8.   Water balance of the Varel II sanitary landfill site.

is reached is very different, depending on the hydraulic conditions of the landfill site. With some landfill sites the total break-down of gas-production can be reported only few years after capping while other sites, like the Mansie landfill, with enough water replenishment from the slopes, as well as from ascendent water from the base, reveal high production rates even 15 years after closure. On the other hand, recently carried out investigations into the behavior of already mechanically-biologically pre-treated waste show clearly that even this almost inert waste will be subject to further biodegradation if the random conditions encourage it. With conventional, untreated waste this must be even more pronounced, as experience from excavations on contaminated sites indicate. So, encapsulation and ensuing mummification must be regarded as preservation of emission potential over generations.

In order to avoid the foregoing 'entombment' situation investigations were carried out into the biodegradation mechanism, taking place in-situ. An investigation was made of many small old household waste tips, ranging in capacity from 20,000 m^3 to 150,000 m^3, which had been provided with sufficient water over a period of 20 to 30 years (Entenmann, 2001). At this stage the gas production left was almost zero and the residual organic substance tended towards a few percent while leachate emission was further declining from an already low level.

This was not to be expected with big central landfills of this age. Investigations of the Varel I landfill, comprising approximately 1.0 million m^3, took place 25 years after the begin of landfilling and 8 years after closure. On the occasion of a drilling campaign for gas wells, samples were taken from the waste and analyzed. The TOC value of those samples, taken from the eluate, reveal with approximately 250 ppm only half the value of the leachate, which is already only half the value of the Varel II landfill still in operation. But the contents of the ammonia are in a much higher range than in the leachate. This was not expected, as the retardation capacity for ammonia is low and immediate flushing out of this end-product of biodegradation was thought to lower the concentration in the pore-space significantly, while residual organic substances would persist there.

The biological degradability of the dry residue in original substance was also measured through-out the profile. Determined as breathing activity it revealed a medium value of 7.7 mg O_2/g (AT$_4$). Compared with MBP as anticipated biodegradation, this value reaches almost the threshold value for MBP-waste of 5.0 mg O_2/g (AT$_4$) according to the ordinance (AbfAblV, 2001). Therefore, it can be stated that the time-span of 8 years (youngest sample) to 20 years (oldest sample) with unrestricted admission of water into the landfill is almost sufficient to reduce the organic content to an extent that is provided with modern pre-treatment technology. On the other side, just in those layers of the landfill that comprise optimum water contents for biodegradation, the TOC-values are significantly higher than 250 ppm, the threshold value for MBP-waste. These findings seem

24

to be contradictory. But the qualitative description of the bore-profile gives an explanation – the rotten household waste is of special consistency. The components which have zero or very low degradability (gravel and boulder fraction in terms of soil mechanics) give a mechanically stable structure with a very big pore-space. The pore-space itself is at least partially filled with sludge from the degradation of the organic fraction of fluid consistency. The proportion of sludge in the samples increases with increasing depth. But it is almost impossible to use drilling to gain samples that reveal the original proportion of sludge. Thus, the above mentioned value for the remaining biological degradability must be considered as too optimistic. The actual value must be higher.

The preceding observations lead to the conclusion that the degree of biodegradation in the landfill is high. If the residual waste were "washed", waste of a very low emission potential would result. Without this washing out, the sludge in the pore-space serves as reservoir of contaminants for future emission. At which degree biodegradation activity in the sludge continues could not be revealed. Gas production eight years after closure is still significant (at 160,000 m^3 per year), but decreasing.

5   CONCLUSIONS AND RECOMMENDATIONS

On the basis of the reported investigations and long-term experience from the operating institutions of 15 landfill sites, conclusions may be drawn about the reclamation planning and after-care with respect to top sealing, internal barriers and monitoring.

5.1   *Top sealing*

- *Landfill sites with base sealing*: For landfills containing non-inert waste, the standard capping (consisting of mineral sealing layer, geosynthetic liner, drainage and cultivated top soil, carried out immediately after completion of filling) is counterproductive in terms of sustainability – conservation of the emission potential actually results and is assumed to be preserved over the lifetime of the capping. After possible failure, the base sealing system is the sole protective measure against leachate emission. But limitations of the performance of the base drainage system mainly from incrustations can be reported from many sites not long after closure (Entenmann, 1996) despite their usual generous dimensioning. With progressive insufficiency of dewatering, hydraulic gradients rise and affect the base sealing. Additionally as there are usually no or only small retention basins for leachate storage. Problems with the leachate treatment plant will result in an immediate storage of leachate in the landfill which is unlawful but not uncommon and finally may even led to road transports in the past.

    Long-term investigations into the emission behavior of only nominally-covered landfills reveal a significant improvement of seepage water quality and continuous decrease of the total emission potential as a result of biodegradation and increasing retention potential of the residual waste body especially for heavy metals. With respect to reclamation planning the necessity of dimensioning water retarding, but permeable, covering systems is therefore emphasised. These require monitoring plans that imply repeatedly drawn up water balances and analyses of the chemical composition of waste, seepage water and residues.

    With landfill sites containing MBP-waste or incineration slag, a standard combination capping has no adverse effect but it is questionable if the costs are justified in relation to the reduction of risk.
- *Landfills without artificial base sealing on natural subsoil of low permeability*: For these closed landfills the main objective must be to either improve an existing base drainage system or to install additional facilities e.g. a deep-reaching ring-drainage system. This is much more effective than a perfection of the capping system, as the necessary time-span of drying out of the landfill body by capping often exceeds 10 years.
- *Landfills without artificial base sealing on natural subsoil of high permeability*: There is no alternative to the installation of high-quality capping of long durability. Results of monitoring

reveal a significant reduction of the concentrations of pollutants in the downstream plume, but not yet in its extension in the first years after implementation.

- *Reclaimed landfills*: Landfills with base sealing, either natural or artificial, and an effective base drainage which were capped in the past solely by a mineral sealing liner without overlying drainage, often provide problems with the sewage water treatment, as the base drainage discharge in wet years is too high. In order to improve the dewatering capacity, measures can be favorable. We are conducting investigations how this can be done by improvement of surface or subsurface water runoff over the existing top sealing liner or enhancement of evapotranspiration.

With respect to old landfills with simple capping that do not have a base sealing and dewatering installations, emission equals infiltration. There is no legal obligation to do anything unless there is an imminent danger of groundwater pollution. With the improved knowledge of the real emission potential from municipal landfill this must be considered as a rare exception.

5.2 *Additional measures for emission reduction*

Contemporary landfilling is based on the multi-barrier principle (Stief, 1986). As depicted before, individual solutions for each site should be favored. This does also mean thinking about alternatives. From the view of the practician there are only two innovative methods just now – Aeration and Ecolandfill. The first of these has to be applied to closed landfills while the second must be applied on landfills in operation or at least some time before closure. Beside these new methods, in some cases, a flushing bioreactor approach is possible (Powrie et al., 2000). This can be suitable for old landfill sites. With regard to newly established sites it should not be applied, as today modern pre-treatment technology and Ecolandfill is available, which is preferable to remediation which flushing in fact means.

- *Aeration*: The random conditions in old landfills are often not suitable for flushing. As an alternative, the aeration method (Heyer et al., 2003) could be applied in order to enhance biodegradation. Although this method is reported to produce significant additional biodegradation, it is questionable whether necessity of its application can be derived from risk assessment.
- *Eco-landfill*: Previous investigations into the hydraulic system of landfills, containing untreated waste, revealed overall hydraulic conductivities of significantly more than 10^{-5} m/s derived from water balances (Entenmann, 1998). Large scale investigations which are being carried out show clearly that, as for bottom ashes from incineration, as well as for mechanically-biologically pre-treated waste, low permeability can not be achieved at all. Therefore, we are far from any internal barrier, except for the retention potential of degraded organic substances that prevents heavy metals from immediate emission.

The principle of Eco-landfill, developed in Austria and proved with the Hehenberg landfill, emphasizes internal barriers. It is based on the addition of mineral fines which are mixed thoroughly with the MBP-waste (Riehl, 1993; Entenmann & Riehl, 2002). This results in an achievable compaction much higher than with conventional MBP-waste, almost impermeability and a fixation of contaminants as a result of chemical and mineralogical processes that reveal long-term stability. If applied from the beginning of landfilling, a leachate treatment plant as well as top sealing liners can be dispensed with, as infiltration (and therefore exfiltration) does not appear and even the dimensioning of the base sealing system can be significantly reduced. If applied to the top layers of the waste body, it results in a significant reduction of leachate. The principle of Eco-landfill compared with conventional MBP-landfill is shown in Figure 9.

- *Flushing bioreactor*: Biodegradation processes in landfills that contain untreated municipal waste, are either limited as a result of insufficient removal of decomposition residues or the emission potential is preserved in form of sludge in the pore-space. This problem can be solved if installations are present to enhance flushing, as could be shown with the Hahn-Lehmden landfill mentioned previously.

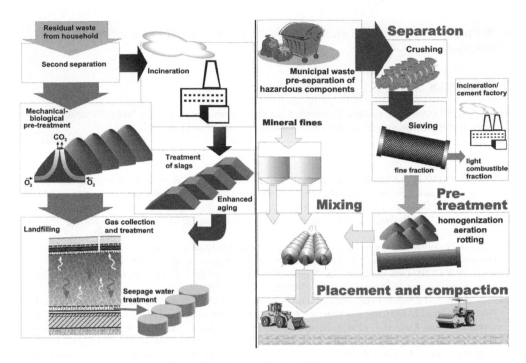

Figure 9.   Contemporary waste disposal (left) versus Eco-landfill.

5.3   *Monitoring*

Monitoring of landfills in the phase of after-care according to the ordinance (WÜ, 1998) means a collection of numerous groundwater, surface-water and leachate analyses, carried out 4 times a year. The benefit of those analyses is very small. In practice, the commitment of threshold values, just recently directed, is very difficult in view of other influence and background contamination. With unsealed landfills, comprising propagating contamination plumes, it is impossible. So in practice, if at all, usually simple qualitative comparisons are made between subsequent values.

The high quantity of analyses should be reduced and be replaced by regularly elaborated assessment of both the hydraulic and the hydrochemical situation considering the hydrogeological background. Nowadays it is possible to utilize the experience of time periods of 30 years and more and it is apparent that the hydrochemical behavior in the groundwater does not change very fast with most landfills. One analysis per year should be sufficient. In order to change the season consecutively, a 15 months time-span is favorable.

The drawing up of water balances is a much more efficient way of testing the effect of top-sealing measures or improvement of runoff than evaluation of groundwater data. The water balance responds immediately to improvement measures whilst groundwater data does not show effects for several years.

In all cases where measures are intended to improve biodegradation, monitoring poses the most demanding task. Measuring the residual biodegradability of waste samples, drilled from the landfill in subsequent series is not practical and by far too expensive. Therefore, indirect measuring of leachate and gas parameters and thoroughly carried out evaluations in distinct periods must be done. As a pre-requisite therefore, additional scientific research should be done.

ACKNOWLEDGEMENTS

Data from investigations, carried out on behalf of Ammerland county, AWV Vechta, AWZ Wiefels, Friesland county, GbA Brake, Vechta county and Wesermarsch county form the basis of this paper. The Authors are grateful for the permission to publish the information from the investigations.

REFERENCES

AbfAblV (2001) *Ordinance on Environmentally Compatible Storage of Waste from Human Settlements and on Biological Waste-Treatment Facilities (Abfallablagerungsverordnung)*, BGBl I, 305p

DepV (2002) *Ordinance on Landfills and Long-Term Storage Facilities and Amending the Ordinance on Environmentally Compatible Storage of Waste from Human Settlements and on Biological Waste-Treatment Facilities*

Entenmann W (1996) *Schädigung einer Deponiebasisdichtung und Stofftransport infolge von Sickerwasser-einstau. – Von den Ressourcen zum Recycling,* Geowissenschaften und Geotechnik im Spannungsfeld von Ökologie und Ökonomie, Ernst, Berlin, pp 61–82

Entenmann W (1998) *Hydrogeologische Untersuchungsmethoden von Altlasten,* Berlin-Heidelberg-New York, Springer, 373p

Entenmann W (1999) *Wasser- und Stoffbilanzen bei der Gefährdungsabschätzung von Altlasten und Deponien,* Geowissenschaften & Umwelt 3, Ressourcen-Umwelt-Management, Springer, Berlin, pp 37–54

Entenmann W (2001) *Untersuchungen zur Wirksamkeit von Deponieabdichtungssystemen im Hinblick auf den Grundwasserschutz,* Barbara-Gespräche 1998, Bd. 5a, Wien, 33p

Entenmann W and Rappert J (1998) *Estimation of seepage discharge from polluted sites and landfills,* Proc 8th International IAEG Congress Vancouver IV, Balkema, Rotterdam, pp 2299–2305

Entenmann W and Riehl-H G (2002) *Diagenetic Inertization – Municipal waste deposition with minimum environmental impact,* Proc 9th JAEG Conf, Durban, 9p

Entenmann W and Schwinn K H (1997) *Emission of pollutants from different contaminated sites into groundwater – a comparative study from Northern German quaternary sediments,* Proc Int Conf IAEG, Athens, Balkema, Rotterdam, pp 1817–1822

EU (1999) *Council directive 1999/31/EC on the landfill of waste,* Official Journal L 182, 16/07/1999, pp 0001–0019

Heyer K-U, Hupe K, Koop A, Ritzkowski M and Stegmann R (2003) *The low pressure aeration of landfills – experience, operation and costs,* Proc Sardinia 2003, CISA, Cagliari, Italy, Chapter 570

Powrie W, Hudson A P and Beaven R P (2000) *Development of sustainable landfill practices and engineering landfill technology,* Final report to EPSRC, Southampton

Riehl-H G (1993) *Die "Diagenetische Inertisierung" – eine umweltneutrale Rückeinbindung von Abfall in den natürlichen Stoffkreislauf,* Restmüll Enquete, Informationsreihe Abfallwirtschaft des Landes Steiermark, Vol. 1, Graz, pp 69–72

Stief K (1986) *Das Multibarrierenkonzept als Grundlage von Bau, Betrieb und Nutzung von Deponien,* Müll und Abfall vol 18(1), Berlin, pp 15–20

TASi (1993) *Technische Anleitung zur Verwertung, Behandlung und sonstigen Entsorgung von Siedlungsabfällen,* Dritte allgemeine Verwaltungsvorschrift zum Abfallgesetz. – Bundesanzeiger Hensfelder – Ludwig (Bearb.), Köln, 1993, 117p

WÜ (1998) *Ordinance on monitoring of landfills*

Geotechnical and Environmental Aspects of Waste Disposal Sites – Sarsby & Felton (eds)
© 2007 Taylor & Francis Group, London, ISBN 978-0-415-42595-7

Leachate leakage from landfill: Causes and mechanisms

H.-Y. Fang
Global Institute for Energy and Environmental Systems, University of North Carolina, Charlotte, USA

A. Kaya
URS, Honolulu, USA

T.-H. Kim
Civil and Environmental System Engineering, Korea Maritime University, Busan, Korea

ABSTRACT: Liquid waste (leachate) leaking from landfill sites is a major problem relating to ground pollution. At present time there is no effective method to control leaking due to the complex soil-liquid-gas interaction of the leachate in the environment. There are three major locations where leakage occurs; the bottom seal (liner) of the landfill, the hydraulic barrier wall and the natural impervious soil layers. The leaking causes, routes and mechanisms at these locations are presented with emphasis on bacterial and chemical corrosions. The factors affecting leachate properties and their characteristics within the soil mass are discussed with emphasis on mass transport phenomena. Leachate leaking-induced instability of landfill controlling facilities is also discussed. All explanations of leaking mechanisms are based on an environmental geotechnical viewpoint.

1   INTRODUCTION

1.1   *General*

Leachate is contaminated liquid from decomposed garbage in a landfill site. It frequently leaks from landfill into aquifers or drinking water wells. The property of leachate is just like garbage, i.e. it varies from country to country, municipality to municipality, site to site, as well as season to season. Leachate is a slow moving flow, which is unsteady, non-uniform, or sometimes discontinuous and also contains hazardous/toxic substances in solid or gaseous forms and may be referred to as creeping flow (Fang, 1986). In order to examine the characteristics of leachate, one must start from the characteristics of fresh garbage itself and the decomposition processes.

There are several ways to dispose of urban refuse (Meegoda et al, 2003) – dumping into landfill is one common method. Urban refuse is called Municipal Solid Waste (MSW) and is also known as garbage or trash. Dumping garbage into a landfill site looks like a simple procedure, but it has a very complex process. The characteristics of garbage in a landfill have a complex life, there are three basic stages between fresh and aged garbage as illustrated in Figure 1. The decomposition (geomorphic) process involves short- and long-term processes, which include mechanical, chemical, physicochemical, and geomicrobiological processes. Figure 1 illustrates that garbage has three distinct stages; the fresh garbage, the aged garbage and the decomposition process between these two stages. These three distinct stages have their own characteristics but are closely interrelated. The fresh garbage is entirely solid state, after the decomposition process, 90% of the fresh garbage becomes a liquid state, and the remaining part of aged garbage becomes organic soil.

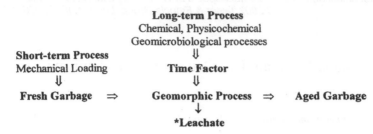

Figure 1.   Decomposition (geomorphic) process: from fresh garbage to aged garbage.

Table 1.   Classification of fresh garbage.

Material category	Garbage components
Fast degradable	Kitchen trash, garden waste, dead animals and manure, papers and paper products
Slow to degrade	Textiles, toys, rugs, glass, ceramics, plastic, rubber, leather goods
Relatively non-degradable	Metals, household appliances, demolition and construction materials, soils, rocks

1.2   *Characteristics of fresh garbage*

MSW consists of anything that cannot be further used or recycled economically. Fresh garbage can be grouped into three categories, i.e. fast, slow and relatively non-degradable material, as indicated in Table 1 which presents the general classification of fresh garbage abstracted from Fang (1995a).

1.3   *Landfill compaction and stability*

Any waste controlling facility has the major purpose of preventing or controlling the hazardous and toxic leachate, which is the liquid waste leaking from a landfill site into the groundwater aquifers or drinking water wells. There are two general types of landfill site; below the ground surface, above the ground surface. The general facilities for controlling the landfill are the top seal (landfill cover) and bottom seal (landfill liner). During the active operation stage of landfilling the top seal does not exist. However, the bottom seal always exists. The compaction process and slope stability for the landfill are the main concerns for fresh garbage.

When garbage is delivered daily by truck and dumped into the landfill site, in most cases this garbage is spread into a thin layer, mixed with locally available soil and compacted by conventional compaction equipment used in construction operations. The main purpose is to cover up an unattractive landfill site or to minimize odour or to prevent animal and bird infestation. There is no standard rule or regulation how garbage should be dumped or compacted at the present time. Fang (1995a) suggested that for compaction in a landfill site;

(a) garbage comes in all types, shapes and sizes and cannot be uniformly distributed in the landfill however, within limits, it can be distributed uniformly within a layer,
(b) heavier items of garbage should be placed close to the centre of the landfill for the purpose of controlling the stability of the fill, avoid dumping these items around the edges of a landfill,
(c) spread the newly dumped garbage as thinly as possible,
(d) mix in locally-available soil,
(e) use heavy compaction equipment (such as dynamic consolidation) to control the compaction.

At present, numerous refined approaches are provided as described in various standard textbooks (Bagchi, 1994; Oweis and Khera, 1998; Sarsby, 2000; Qian et al, 2002).

The compaction procedure is closely related to the stability of landfill. The better the compaction of the loosely-dumped garbage the better is the stability of the landfill slopes, as discussed by Fang (1995a), Saarela (1997) and many others.

1.4 *Waste interaction with the environment*

Within degradable waste, some is hazardous/toxic and some is not. There are various states of matter, solid, liquid and gas forms depending on local environmental conditions. Many hazardous wastes, when mixed with other waste or materials can produce effects that are harmful to human health and environment such as;

- heat or pressure,
- fire or explosion,
- violent reaction,
- toxic dusts, mists, or gases,
- flammable fumes or gases.

All generators of waste materials are required to determine whether the waste is hazardous in one of two ways. Firstly it is either waste and spent material that is hazardous by definition and contained in specific lists such as those issued by USEPA (EPA, 1996), or secondly it exhibits one of four hazardous characteristics such as inflammability, reactivity, corrosiveness, toxicity.

When two types of waste are mixed together they may produce heat or pressure or fire or explosion. Burning in landfill frequently occurs causing internal burning. A well-publicized disaster of an underground mine fire at the mining town of Centralia (Pennsylvania) started from a small garbage dump fire accident.

Methane (CH_4), is a colourless, odourless gas. It is formed during the decomposition of vegetable matter and is found in landfill. Methane gas can seep into subsurface soil layers and can travel as far as 8 km from a landfill site.

2 DECOMPOSITION OF FRESH GARBAGE

2.1 *Decomposition*

The decomposition (geomorphic) process is the second stage of the garbage process as indicated in Figure 1. This stage covers the complex decomposition process including mechanical, chemical, physico-chemical and geo-microbiological processes. Any materials in a given location, due to various environmental factors after a certain time gradually change form and properties. Figure 1 shows the concept of geomorphism (Fang, 1986, 1997) of solid municipal waste. Mechanical alteration is considered a short-term process, while the other alterations are classified as long-term processes.

2.2 *Factors affecting decomposition of fresh garbage*

As indicated in Figure 1, the short-term geomorphic process includes compaction, kneading, shearing. The mechanical process itself is not included in the decomposition process. The larger amount of volume change reflected as subsidence or settlement in the landfill site is contributed by mechanical load such as surcharge, snow, rainwater, etc. The mechanical process involves only reducing the voids in the fresh garbage.

The long-term process includes three major complex processes; chemical, physico-chemical, and geo-microbiological:

- The chemical process includes carbonation, hydration, hydrolysis, carbonation and hydrolysis, oxidation. Detailed discussions of these chemical processes are given in many standard textbooks on soil sciences, soil chemistry and others (Yong and Mulligan, 2003).

31

- Ion exchange reaction is the major process in the physico-chemical process. From an environmental geotechnology viewpoint the ion exchange reaction can provide two important reactions;
  – ion exchange reaction changes water properties from soft water to hard water and vice versa,
  – may change soil structure from dispersive to flocculate structures and vice versa.
  Hard water is corrosive and flocculate soil structures are more permeable than dispersive structures.
- Bacterial activity is a major part of the geomicrobiological process. In general, there are two basic types of bacteria, i.e. heterotrophic and autotrophic. The heterotrophic type of bacteria acquires energy and carbon directly from organic matter, however, the autotrophic type acquires energy from oxidation of its mineral constituents. In the landfill, it is the heterotrophic type of bacteria that most likely exists. Decomposition by bacteria has five stages (as proposed by Wardwell et al, 1982) namely;
  – aerobic,
  – anaerobic (non-methanogenic),
  – anaerobic (methanogenic),
  – anaerobic decline,
  – regain of aerobic growth.

The decomposition process will change the properties of leachate and also will cause cumulative settlement (subsidence) in landfill, which can break the top seals (covers). As a consequence rainwater, floodwater and surface runoff will intrude into the landfill sites. A typical example of the effect of bacteria on pore fluid is illustrated in Equation 1, whereby bacteria can cause a change from a weak acid (H_2S) to a strong acid (H_2SO_4).

$$H_2S + 2O_2 \xrightarrow{\text{Bacteria}} H_2SO_4 \qquad (1)$$

3 LEACHATE

Leachate from the decomposed garbage is the second stage of garbage ageing process as indicated in Figure 1. The ageing garbage contains both solid and liquid wastes. The liquid phases are more complex than the solid phases.

3.1 Factors affecting quantity of leachate

- Composition of garbage. Leachate is decomposed liquid from fresh garbage. The composition of fresh garbage directly relates to the properties of leachate. Since the garbage varies from site to site and from time to time the composition of the leachate will follows the same pattern.
- Type of landfill. A landfill will be influenced by hydrogeological and geohydrological conditions according to whether it is located above or below ground.
- Age of the landfill. Time plays an important role because the degree of decomposition relates to time (the age of the landfill).
- Weather conditions. Conditions such as rainfall, snow, flood, will influence the stability of a landfill site and consequently will affect the quantity of leachate inside the landfill.
- Hydrogeological and geohydrological conditions. These characteristics will affect the subsurface soil conditions, groundwater fluctuation, characteristics of aquifers and geological formations. These factors are directly related to the characteristics of a landfill site and consequently affect the properties of leachate.
- Bacterial and chemical breakdown. These activities will directly change the properties of leachate.

3.2 Properties of leachate

As indicated in Table 1 a landfill site contains various types of solid waste. However, after a certain time, due to various natural and man-made decomposition processes, 90% of the solid waste

32

Table 2. Ranges of leachate properties in USA (USEPA, 1988).

Component	Range (ppm)
PH	3.7–8.5
Hardness, CaCO$_3$	200–7,600
Alkalinity, CaCO$_3$	720–9,500
Ca	240–2,400
Mg	64–410
Na	85–3,800
K	28–1,700
Fe	0.15–1,640
Ferrous ion	8–9
Chloride	50–2,400
Sulfate	20–750
Phosphate	0.5–130
Organic-N	3.0–490
NH$_4$-N	0.3–480
BOD	22,000–30,000
COD	800–50,000
Zn	0.02–130
Ni	0.15–0.9
Suspended solids	13–27,000

Table 3. Properties of leachate from central Pennsylvania (Fang and Evans, 1988).

Parameter	Concentration (mg/L)
Aluminum (Al)	2.1
Biochemical oxygen demand (BOD)	7,300
Chemical oxygen demand (COD)	11,000
Chloride	898
Fluoride	0.41
Manganese (Mn)	300
Nitrite (NO$_3$)	0.89
Oils and greases	43
Sulfate (S)	400
Total organic carbon (TOC)	3,000
Total organic nitrogen	220

becomes liquid waste. The major purpose of controlling the landfill facilities is not only to deal with the solid waste but to control the liquid waste as well. In many cases, controlling the liquid waste is more important than controlling the solid waste.

The range of leachate properties is indicated in Table 2.

General physico-chemical properties of leachate from a landfill site located in central Pennsylvania are summarized in Table 3.

The colour of leachate is generally dark brown. Characteristics of leachate also produce higher temperature, higher ion concentration and exchangeable ion reaction, redox reaction, and higher bacteria activity. Because the properties have a high abrasive erosion capability, the bottom seal (liner) hydraulic barrier wall and impervious natural soil layers are attacked.

Further information on leachate processes and properties has been provided by many investigators, e.g. Lu et al (1985); Qasim and Chiang (1994); McBean et al (1995); Qian et al (2002).

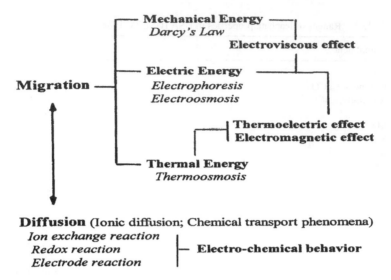

Figure 2. Diffusion and migration characteristics in soil-water systems.

3.3 Leachate flow

The movement of leachate in the landfill site is a complex process. It does not follow Darcy's Law as commonly used in geotechnical engineering, it follows various potential heads depending on the local environmental conditions. Leachate contamination is a non-uniform, unsteady liquid flow, possible components are:

- *Advection* is movement of leachate in a porous medium under a hydraulic gradient following Darcy's Law.
- *Diffusion* is the process whereby ionic or molecular particles move in the direction of their concentration gradient under the influence of their kinetic energy. The process of diffusion is often referred to as self-diffusion, molecular diffusion, or ionic diffusion. In geotechnical engineering these are referred to as chemical transport phenomena. The mass of the diffusing substance passing through a given cross-section per unit time is proportional to the concentration gradient (Fick's First Law). Some investigators have used this Law for predicting the leachate movement in landfill sites. However, Fick's Law is valid for low ion concentration in the groundwater aquifer not for contaminated high concentration of various types of ions existing in the leachate, which is a nonlinear and unsteady flow. Therefore, using Fick's Law for predicting leachate movement in landfill is questionable.
- *Migration* phenomena in soil are part of the dynamic behaviour of soil moisture movement and dynamic equilibrium between water in the various states. Within this individual water molecules will pass back and forth at stabilized rates, between the vapour and the liquid states, between the strata of capillary and hydroscopic water. Migration is a transient redistribution of water rather than a continuing flow. There are many possible factors such as capillarity, vapour diffusion, moisture film transfer, electromotive force, electrokinetic force, etc. The relationship between diffusion and migration in a soil-water system is presented in Figure 2.

3.4 Leachate movement in landfill

In general, soil is composed of electrically negative mineral surfaces while water is composed of electric water dipoles and predominantly positively charged ions. It follows that the soil-water system possesses a highly electrical character and hence, will respond to the application of an electric potential. The interaction of these electrically-charged components of the system is a

34

(a) Leachate (Liquid waste) Through Soil Deposit

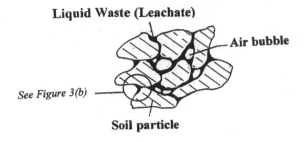

(b) Possible Leachate Movement in Landfill Site

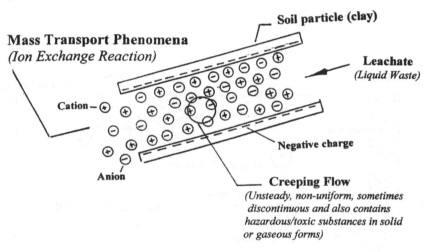

Figure 3.   Potential micro-scale flow in a contaminated soil-water system.

function of temperature. Therefore, the thermal-electrical properties are closed and exist in the natural environment.

Water in a soil system may simultaneously be present as vapour, as a liquid of varying viscosity and as a solid of varying plasticity. Water possesses a well-developed structure of highly electric character due to its own polar molecular nature, as well as being under the influence of the electrically charged surfaces and ions in solution. In view of the manifold properties of water as a substance and modifications in the soil environments, water in soil responds to the imposition of an energy gradient be it; mechanical (Darcy's Law), thermal (Laws of thermodynamics), electrical (Ohm's Law), magnetic (Faraday's Law), or other. This water responds to, or exhibits, coupling effects indicative of the disturbance of the other energy fields. Such response usually results in mass transport phenomena and therefore leachate flow through contaminated landfill site is not simply a function of hydraulic conductivity as described by Darcy's Law, as commonly assumed in geotechnical engineering. If mass transport phenomena exist in such a case then the leachate flow in such a system is controlled by thermal-electric-magnetic energies and not by mechanical energy alone (Fang, 1997).

Figure 3 is a schematic diagram illustrating the possible flow movement in a contaminated soil layer at micro-scale. Water movement between soil particles in a contaminated soil-water system is influenced by electro-viscous effects (Street, 1959) and types and concentrations of the exchangeable ions play an important role.

35

Table 4. Factors causing leachate leakage from landfills.

Factor	Specific items
Errors in assumptions	Surface overburden load, snow loading
Analytical approach	Inappropriate methodology for selecting the analytical procedures for the control facility
Design concepts	Slope stability and settlement analysis
	Permeability data and test procedures
Construction procedures	Weathering effects
	Lack of proper field inspection and supervision
	Incorrect location of impervious soil layer
	Poor workmanship
Maintenance and monitoring	Inadequate routine maintenance
	Inappropriate monitoring system

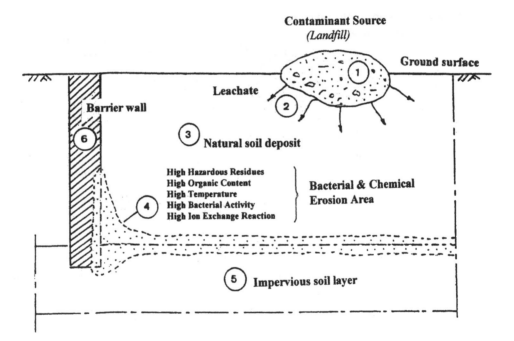

Figure 4. Relative positions of components of an engineered landfill.

4 LEACHATE LEAKAGE FROM LANDFILL

4.1 *Causes and routes*

From a theoretical viewpoint leakage from landfills is common – it cannot be stopped or eliminated. However, it can be controlled or reduced to a manageable level. There are numerous factors causing leachate leakage from landfills as indicated in Table 4.

In general there are three major locations for leakage routes as indicated in Figure 4 – area no.4 is the most critical area because it is where most leachate collects, area no.5 is a naturally-impermeable soil layer and area no.6 is a barrier wall.

36

4.2 Leakage through the liner

Leachate leakage through an engineered landfill liner occurs due to a variety of reasons:

(1) *Leakage due to puncture.* Waste contains everything from sharp metal scraps, broken appliances, nails, broken glass, metal container, etc. – all of these are particularly hazardous for the synthetic liners during the early stage of filling when a compaction process is applied. Numerous reports indicate that synthetic liners suffer punching failure.

(2) *Shrinkage, swelling and cracking.* These are natural processes that occur frequently in earthen structures due to internal energy imbalance in the soil mass. The two common causes of energy imbalance in soil layers are non-uniform moisture and temperature distribution. These phenomena are significant around many hazardous/toxic waste sites. Significant volume changes and cracking may be associated with fluid transport in such a complex system. When water is removed from soil it will shrink while the addition of water will increase the soil volume. Swelling and shrinkage is not a reversible process meaning that swelling is not equal to the shrinkage. The shrinkage process is associated with the thermal energy field whilst swelling relates to the multimedia energy field. Swelling-shrinkage cycles will lead to cracking and fracture of the soil.

(3) *Leakage by horizontal capillarity.* Many earth structures such as clay liners or hydraulic barrier walls are constructed from locally available soil. Laboratory experiments (Fang et al, 1997) have demonstrated how horizontal capillary action can cause slow-moving pore fluid to pass through soil layers with non-uniform density of soil material and non-uniform compaction from construction operations.

(4) *Thermal-osmosis forces.* The thermo-osmotic phenomenon is one of the important factors relating to the landfill failure mechanism. The thermal gradient on the inside and outside of landfill sites is significantly different. The inside of the landfill is at much higher temperatures due to the decomposition process. In some cases the radioactive nuclear waste in landfills creates very high temperatures (Fang, 2002). In many cases leachate movement in the landfill includes covers and liner upon application of the thermal gradient. Thermo-osmosis force has been estimated and limited experimental data derived (Winterkorn, 1958, 1963).

(5) *Leakage through vegetation.* Landfill areas are often planted with trees/vegetation for the purpose of improving the appearance of unwanted landfill sites. To plant trees or vegetation also has detrimental side effects and two important ways that tree roots attack liners are:
    (a) through tree roots looking for nutrition (food) – the landfill area is an ideal place and roots are capable of, and will, penetrate liners,
    (b) by providing conduits for fluid flow as landfill leachate seeps out through root hairs subject to suction forces.
Roots are always looking for food and water and landfill sites provide all these necessities for growing trees/vegetation. The force produced by tree roots is strong enough to penetrate the clay liners or synthetic membrane. For example, bamboo roots can penetrate a 15 cm thick concrete wall. Other common trees planted around landfills areas are silver maple, willow and many others. Some vegetation such as squashes, spread their roots laterally and widely within the 1 ft (0.305 m) surface zone. Some root systems are coarse and open, whereas others consist of masses of fine rootlets. Larger trees and their roots can reach great deeper depths. Detailed discussion on tree roots related to landfill sites is given by Fang (1995b, 1997).

(6) *Geomembrane defects.* These defects are a major problem as reported by Giroud and Badu-Tweneboah (1992) and Qian et al (2002). Soil-pollution interaction effects on the stability of geosynthetic composite walls and geosynthetic liners have been reported by Fang et al (1992).

Leakage from landfill cannot be completely stopped or eliminated. Most investigators focus on leakage from landfill as caused by hydraulic gradient (Darcy's Law). Unfortunately, this approach is not the whole truth, there are numerous factors affecting liner leakage. Individually each of these factors produces only small quantities of leakage and would generally go unnoticed. However, the accumulations of these small amounts can be significant enough to cause groundwater aquifer pollution, thereby, affecting the stability of all waste control systems. For example, the velocity head in soil is commonly negligible, however, for estimation of pollution migration purposes, this small item must be also considered.

4.3 *Leakage through an impermeable soil layer*

Soil is the cumulative result of geological ages under changing environments, combined with the physical effects of weathering and biological processes – Fang (1996, 1997). Soils are creations of climatic forces and influenced by local environmental factors. As a result of these factors, the soil's in-situ (undisturbed) condition shares many essential properties with living systems. Furthermore, microfauna that are dispersed in soils render soils actual living systems. No soil is truly impermeable. Boring logs often show fungi, three roots, and bacteria distributions in all the soil profiles and horizons (Figure 4, area no.4). Little information is available on the characteristics of the impermeable soil layer and degree of safety with respect to the waste control facility.

4.4 *Subsidence of landfill*

Subsidence is ground movement with a major component in the vertical direction. In a landfill site most of the ground movement phenomenon is subsidence and results from the volume change of the landfill caused mainly by decomposition of fresh garbage and leachate leaks from the landfill. Subsidence is an important factor relating to the landfill stability as well as it controlling facilities. There are three general types of subsidence namely, initial, primary and secondary subsidence:

- Initial subsidence results from the fact that fresh garbage is loose and contains a large volume of air. The comparatively sudden reduction in volume of garbage mass (matrix) under an applied load such as a compaction process, as well as under surcharge weight, is due principally to expulsion and compression of gas or air in the garbage voids.
- Primary subsidence results from the reduction in volume of a garbage mass caused by the application of a sustained load to the garbage mass (matrix) and is due principally to a squeezing out of liquid from the void spaces of the garbage mass accompanied by a transfer of load from garbage-water to garbage-solids.
- Secondary subsidence results from the reduction in volume of a garbage mass caused by application of a sustained load to the garbage mass and is due principally to the adjustment of the internal structure of the garbage mass after most of the load has been transferred from decomposed garbage to the stable garbage.

Other causes of subsidence are freeze-thawing, shrinkage-swelling, tree root intrusion. In general subsidence will cause the clay top seal covers and geosynthetic liners to be significantly distorted so that they may crack and break, cave-in of covers and liners.

4.5 *Design and construction deficiencies*

Basic considerations for analysis and design of landfill facilities consider also the design and construction deficiencies. Two important items to point out from an environmental geotechnology viewpoint are; the unbalanced lateral earth pressure around the landfill, the stability of the hydraulic barrier wall.

There is an imbalance between the lateral earth pressures inside and outside of the landfill liners because of the decomposition of the garbage. When fresh garbage decomposes, the density of the garbage changes from solid state into liquid state and volume of garbage in the landfill reduces significantly (Figure 5a). The unit weight of soil outside of the landfill remains relatively constant (at about 20 to 22 kN/m^3) and thus an unbalanced lateral earth pressure is produced. This pressure can cause clay liner cracks and geotextile deformation which moves it from the originally designed position (Figure 5b).

(a) **Dotted Line Indicates the Subsidence or Volume Change of Landfill**

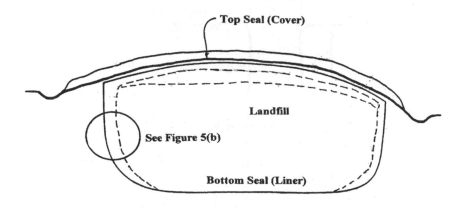

Top Seal (Cover)

Landfill

See Figure 5(b)

Bottom Seal (Liner)

(b) Unbalanced Earth Pressure Between Inside and Outside of Liner

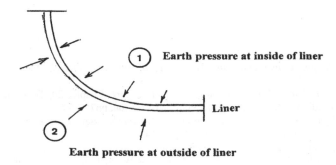

① **Earth pressure at inside of liner**

Liner

②

Earth pressure at outside of liner

Figure 5.   Volume change of landfill due to garbage decomposition.

The purpose of the hydraulic barrier wall is to prevent hazardous/toxic liquid waste (leachate) from seeping into drinking water wells. In general, this barrier wall is made from bentonite slurry mixtures (Evans, 1991; Inyang and Tumay, 1995). The width (W) of the trench/wall is generally three feet (1 m), because of the design of the digger/backhoe machine (Figure 6). The depth (D) to the impermeable layer varies from site to site. Therefore the height of hydraulic barrier wall and width-depth ratio (W/D) varies from case to case. The larger the value of (W/D) the more unstable the wall becomes. The nature of the barrier wall is considered a flexible retaining wall. The stability of the wall at its extreme bottom part is most critical.

5   CONCLUDING REMARKS

Natural decomposition of garbage (solid municipal waste) can be considered as a three-stage process, namely; fresh garbage, aged garbage, the decomposition process (both short-long-term). Ground subsidence, due to garbage decomposition (geomorphism), chemical and bacterial erosion, detrimentally affects the stability of the waste controlling facility.

Leaking causes, routes and mechanisms are functions of the landfill itself, the nature of any underlying impermeable soil layers, design errors (both natural and man-made).

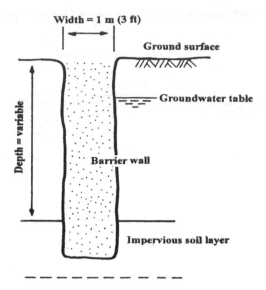

Figure 6.   Typical cross-section of hydraulic barrier wall.

REFERENCES

Bagchi A (1994) *Design, Construction and Monitoring of Sanitary Landfill,* John Wiley and Sons, New York.

Evans J C (1991) *Geotechnics of hazardous waste control systems,* Chapter 20, Foundation Engineering Handbook (Second edition), Kluwer Academic Publishers, Boston, pp750–777.

Fang H-Y (1986) *Introductory remarks on environmental geotechnology,* Proceedings 1st International Symposium on Environmental Geotechnology, v 1, pp1–14.

Fang H-Y (1995a) *Engineering behaviour of urban refuse, compaction control and slope stability analysis,* Proceedings GREEN 93, Waste Disposal by Landfill (Ed R W Sarsby), Publ A A Balkema, Rotterdam, pp47–72.

Fang H-Y (1995b) *Bacteria and tree root attack landfill liners,* Proceedings GREEN93, Waste Disposal by Landfill (Ed R W Sarsby), Publ A A Balkema, Rotterdam, pp419–426.

Fang H-Y (1996) *Leaking mechanism from landfill through impervious soil layer,* Proceedings International Congress on Environmental Geotechnics IS–Osaka96, Japan, v1, pp491–495.

Fang H-Y (1997) *Introduction to Environmental Geotechnology,* CRC Press, Boca Raton, Florida.

Fang H-Y (2002) *Radioactive nuclear wastes,* ASCE Practice Periodical of Hazardous, Toxic and Radioactive Waste Management, v6, no.2, pp102–111.

Fang H-Y and Evans J C (1988) *Long-term permeability tests using leachate on a compacted clay-liner material,* ASTM Special Technical Publication STP963, pp397–404.

Fang H-Y, Pamukcu S and Chaney R C (1992) *Soil-pollution effects on geotextile composite walls,* Slurry Walls Design, Construction, and Quality Control (Eds Paul, Davidson and Cavalli), ASTM Special Technical Publication STP 1129, pp103–116.

Fang H-Y, Daniels J L and Inyang H I (1997) *Envirogeotechnical considerations in waste containment system design and analysis,* Proceedings, 1997 International Containment Technology Conference, St. Petersburg, Florida, pp414–420.

Giroud J P and Badu-Tweneboah K (1992) *Rate of leakage through a composite liner due to geomembrane defects,* Geotextiles and Geomembrances, v11(1), Elsevier Science Publishers, London, pp1–28.

Inyang H I and Tumay M T (1995) *Containment systems for contaminants in the subsurface,* Encyclopedia of Environmental Control Technology, Gulf Publishing Company, pp175–215.

Lu J C S, Eichenberger B and Steams R J (1985) *Leachate from Municipal Landfills, Production and Management,* Publ Noyes, New Jersey, pp109–121.

Meegoda J N, Ezeldin A S, Fang H-Y and Inyang H I (2003) *Waste immobilization technologies,* ASCE Practice Periodical of Hazardous, Toxic and Radioactive Waste Management, v7, no.2, January, pp46–58.

McBean E A, Rovers F A and Farquhar G J (1995) *Solid Waste Landfill Engineering and Design,* Prentice Hall, New Jersey.

Oweis I S and Khera R P (1998) *Geotechnology of Waste Management (2nd edition),* PWS Publishing Co, Boston.

Qian X, Koerner R M and Gray D H (2002) *Geotechnical Aspects of Landfill Design and Construction,* Prentice Hall, New Jersey.

Qasim S R and Chiang W (1994) *Sanitary Landfill Leachate,* Technomic Publishing Co, Pennsylvania.

Saarela J (1997) *Hydraulic approximation of infiltration characteristics of surface structures on closed landfills,* Monographs of the Boreal Environment Research, Finnish Environment Institute, Helsinki.

Sarsby R W (2000) *Environmental Geotechnics,* Publ Thomas Telford, London.

Street N (1959) *Electrokinetics II, electroviscosity and the flow of reservoir fluids,* Illinois Geological Survey Circular no. 263.

USEPA (1988) *Summary of data on Municipal Solid Waste Landfill Characteristics – Criteria for municipal solid waste landfills (40CFR Part 258),* OSWER, US Environmental Protection Agency, EPA/530-SW-88-038.

USEPA (1996) Hazardous/toxic waste lists.

Wardwell R E, Charlie W A and Doxtader K A (1982) *Test method for determining the potential for decomposition in organic soils,* ASTM Special Technical Publication STP820, pp218–229.

Winterkorn H F (1958) *Mass transport phenomena in moist porous systems as viewed from the thermodynamics of irreversible processes,* Highway Research Board Special Report 40, pp324–335.

Winterkorn H F (1963) *Soil water interaction and its bearing on water conduction in soils,* Engineering and World Water Resources, Princeton University Conference, pp15–34.

Yong R N and Mulligan C N (2003) *Natural Attenuation of Contaminants in Soils,* CRC Press, Boca Raton, Florida.

Vegetable fibre degradation in polluted water

R.S. Karri, R.W. Sarsby & M.A. Fullen
SEBE, University of Wolverhampton, UK

ABSTRACT: Over the past 25 years large quantities of vegetative matter (particularly gardening waste) have been deposited in engineered landfills. The fibrous nature of this type of waste initially creates a form of 'soil reinforcement' within the refuse mass. With time the fibres will degrade and the reinforcing effect will be lost and this could have a serious effect on the stability of refuse slopes. Laboratory tests have been conducted to investigate the effect of pore water composition on the strength properties of fibrous vegetable matter and individual vegetable fibres. This preliminary assessment of whether the stability of 'as-constructed' landfill slopes is likely to be affected significantly by decomposition of vegetable matter within the refuse was conducted using a 'typical' vegetable fibre.

1 INTRODUCTION

Waste comprises a heterogeneous mass of material that varies widely according to source and time particularly. The loss of the inert, denser constituent and the increase in volume of paper, vegetable matter including fibres, rag and plastics has had a significant effect on the biological and geotechnical properties of refuse fill. Furthermore, the elimination of domestic burning of refuse led to a drastic increase in the volume of household refuse, or Municipal Solid Waste (MSW), which needs to be disposed of in a controlled manner. Recent domestic refuse contains a high proportion (around 55%) of organic material. Commercial and mixed industrial wastes can contain even higher proportions of organic material this categorization includes both vegetable and putrescible matter including fibres, paper-based waste.

Soil reinforcement is an established engineering technique for improving the strength and stability of soils. The 'traditional' system of soil reinforcement employs regularly-placed horizontal reinforcing elements. However, a system of using randomly-distributed, discrete fibres within soil has also been developed and this arrangement is very similar to the situation within landfilled, mixed refuse (Karri, 1997a, 1997b). Vegetable fibres present in the domestic refuse may act as reinforcement in slopes comprised of refuse. Since the side slopes of most landfills are constructed as steeply as possible during waste placement, fibre degradation, variation of moisture content and settlement will affect the Factor of Safety against shear failure of the refuse slopes. The justification for initially steep slopes is often the apparent possession of effective cohesion by the waste mass (Sarsby, 2000), particularly MSW with a high vegetable fibre content. The effect of various liquids on the mechanical properties of vegetable fibres was assessed using coconut fibres (coir is a typical vegetable fibre, although possibly more durable than many fibres encountered in domestic refuse, which would provide a picture of the performance which might be expected from vegetable fibres acting as natural reinforcement for landfilled refuse).

2 EXPERIMENTS

2.1 *Methodology*

Coconut husks and fibres were collected from different sources and were placed in water and several other aqueous solutions containing different concentrations (1.0 N and 0.5 N) of NaCl, NH$_4$OH,

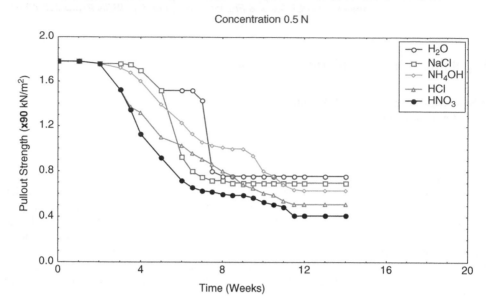

Figure 1. Separation strength of fibres in 0.5 N solutions.

HCl and HNO$_3$. The husks and fibres were submerged in the solutions. Each week one husk and five fibres were taken at random and were subjected to separation/pull-off tests (to indicate how the 'reinforcing' elements would disaggregate) and tensile tests (to indicate how the fibres themselves would disintegrate).

2.2 *Separation tests*

The ends of five fibres attached to a husk were peeled away from the husk so that each fibre had a free end about 2 cm long. The husk was then attached to a load frame and the free ends of the fibres were attached to a constant-rate-of-loading device. The load applied to the fibres was increased until the fibres separated from the husk. The average separation load per fibre was calculated on the assumption that each fibre carried the same load.

2.3 *Tensile tests*

The tensile strength of fibres was determined using a tensometer. A group of five fibres was taken and their ends were clamped in the tensometer so that an effective length of 7 cm was left between the grips. The fibres were then subjected to an increasing tensile load which was applied at a constant rate of strain until the fibres failed in tension. It was assumed that the maximum tensile load was carried equally by all of the fibres. This process was repeated when the fibres had been soaked for various times.

3 RESULTS

3.1 *Separation strength*

Figure 1 shows the variation of separation strength for fibres immersed in all 0.5 N aqueous solutions for a period of approximately 4 months. It is seen that the resistance to separation of the fibres from their parent material starts very shortly (within 2–4 weeks) of their immersion in the various

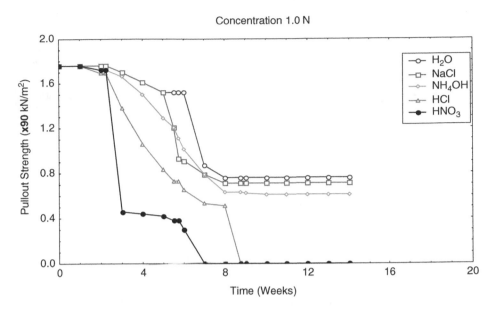

Figure 2.　Separation strength of fibres in 1.0 N solutions.

solutions. This behaviour is to be expected since the traditional method of separating such vegetable fibres for commercial purposes, i.e. retting, involves soaking of the vegetation in water tanks followed by mild beating.

After 2 to 3 months of soaking in the 0.5 N solutions the separation strength fell to its minimum value – between 20 and 40% of the original strength approximately. Acidic soaking conditions initiated loss of separation strength slightly sooner than neutral or alkaline conditions. However the rate of strength loss and actual size of the loss was significantly higher with acidic conditions.

The foregoing picture is more exaggerated when the solutions are more concentrated as shown for the 1.0 N solutions in Figure 2. The fibres immersed in neutral or alkaline solutions exhibited a very similar strength loss pattern to that observed for the 0.5 N solution. However, with acidic solutions the strength loss was rapid and total within a period of around 2 months. Within landfilled domestic refuse the pH of the leachate falls markedly during the first stage of anaerobic decomposition due to acid fermentation (acetogenesis). The leachate becomes strongly acidic (pH values in the range 4 to 5 are common) and then takes several years to rise towards a neutral value (Sarsby, 2000). Thus, it can be assumed that within the refuse body there will be rapid disaggregation of fibres within vegetable matter. Whilst the fibres may still provide tensile reinforcement of the waste mass they will lose their collective ability to provide stiffness within the waste so that compression and settlement of the mass will not be inhibited by the vegetable material.

3.2　Fibre tensile strength

The variation of fibre tensile strength with soaking time (in a 0.5 N solution) is shown in Figure 3. It is readily apparent that neutral water had virtually no effect on the tensile strength of the fibres – hence the success of the traditional retting process for separation of fibres for use in the manufacture of ropes, mats, etc. The fibres were also relatively resistant to saline water (coir ropes have been widely used in seawater environments for many years) and still retained approximately 80% of their tensile strength after about 4 months immersion. However, acidic water conditions had a very detrimental effect on the tensile strength of the fibres. The fibres exhibited abrupt and very rapid loss of strength after being in acidic water for only 2 weeks and after 3 months the fibres only retained between 10 and 20% of their original strength.

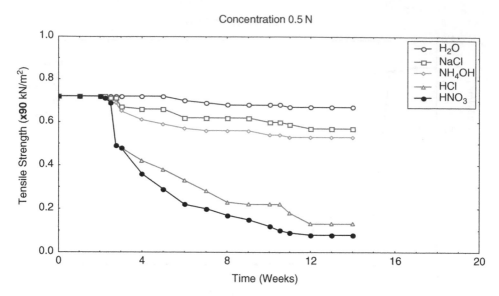

Figure 3. Tensile strength of fibres in 0.5 N solutions.

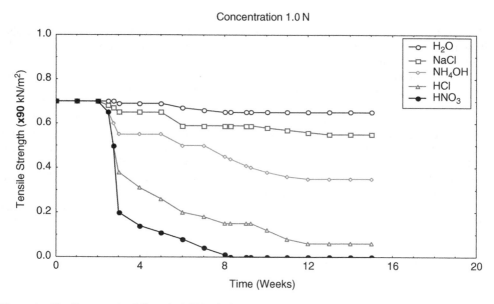

Figure 4. Tensile strength of fibres in 1.0 N solutions.

The detrimental effect of acidic solutions on tensile strength was even more dramatic for the 1 N solutions as shown in Figure 4. It only took between 2 and 4 months immersion in the HCl and HNO$_3$ solutions for the fibres to effectively lose all tensile strength. Bearing in mind the previous comment, that it is the normal state for landfill leachate to exhibit low pH values in the 'early', acetogenic stage of refuse decomposition, it would appear that any form of 'soil reinforcement' effect from vegetable matter within the landfilled refuse is likely to decline rapidly after infilling. This finding is in keeping with the reporting of zero effective cohesion by many researchers who have studied the shear strength characteristics of landfilled refuse (Sarsby, 2000). However, the results also indicate that when designing the side slopes of landfills, and refuse landraises especially, it is unwise to assume that MSW will exhibit any reliable effective cohesion, even for short-term conditions.

4 CONCLUSIONS

Immersion of coir fibres within non-neutral aqueous solutions reduced both the force needed to separate individual fibres and the tensile strength of individual fibres themselves. The effects were particularly severe when the solutions were acidic and within a few months the fibres had effectively lost all of their strength.

Since acidic conditions will exist within landfill leachate in the early stages of anaerobic decomposition of the refuse it is considered that any 'soil reinforcement' created by landfilled vegetative matter will be short-lived. Consequently it would not be appropriate to assign any effective cohesion to landfilled domestic refuse when undertaking slope stability analyses, even for the short-term condition.

REFERENCES

Karri R S (1997a) *Improvement of CBR value using coir fibres,* Project report, College of Engineering, GITAM, Visakhapatnam.

Karri R S (1997b) *Strength improvement of cohesive soil by coir fibres,* Project report, College of Engineering, GITAM, Visakhapatnam.

Sarsby R W (2000) *Environmental Geotechnics,* Thomas Telford Publishers, London.

Geotechnical and Environmental Aspects of Waste Disposal Sites – Sarsby & Felton (eds)
© *2007 Taylor & Francis Group, London, ISBN 978-0-415-42595-7*

Evaluation of geotechnical parameters for effective landfill design and risk assessment

T. Koliopoulos
Department of Civil Engineering, University of Strathclyde, UK

V. Kollias
University of Thessaly, Volos, Greece

P. Kollias
Sanitary Civil Engineer, Athens, Greece

G. Koliopoulou
Department of Medicine, University of Ioannina, Greece

S. Kollias
Department of Mathematics, University of Athens, Greece

ABSTRACT: Sanitary landfill remains an attractive disposal route for household, commercial and industrial waste, as it is more economical than alternative solutions. The landfill biodegradation processes are complex, including many factors that control the progression of waste mass to final stage. This paper examines several geotechnical parameters in landfill emissions, as a result of change in waste composition and management. The purpose of the research is to investigate and estimate the design and treatment of different solid waste properties in different landfill geotechnical behavior. Projections are made for landfill settlements in relation to different landfill mass's geotechnical properties. Also projections of landfill emissions' risk are made for several Greek sites. In the end, SIMGASRISK risk assessment model is applied to heat transfer in landfill mass taking into account the particular geotechnical parameters of waste materials.

1 INTRODUCTION

Sanitary landfill remains an attractive disposal route for municipal solid waste, as it is more economical method than other alternative waste disposal systems, ie incineration method. The landfill biodegradation processes are complex, including many factors that control the progression of the waste mass to final stage (Fleming, 1996; Kollias, 2004; Koliopoulos et al., 1999, 2002, 2004; Skordilis, 2001; Tchobanoglous et al., 1993).

Efficiently managed sustainable landfill sites can generate considerable volumes of methane gas (CH_4), which can be exploited by landfill gas recovery installations to produce electricity. According to the EU waste management policy and strategy separate collections will influence rates, yields, development of renewable resources technologies and exploitation of global amounts of landfill gas for several land uses. The increasing of the SWM recycling rates will influence the waste management systems; waste composition streams; costs and emissions from waste treatment and disposal activities.

A plethoric flow and use of resources characterise our society in an unsustainable way. Waste management is the discipline that is concerned with resources once society no longer requires them. A successful socioeconomic sustainable development requires a continuous change and harmonisation to the products' life cycle of our society, bearing in mind its current-future necessities in goods and waste management. Extensive sustainable programs in education and information to the public for active participation in recycling and waste minimisation projects are necessary. Therefore, the problem is transferred to the dilemma on how we can manage our waste better so as to avoid environmental pollution and associated hazards in public health (Koliopoulos, 1999, 2000). However, the effectiveness of all bioremediation techniques is depended on the degree by which the risk potential of a contaminated site is reduced. Hence, the selection of the particular bioremediation action technique is based on the type of the contaminant risk. Hazard is a property or situation that in particular circumstances could lead to harm. Risk assessment is an analysis of the potential for adverse health effects (Fleming, 1991, 1996; Koliopoulos et al., 2002, 2003, 2004).

2 SETTLEMENTS vs WASTE MASS PROPERTIES – SIMULATION RESULTS

In landfill design the knowledge of the long-term behaviour of landfill covers after closure is important for the performance of the capping system and future rehabilitation, development of the site. The majority of immediate settlement is due to mechanical mechanisms. Sowers (1973) estimated that these mechanisms are completed in 1 month. This can be explained due to the rapid dissipation of landfill gas and pore fluid through the voids while the waste permeability is high. This settlement is characterized as primary compression.

Primary compression is followed by long-term settlement which is due to volume reduction, mechanical mechanisms, creep phenomena, ravelling, physico-chemical changes and bio-chemical decomposition during the waste biodegradation stages. Decomposition depends on temperature, waste biodegradation and moisture conditions within the landfill (Fleming, 1990, 1996; Koliopoulos, 2000; Kollias, 2004; Skordilis, 2001; Tchobanoglous et al., 1993).

Sowers (1973) attributed the long-term settlement of refuse fills to secondary compression caused by decaying waste mass as a result of the physico-chemical and bio-chemical decomposition mechanisms which continue until the end of waste biodegradation. A primary factor in predicting landfill capacity is the refuse density. The prediction of final refuse density is difficult since the density depends on the waste type, water content and time since placement. The final refuse density influences the degree of settlement. However, apart from the change in refuse density, as the strain in a porous media increase, the permeability decreases affecting the waste mass hydraulic conductivity in time and related moisture regime, waste biodegradation, produced landfill gases and leachates.

An effective and simple approach for predicting the attenuation of landfill long-term settlements rates under self-weight has been developed by Coumoulos and Koryalos (1997). It is easy to be applied, based on monitoring of landfill surface settlements after construction due to secondary compression. The basic assumption of the examining approach is that long-term settlement of solid waste under self-weight can be approximated by a straight line when plotted against logarithm of time. If this assumption is valid for a site, then the vertical strain due to secondary compression under self weight, can be expressed by the following well known equation in geotechnical engineering:

$$\frac{\Delta H}{H} = C_a \log \frac{t}{t_1} \tag{1}$$

where:

ΔH settlement of the waste column, which occurs between t and t_1,

H   height of the waste mass in the landfill,

C_a coefficient of secondary compression of the waste, it is expressed as vertical strain per cycle of log time,

t, t_1 elapsed times on the secondary settlement curve after closure.

The slope of the curve expressed by equation (1) provides the vertical strain rate due to long term compression. Therefore:

$$\frac{d\frac{\Delta H}{H}}{dt} = \frac{0.434C_a}{t} \tag{2}$$

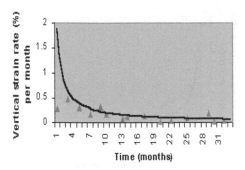

Figure 1.   Vertical strain rate v elapsed time since closure of MACH cell 1 (from Koliopoulos et al., 2002).

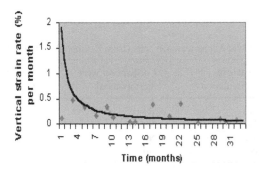

Figure 2.   Vertical strain rate v elapsed time since closure of MACH cell 2 (from Koliopoulos et al., 2002).

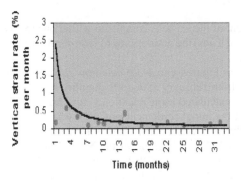

Figure 3.   Vertical strain rate v elapsed time since closure of MACH cell 3 (from Koliopoulos et al., 2002).

Equation (2) belongs to the category of a rectangular hyperbola which is asymptotic to the axis of time. The latter equation expresses the attenuation of vertical strain rate versus time, reflecting the settlement behaviour of the waste. The above formula has been applied for Mid Auchencarroch (MACH) experimental landfill cells (Koliopoulos et al., 2002). MACH is a UK Environment Agency (EA) and industry funded research facility. It has been capped since 1995. The experimental variables are waste pretreatment, leachate recirculation and co-disposal with inert material. The project consists of four cells each of nominal plan dimensions 28 m × 30 m and 5 m deep, giving a nominal volume of 4200 m³.

Mid Auchencarroch project attempts to develop and assess techniques to enhance the degradation, and pollutant removal processes for Municipal Solid Waste (MSW) landfill. The wet-flushing bioreactor batch landfill model is seen as the method of achieving the goal of sustainability. Figures 1 to 4 contain the attenuation curves of vertical strain rates of the four Mid Auchencarroch cells.

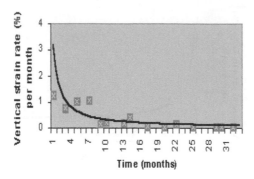

Figure 4.   Vertical strain rate v elapsed time since closure of MACH cell 4 (from Koliopoulos et al., 2002).

Table 1.   Data of Mid Auchencarroch cells (from Koliopoulos et al., 2002).

Site	Leachate recirculation	Co-disposal with inert material	Density t/m^3	C_a
Cell 1	Yes	Yes	1.19	0.034
Cell 2	No	No	0.74	0.026
Cell 3	Yes	No	1.03	0.033
Cell 4	No	No	0.73	0.044

Since primary settlement of the waste is completed within one to five months of the fill (either self weight or superimposed loads) the main source of long-term settlement would be the secondary compression of the waste.

The attenuation curves of Figures 1 to 4 use equation 2, based on the settlement field data of MACH experimental site from February 1996 to September 1997 (Koliopoulos, 2000). However, the values of C_a, which are calculated using equation 1, for Mid Auchencarroch cells are presented in Table 1. Assuming that the initial settlement of Cell 4 was indeed primary compression, then it is clear that the two pulverised waste cells, 1 and 3, have settled more than the untreated waste cells. However, NAVFAC DM7.3 (U.S. Dept. of the Navy, 1983) recommends values for C_a for refuse, ranging from 0.02 to 0.07 for landfills between 10 and 15 years. Oweis and Khera (1990) recommend values between 0.01 and 0.04. In this paper, the calculated C_a values, are consistent with C_a values cited in the literature (Sowers, 1973; NAVFAC, 1983; Oweis and Khera, 1990; Sharma and Lewis, 1994; Phillips et al., 1993; Jessberger and Kockel, 1993).

Continued settlement of the site is expected, however, the long-term secondary settlements of the fill can be accurately predicted based on the data developed from field investigation. A remedial crack repair work, in concern with future maintenance, will mitigate against future cracking, potential erosion of the fill and risk of landfill emissions' migration. A neural network analysis, taking into account particular geotechnical parameters, can be used for accurate prediction of settlements and associated maintenance. An efficient landfill design and the confrontation of landfill emissions' hazards and associated risks have to evaluate waste input synthesis's geotechnical properties, geological strata properties next to landfill boundaries, return period of floods, and seismicity of landfill region. An extended risk assessment in time is necessary, evaluating site's behaviour, so as to control and confrontate the migration of particular pollutants, protecting public health from hazardous emissions' toxic concentrations (Koliopoulos et al., 2003, 2004). In Figure 5, are presented risks of hazard at twelve Greek sites related to the seismicity of landfills' regions and respective disposed biodegradable waste fractions into examining sites.

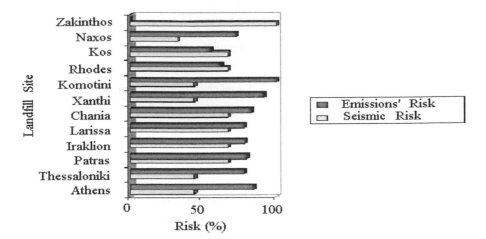

Figure 5. Risks of hazard at Greek sites.

3 HEAT TRANSFER AND GAS MIGRATION vs GEOTECHNICAL PROPERTIES – RISK ASSESSMENT

In landfill design, during the construction of geotechnical barriers or the installation of liners, account should be taken of magnitudes of landfill mass temperature, produced gas pressures and advection velocities of produced gas so as to avoid landfill gas migration and the associated harms. Landfill gas migration is related to the gas production; waste mass heat generation, heat transfer, waste biodegradation, geotechnical properties of the waste mass and the geological strata of the surrounded landfill area.

A risk assessment model of landfill gas emissions has been developed named SIMGASRISK, taking into account heat transfer, waste mass properties and geotechnical properties. Heat transfer modules in one dimension (1-D) and two dimensions (2-D) have been developed in SIMGASRISK and have been applied for different landfill conditions, presenting satisfactory results (Koliopoulos et al., 2004). Comparisons have been made between numerical outputs and field data, showing that SIMGASRISK is a robust and useful model for landfill design and diagnosis. The waste mass's porous medium is assumed as homogenous and isotropic. The general governing equation for the heat transfer (1D) in the waste mass is the following:

$$\frac{\partial U(y,t)}{\partial t} - \beta \frac{\partial}{\partial y}\left(\frac{\partial U(y,t)}{\partial y}\right) = \alpha \tag{3}$$

where
$\beta = k/\rho C_\upsilon$
k   thermal conductivity (kcal/day m °C)
ρ density (kg/m^3)
C_υ heat capacity (kcal/kg °C)
U   temperature on particular node of the grid in vertical location (°C)
t   time (day)
y   vertical distance in shallow landfill depth (m)
α source term (kcal/m^3 day)

Moreover, the lateral heat transfer in landfill boundaries is a 2-D problem and it is described by the following differential equation:

$$\frac{\partial U(x,y,t)}{\partial t} - \beta \frac{\partial}{\partial x}\left(\frac{\partial U(x,y,t)}{\partial x}\right) - \beta \frac{\partial}{\partial y}\left(\frac{\partial U(x,y,t)}{\partial y}\right) = \alpha \tag{4}$$

where

$\beta = k/\rho C_\upsilon$

k	thermal conductivity (kcal/day m °C)
ρ	density (kg/m^3)
C_υ	heat capacity (kcal/kg °C)
U	temperature on the particular node of the grid in particular x, y location (°C)
t	time in days
y	vertical distance in depth (m)
α	source term (kcal/m^3 day)

The latter differential equation is solved numerically in SIMGASRISK so as to calculate the tie-temperature regime of geotechnical barriers (i.e. clay, bentonite etc.) next to landfill boundary. This equation has been applied for shallow sequential batch bioreactors. Equation (4) has been applied for 1 m width (x axis) in clay porous medium properties and at height equals to landfill depth (y axis). The porous medium of clay or other examining geotechnical barriers is assumed as homogenous and isotropic. The latter assumption explains the reason for the application, in the heat transfer module, of a two-dimension numerical solution in SIMGASRISK and not a three-dimension one. Numerical solution of equations (3) and (4) is made by SIMGASRISK modules so as to proceed in the calculation of landfill gas advection velocities in time. The next stage follows, covering landfill's associated risk assessment.

SIMGASRISK develops a primary risk assessment for lateral LFG migration based on the produced LFG pressure taking into account the mid-depth waste mass temperature and biogas generation. Also SIMGASRISK develops a secondary risk assessment for lateral LFG migration from landfill boundaries. It is focused on the calculation of biogas migration advection velocity taking into account the particular source and pathway risk factors like heat generation-transfer, permeability and porosity of the porous medium. SIMGASRISK can be applied easily to particular landfill cases, giving satisfactory results as a diagnostic tool for future effective designs. The most important parameters for LFG generation and heat generation were selected on the basis of the waste input materials' characteristics (biodegradation, moisture content, thermal properties of each waste material). In Figure 6 is presented the flow chart of SIMGASRISK model.

SIMGASRISK is a dynamic numerical model as it calculates the followings: (i) LFG peak production for wet/dry site conditions; (ii) Peak Temperatures for different waste inputs; (iii) Risk assessment of lateral landfill gas migration based on the source and pathway risk factors of the examining site; (iv) Heat transfer vertical & lateral distribution at shallow landfill bioreactor during its life cycle; (v) Calculation of LFG advection velocity; (vi) Threshold distances vs methane explosive levels in time; (vii) Risk assessment of different waste inputs-management vs heat emissions-LFG emissions. SIMGASRISK uses several analytical and numerical solutions so as to solve the particular differential equations of heat transfer and lateral LFG migration from landfill boundaries. The heat generation source term α has been found that it depends on waste density, landfill gas density and biodegradation parameters as it is presented in equation (5) (Koliopoulos et al., 2004).

$$\alpha = D_{LFG} D_{waste} Gt\Omega \tag{5}$$

where;

α		heat generation source term at landfill mid-depth (kcal/m^3 day)
$D_{LFG} = 1.18$		average LFG density (kg LFG/m^3 LFG)
D_{waste}		waste density (kg/m^3)
Gt		LFG production in time (m^3 LFG/1,000 kg waste day)
$\Omega = \mu e^{-lt}$		μ heat generation (kcal/kg LFG), l biodegradation rate (day^{-1}), t (day)

Long-term exponential heat generation curves, have been developed in SIMGASRISK providing satisfactory results with field data (Koliopoulos et al., 2004). The development and calibration of heat generation curves were based on Mid Auchencarroch experimental four cells and Tokyo

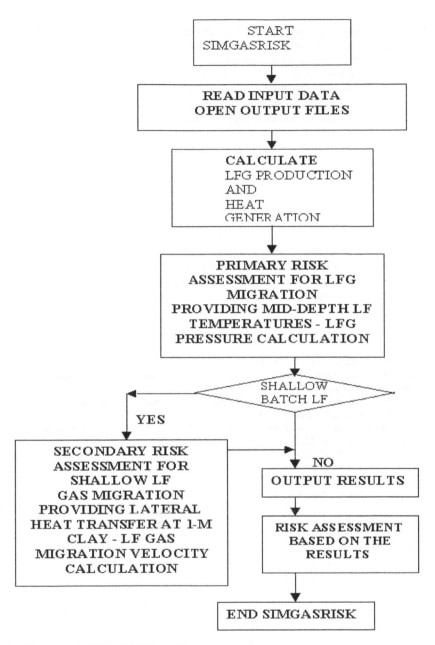

Figure 6.   Flow chart of SIMGASRISK model.

Metropolitan landfill field data, taking into account different waste input material characteristics, waste management techniques and geotechnical properties.

4   CONCLUSIONS

Values of C_a parameter for the attenuation curve of landfill settlements for the four Mid Auchen-carroch cells have been presented and were generally within the ranges reported in the literature.

Whilst these parameters can be used to simulate settlement rates at other sites with similar characteristics, it is important to validate model simulations with field observation data. Future efficient landfill designs should take into account all the geotechnical parameters and waste material properties which were analysed in this paper. Factors and records about waste composition, unit weight, water content, waste mass temperature, settlements and fluctuations of leachate level in time are necessary for the improvement of mathematical modelling-prediction of landfill emissions, life cycle analysis of waste material and efficient confrontation of landfill behaviour and exploitation of landfill emissions for several land uses.

Risk assessment estimations of several environmental pollution subjects, must be site specific taking into account the geotechnical parameters of waste input properties and geological strata regime next to landfill boundaries, yet no single preferred method is available. SIMGASRISK can be used as a risk assessment diagnostic tool for site specific case studies. Both primary and secondary risk assessments in SIMGASRISK are focused on the quantification of source and pathway risk factors. The evaluation of risk factors is necessary so as to take additional measures, like timing of risk assessment and communication of results. The timing of risk assessment is promoted by recommending that it should be undertaken in parallel with frequent in situ measurements in time, landfill emission's monitoring and have to prevent excessive or repeated data collection. A monitoring network would be useful to provide available data for a risk assessment and communication with the responsible authorities.

REFERENCES

Coumoulos, D.G., Koryalos, T.P. (1997). *Prediction of attenuation of landfill settlement rates with time.* Proceedings XIV International Conference on Soil Mechanics and Foundation Engineering, Hamburg, 6–12 September, Vol. 3, pp. 1807–1811.

Fleming, G. (1991). *The marginal and derelict land problem*, in Recycling derelict land, Thomas Telford, Institution of Civil Engineers, London, UK.

Fleming, G. (1996). *Hydrogeochemical Engineering in Landfills*, Geotechnical Approaches to Environmental Engineering of Metals, Rudolf, R. (ed.), Springer, 183–212.

Jessberger, H.L., Kockel, R. (1993). *Determination and assessment of the mechanical properties of waste materials.* Proceedings Symposium Green 93, Geotechnics Related to the Environment of waste materials, Bolton, U.K., pp. 313–322.

Koliopoulos, T., Fleming, G., Skordilis, A. (1999) *Evaluation of the Long Term Behaviour of Three Different Landfills in the UK and in Greece*, in Proc. Sardinia 99, 7th Int. Waste Management and Landfill Symposium, S.Margherita di Pula, Cagliari, Eds. Christensen, T., Cossu, R., Stegmann, R., Italy, vol. I, 19–26.

Koliopoulos, T. (1999). *Sustainable Solutions for the Most Pressing Problem within Solid Waste Management*, International Solid Waste Association Times Journal, Copenhagen, Denmark, **3**, pp. 21–24.

Koliopoulos, T. (2000). *Management and Risk Assessment of Mid Auchencarroch Landfill, Scotland*, Researchers' Conference, E.A., International Water Association, Environment Agency Headquarters, Trentside, Nottingham, England, U.K.

Koliopoulos, T. (2000). *Numerical Modelling of Landfill Gas and Associated Risk Assessment*, PhD Thesis, University of Strathclyde, Dept. of Civil Engineering, C.E.M.R., Glasgow, Scotland, U.K.

Koliopoulos, T., Fleming, G. (2002). *Prediction of long-term settlement behaviour of M.S.W. landfills – Mid Auchencarroch*, I.S.W.A Congress, Proceedings vol. 2, pp. 1067–1076 (Ed. G. Kocasoy, T. Atabarut, I. Nuhoglu), Istanbul, Turkey.

Koliopoulos, T., Kollias, V., Kollias, P. (2003). *Modelling the risk assessment of groundwater pollution by leachates and landfill gases*, U.K. Wessex Institute of Technology, (Ed. C.A. Brebbia, D. Almorza, D. Sales), pp. 159–169, W.I.T. Press.

Koliopoulos, T., Fleming, G. (2004). *Modelling the biodegradation of treated and untreated waste and risk assessment of landfill gas emissions – SIMGASRISK*, Proceedings of the European Symposium on Environmental Biotechnology, ESEB 2004 (Ed. W. Verstraete), pp. 625–628, Oostende, Belgium, Balkema publishers.

Kollias, P. (2004). *Solid Waste*, Athens, Greece.

Skordilis, A. (2001) *Controlled Non-Hazardous Waste Disposal Technologies*, Athens, Greece.

NAVFAC (1983). *U.S. Dept. of Navy, Soil Mechanics Design Manual 7*, Naval Facilities Engineering Command, Alexandria, Va, U.S.A.

Oweis, I.S., Khera, R.P. (1990). *Geotechnology of Waste Management*, Butterworths and Co Publ. Ltd, London, chapter 10, pp. 177–213.

Phillips A.B., Walace K., Chan K.F. (1993). *Foundations for reclaiment landfill sites*. Proceedings Conference on Geotechnical Management of Waste and Contamination, Sydney, NSW, R. Fell, T. Phillips, G. Genard (eds.), pp. 185–208.

Sharma, H.D., Lewis, S.P. (1994). *Waste Containment Systems*, Waste Stabilization and Landfills,John Wiley and Sons, Inc., New York, 1994.

Sowers, G.F. (1973) *Settlement of waste disposal fills*. Proceedings 8th International Conference on Soil Mechanics and Foundation Engineering, Moscow, Vol II, pp. 207–210.

Tchobanoglous, G., Theisen, H., Vigil, S. (1993). *Integrated Solid Waste Management*, McGraw-Hill Book Company, New York, USA.

Hydraulic design of capillary covers to accommodate scatter in infiltration volumes

S.A. Reyes
R & A Consulting Engineers Ltd, Bahia Blanca City, Argentina

ABSTRACT: Capillary barriers acting as covers have been used to isolate different kind of wastes. Depending on the waste nature they may give more or less successful performance. Some low level radioactive nuclear waste disposal sites (in desert areas) have been provided with capillary cover barriers as final ones. However, unsaturated conditions (cover desiccation) rendered them unable to avoid vapour emissions (Richardson, 2003; Porro and Keck, 1998). Household waste landfills may benefit from capillary layers (CLs) in the early stages of disposal and some time after landfill closure. They may be designed to provide a certain volume of water infiltration into the waste mass thus enhancing organic fraction biodegradation. Protective layers (PLs) have been argued to provide water retention capacity on top of the capillary one (Jelinek and Amann, 2001) and loamy soils have been stated as a suitable material type for such a purpose. However, protective layer desiccation (fissuring) or severely unsaturated conditions (also at the CL) allow methane and carbon dioxide venting through them (Reyes, 2003). Hydraulic design of capillary layers usually relies on their diversion capacity and, provided the infiltration volume is correctly estimated, values of maximum CL (allowable) lengths may be calculated (Reyes, 2003). This paper discusses some aspects of hydraulic cover behaviour that may lead under certain climatic conditions to avoid using loamy soils as a PL material. Furthermore PLs may be avoided altogether except when placed in regions which flood regularly and after landfill closure.

1 INFILTRATION PATTERNS

Loamy soils invariably consist of a varying content of plastic fabric. They desiccate and fissure during relatively rigorous evapo-transpiration periods. They thus allow large volumes of water into the CL and possibly past it into the capillary block (CB). On the other hand, designing against finger flow patterns becomes extremely difficult, since fissures concentrate flow into preferred paths. It should be remembered that each time the infiltration rate becomes smaller than the hydraulic conductivity of the CL (at the water entry value of the capillary one), flow becomes unstable, initiating finger flow. This finger flow, if regular, may lead to significant leachate volume generation in the long run (even under relatively low precipitation regimes). Fissuring may promote finger flow even under infiltration rates not smaller than the hydraulic conductivity of the sub-layer, due to already existing preferred flow paths (Reyes, 2003).

A major issue is the unusual scatter in precipitation values (stormy events) that has been taking place during the last decade which makes it very difficult to design PLs composed of loamy (or plastic) soils. In fact, an efficient hydraulic design for such opposite problems as excessive (rain) water surplus (hence infiltration into the CB) and finger flow is almost impossible for a single layer (whether a PL or not). Great scatter in precipitation values and most of all, their concentration in very short stormy events, cannot be efficiently accommodated with a single layer. Moreover plastic soils' fabrics are highly affected by evapo-transpiration.

Three or more non-plastic soil layers (CLs) one on top of the other may be designed to accommodate a maximum amount of infiltration each. The efficiency of such cover systems may be acceptable for regions where acute storm events do not regularly take place. Otherwise, large volumes of water may infiltrate in very short time periods. However, site-scale experimental work (Porro and Keck, 1998) has shown that, even under rigorous water infiltration rates, thick Capillary Layers are efficient in retarding infiltration and promoting evapo-transpiration. Even when compared to thick vegetative evapo-transpiration layers CLs behave better in decreasing percolation past themselves (Porro and Keck, 1998). Where the arrangement is a succession of CLs, once the diversion capacity of the top one has been exceeded the following sublayer (acting as a CL now) should start diverting the excessive flow volume. Now the CB function is relocated to the bottom (or next) layer. In this way the effects of precipitation scatter can be better managed. The interesting point is the water volume that may be managed during a storm event. In other words, what is the maximum water volume that, during a storm event, a succession of capillary layers may efficiently accommodate.

2 DESIRED BEHAVIOUR

In household waste landfills, for there to be waste biodegradation a certain amount of water (to keep an optimum value of moisture content) must be continually provided to the waste mass. A balanced water content helps indigenous bacteria to act as carbon receptors and degrader agents. However, too much water can halt this mechanism. The amount of water allowed to freely enter the waste mass is directly related to, bottom drainage system capacity, capillary layer and cover design.

It should be noted that landfill emissions have been stated as the fourth largest source of methane anthropogenic contributions (Stern and Kaufmann, 1996). Therefore landfill covers must play a key role not only in methane generation (related waste biodegradation), but also in controlling methane dispersion. Another important design issue is long term aquifer protection. Each time this is considered a balance between low and high infiltration cover design philosophies should be attempted (Rowe et al, 1997). According to Rowe et al high infiltration into the landfill during the operational phase and some time after waste disposal ceases may be allowed. Then, after the landfill has been largely stabilized, a low infiltration cover should be provided. Thus, the performance of the cover should limit the volume of infiltrated water while distributing it and homogenizing it into the waste mass. Probably the maximum infiltration should correspond to the drainage capacity of the bottom drainage system. In this way bioreactor processes may be enhanced and landfill cells will not be flooded.

A capillary layer constitutes a landfill capping option although as stated only for regions where precipitation follows a regular annual non highly stormy pattern. Some authors (Jelinek and Amann, 2001) propose shielding the capillary layer by a protecting one of loamy material to reduce high peaks of precipitation and promote a water balancing effect. Particle size and layer thickness constitute major design parameters for Capillary Barrier Covers (Shackelford and Nelson, 1996). Both the capillary layer and retention or Protective Layer (top, PL) must be designed not only to divert excessive water but, on the other hand, to retain it during high evapo-transpiration periods. This will minimize the extent of the drying period acting on the CL and thus minimize methane emissions through the top layer (Tremblay et al, 2001). In some cases this may not be naturally achievable because of precipitation intensity and its annual distribution, particularly under low humid to dry climatic conditions (Reyes, 2003).

3 CAPILLARY LAYER DESIGN

Ursino and Cossu (2001) have stressed the need to evaluate soils used for capillary barrier construction against finger flow. The scatter in infiltration (precipitation) values makes it very difficult to avoid finger flow, since it is likely to take place each time infiltration (precipitation) is of relatively low magnitude. However, if the PL on top of the CL is kept free of fissures then infiltration values arriving at the PL-CL interface may be diminished (and homogenized). Needless to say, it has always been very difficult to keep PLs made from plastic materials free of fissures. The main reason being crust desiccation and, depending on the extent of the drying period, fissures may develop over a wide range of depths.

Each time the CL is overflowed by excessive infiltration due to unusual precipitation intensity and concentration, water will travel past the interface into the CB. Thus, the CL must be designed to avoid excessive flow into the CB. Because of this phenomenon there is a threshold value for water infiltration volume over which Capillary Barrier Covers are no longer efficient.

3.1 Diversion capacity

The diversion capacity of a capillary layer can be estimated from the following expression:

$$Q_{MAX} = K_s T_h \tan \theta \qquad (1)$$

where, T_h is the thickness of the capillary layer, θ is the slope angle of the capillary layer, K is the saturated conductivity of the capillary layer.

The ratio between the diversion capacity and the infiltration rate corresponds to the unit area (area per unit width of capillary layer drainable at maximum flow):

$$\frac{Q_{MAX}}{I} = A_{unit} \qquad (2)$$

Hence, for a given value of T_h the A_{unit} corresponds to the maximum drainable surface per meter width that avoids or prevents water from entering the capillary block (by excessive infiltration and not by finger flow). This value of drainable surface (per metre width) also corresponds to the maximum theoretical slope length (MSL) that should avoid water flow into the capillary block. A larger length requires a thicker capillary layer in order to restrict flow to its limits. A parametric analysis can be performed to evaluate the limiting values to the dimensions (thickness and length) of the capillary layer. Then for a given material and slope value, it is possible to establish maximum allowable distances between drainage ditches so as to avoid or minimize flow into CBs.

Usually volumes of daily or monthly infiltration are considered to perform an annual water balance across the PL (on top of the CL). Then, average infiltration daily volumes are obtained and considered to calculate A_{unit} (equation 2). This type of design approach has been already performed (Reyes, 2003). It has been shown that for regions where precipitation does not present regular storm events CLs constitute a valid option. In fact, it has been shown that for cases where an efficient PL can be provided as an attenuating layer, the average annual infiltration rate through it can be efficiently accommodated by the CL. However, more realistic (non-annual averaged) infiltration (precipitation) values must be considered. Instead, peak precipitation patterns during unusual storm events, e.g. 30 mm in half an hour or 100 mm/day for South-Eastern Buenos Aires province (Argentina), may be employed. Under such scenario, Figure 1 states hourly allowable infiltration values, i.e. the volume of drainable water directly infiltrated and laterally diverted by a capillary layer, for a succession of CLs. A succession of three CLs with K_s varying from 1×10^{-2} cm/s to 1×10^{-4} cm/s at 10% inclination with 0.3 metre thickness and 30 metres length has been considered.

It can be observed that the allowable (drainable) volume does not match the worst case scenario (or storm event previously stated). In fact, the accumulated hourly allowable volume for a succession of three CLs amounts to 0.4 mm which corresponds to a daily maximum precipitation of 9.6 mm, clearly insufficient. We then may apply, an increment to the thickness of the three CLs up to half a metre and increase the slope to 50% (26.56°). Under this scenario, the allowable hourly volume of precipitation increases up to 3.3 mm taking the daily allowable volume up to almost 80 mm (Figure 2). Now, the daily volume seems to be high enough for several regions in the country. This precipitation value can still be increased due to consideration of run-off in the above analysis. This would probably add a further small percentage, probably taking the daily maximum precipitation up to nearly 100 mm/day. This last value is fairly reasonable for non-highly rainy regions. However,

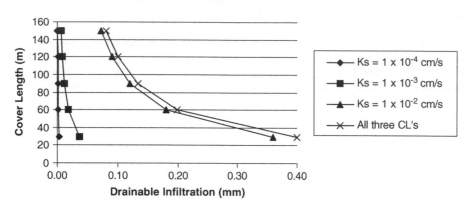

Figure 1. Maximum hourly drainable infiltration (precipitation) height capillary layers 0.3 m thick at 10% inclination.

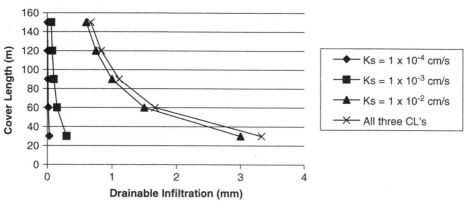

Figure 2. Maximum hourly drainable infiltration (precipitation) height capillary layers 0.5 m thick at 50% inclination (1V:2H).

for other places where storm events can produce up to 30 mm in twenty to thirty minutes, this arrangement of CLs seems inadequate.

In addition, by increasing the thickness of the bottom layer we may reach values of drainable (infiltration) volumes even higher. For instance, we may match the experimental values at Idaho National Engineering and Environmental Laboratory where a scaled capillary block of over 75 cm thickness has been thoroughly instrumented. In fact, this may be a sound choice, as the bottom capillary block will only act either as a CB or in the end as a drainage layer.

If for instance we choose a metre thickness for the lower CB then recalculating drainage capacity of the combined CLs gives a total infiltration value (Figure 3) of 6.33 mm/hour or 152 mm/day (for 30 metres maximum slope length or drainage ditch spacing). Note that cover length may also be interpreted as drainage ditch spacing placed perpendicular to the slope within the CLs.

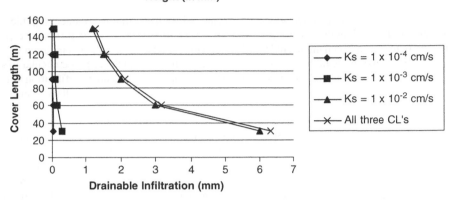

Figure 3. Maximum hourly drainable infiltration (precipitation) height (two capillary layers 0.5 m thick at 1V:2H and bottom capillary block 1.0 m thick).

Table 1. Maximum Drainable Area (m^2) before flow into CB for a single 0.3 m thick capillary layer at 10% slope.

	Saturated conductivity capillary layer K_s (cm/s)	
Annual Total Percolation (mm) (ATP)	1,0E-10^4	1,0E-10^3
13	72,78	727,75
40	23,65	236,52
80	11,83	118,26
120	7,88	78,84
150	6,31	63,07
215	4,40	44,00

4 DRAINAGE DITCH LAYOUT

Table 1 shows Maximum Drainage Area (MDA) values for a 0.30 m thick capillary layer, for two values of hydraulic conductivity and several Annual Total Percolation (ATP) values. The minimum and maximum ATP values are 13 mm and 215 mm respectively (calculated from the last ten years precipitation records in the South-Eastern part of Buenos Aires province). Each value of the MDA corresponds to a 1 m width section of the capillary layer, thus the MDA also states the maximum slope length that avoids flow into the CB.

It is obvious to the reader that, if precipitation records result in unusual values for short time periods, a significant increase of percolation volume into the CL may render the proposed layout inadequate. Drainage ditches (DD) should be provided to diminish the diversion length and thus collect water before it goes into the CB. Yet, in cases where severe storms take place in very short periods placing DDs closer together may not be enough. Under this scenario, the amount of water allowed through the capillary block may be excessive and thus uncovered combined CLs successions may be abandoned as a valid option. If this last is not the case then MDA values and allowable slope lengths for each CL in the proposed succession scheme should be equally calculated and designed. This can also be obtained from Figures 1 to 3 – ordinate values correspond to allowable cover length values (landfills already constructed cases). Clearly, under this scenario, the DD spacing will differ for each capillary layer. Designing for multiple values of DD spacing may render construction and thus side (collection) sump placement more efficient.

5 CONCLUDING REMARKS

A parametric analysis has been performed by varying values of hydraulic conductivity, thickness, slope inclination and length of capillary layers in order to maximize their diversion capacity. An arrangement of combined uncovered CLs has been proposed to cope with precipitation volumes not amounting to extremely high flow volumes. Drainage ditches which are perpendicular to slope direction within the Capillary Barrier Cover have been proposed to attenuate infiltration volumes into capillary blocks. These drainage ditches have proved a sound solution to capillary layers design in regions where precipitation does not exceed values of 150 mm/day. However, there is an important drawback in cases where combined CLs are chosen as a cover system. In fact, designing them for a given storm event, e.g. 100 mm/day, implies that patterns lower than the design value would lead to non-saturated conditions and they may not behave as well with finger flow phenomena. However, total volume due to finger flow is more regularly distributed (due to its low value). Therefore, if it is not excessive during long-term periods (for instance, yearly), it may be assumed to provide the necessary water content to enhance bioreactor type of behaviour (household waste landfills only). Biogas dispersion through the cap in uncovered combined CLs may be of concern. In the author's opinion CLs (single or in multiple layouts) should not be used in the long-term as a final cap unless covered with an efficient non-plastic fabric PL.

PLs seem necessary to avoid direct exposure to precipitation in flooded stormy regions. This is despite the good performance (Porro and Keck, 1998) that scaled, instrumented experimental CLs showed when subjected to very high water infiltration values (of up to 700 mm in few days). In fact, this high value of infiltration would not have been achievable if the CL had not been covered (Porro and Keck, 1998). This covering may be achieved by providing a vegetative non plastic soil layer on top of the Capillary Barrier Covers.

An arrangement of combined CLs sitting one on top of another (three at least) can accommodate precipitation in regions not affected by acute storm events. In this way, persistent precipitation not corresponding to extreme rates may be efficiently managed. By so doing high infiltration rates may be diverted, particularly by placing DDs at convenient positions perpendicular to the slope of the CLs. Simultaneously, a given design value of water flow may be allowed past CLs into the waste mass enhancing bioreactor-type behaviour. This last option allows design for even higher values of allowable infiltration volumes during the operational phase (household waste landfills) – this is the stated approach of Rowe et al (1997) with regard to cover design philosophy. Thus, a capillary barrier may be designed to allow a given value of water infiltration past the CB into the landfill cell (enhancing bio-reactions). This should be done with strict consideration to a maximum allowable height of leachate into landfill cells. To achieve this, we need to know the value of the maximum drainable area that avoids flow into the CB. In this way, we can design the capillary layer to divert a given (design) infiltration volume and allow the rest into waste cells.

Finally, protection between adjacent CLs against finer fraction cross contamination can be achieved by interface geosynthetic placement. Simultaneously, care should be taken not to decrease the reflection effect of capillary phenomena.

REFERENCES

Jelinek D and Amann P (2001) *Design and Construction of Capillary Barriers in Top Cover Systems for Landfills – Field Experience,* Proc Eight International Waste Management and Landfill Symposium, V III, October, pp345–354.

Porro I and Keck K (1998) *Engineered Barriers Testing at the INEEL Barriers Test Facility FY-1997 and FY,* INEEL Report No EXT 98-00964.

Reyes S (1997) *Fissured covers and Consequences in Some Latin American Landfills,* Proc 6th International Symposium on Landfill Design, Sardinia, Vol V, October, pp393–403.

Reyes S (1999) *Common Features in Argentinean Sanitary Landfills and Necessary Steps Towards Their Remediation,* Proc Eight International Waste Management and Landfill Symposium, Sardinia, Vol IV, October, pp617 622.

Reyes S (2003) *Infiltration Control To Enhance Bioreactor Behaviour By Use Of a Capillary Cover,* Proc Ninth International Waste Management and Landfill Symposium, Sardinia, October.

Richardson G (2003) Personal Communication.

Rowe RK, Quigley RM and Booker JR (1997) *Clayey Barrier Systems for Waste Disposal Facilities,* E& F, p389.

Shackelford CD and Nelson JD (1996) *Geoenvironmental Design Considerations for Tailing Dams,* Proc International Symposium on Seismic and Environmental Aspects of Dams Design, pp131–188.

Stern DI and Kaufmann RK (1996) *Estimates of Global Anthropogenic Methane Emissions 1860–1993,* Chemosphere, 33, pp159–176.

Tremblay M, Heroux M and Nastev M (2001) *Monitoring of Three Instrumented Final Covers at the City of Montreal's Landfill Site,* Proc Eight International Waste Management and Landfill Symposium, Vol III, Sardinia, October, pp374–384.

Ursino A and Cossu R (2001) *Combining Local Measurements Modelling and Scaling Theory of Capillary Barriers,* Proc Eight International Waste Management and Landfill Symposium, Vol III, Sardinia, October, pp335–344.

Geotechnical and Environmental Aspects of Waste Disposal Sites – Sarsby & Felton (eds)
© 2007 Taylor & Francis Group, London, ISBN 978-0-415-42595-7

Ground treatment by dynamic compaction in landfill materials

C.J. Serridge
Pennine Vibropiling Limited, Bacup, UK

ABSTRACT: The dynamic compaction (DC) technique has been applied to landfill materials within the UK for some 25–30 years. However, its application is strongly influenced by the age (and composition) of the landfill, together with the nature of the proposed development or end use. For landfills containing a significant domestic refuse content, DC has largely been restricted, in these situations, to highway embankment schemes and surface parking areas. Within continental Europe and the USA, DC of Municipal Solid Waste (MSW) is increasingly being recognised as important in landfill operations, since increasing the density of MSW increases the storage capacity of the landfill, and by inference, the integrity of the final cover system. Some applications and discussion of the DC technique in the above contexts are presented. A novel application of where DC has been used to improve the geotechnical properties of a loose, inert, sand fill deposit in former sand and gravel workings prior to development of a new landfill cell in Suffolk (UK), is also presented.

1 INTRODUCTION

Within the UK, high population density and increased traffic volumes has led to the expansion of urban areas which have encroached on former landfill sites where refuse is present in significant quantity. This has necessitated the use of ground treatment techniques, such as Dynamic Compaction (DC) to improve the bearing capacity and settlement characteristics of these soils prior to site development.

Waste in landfills comprises a heterogeneous mass of material that varies widely according to location, time, provenance (and indeed from country to country, dependent upon socio-economic circumstances and climate). The typical composition, by weight, of household refuse in the UK over the past 50 years is illustrated in Figure 1.

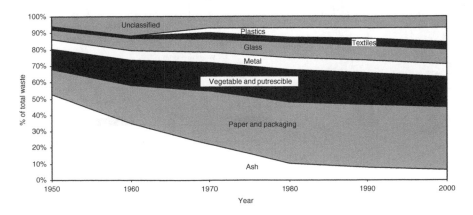

Figure 1. Composition of UK domestic waste by weight (after Sarsby, 2000).

The 1956 Clean Air Act resulted in a major reduction in the use of solid fuel, and hence in the ash content of refuse (it fell from over 50% in 1950 to around 8% in 1990). The loss of this inert, denser constituent and the increase in volume of paper and packaging, vegetable matter, rag/textiles and plastics, has a significant effect on the biological and geotechnical properties of refuse fill. Recent domestic refuse contains a high proportion (around 55%) of organic material. Commercial and mixed wastes can contain even higher proportions of organic material. It is important to recognise, however, that as wastes are progressively minimized, re-used, recycled, processed and recovered, (in response to European Union (EU) directives), the characteristics of waste received at landfill sites in the UK are likely to further change.

Slocombe (1993) distinguishes three main eras of landfill within the UK, which have different suitability for the support of structures. He describes the early (pre 1940's) domestic refuse (typically burnt ash and cinders, before collection and landfilling), as suitable for consideration of the support of structures following ground treatment, e.g. DC. Around the 1940's and 1950's the materials tended to be not as well-burnt and with minor proportions of material still liable to long term degradation. These materials are described as suitable for treatment to buildings that are tolerant to some differential settlement movements such as lightweight industrial units. The more recent refuse is not burnt, but collected in polythene bags. The degradable component is high, and as such there is potentially a high risk for excessive movements occurring as a result of long-term decay for most structures. However it is indicated that road embankments, surface parking areas, services and utilities can benefit from treatment of these younger fills. It is important to recognise that each site should be assessed individually in respect of ground treatment application.

2 DYNAMIC COMPACTION

Dynamic compaction (DC) has been applied to older landfills in the UK for some 25–30 years. The technique employs the controlled repeated application of high-energy impacts on the ground using steel or concrete tampers (weighing typically 8–15 tonnes), from heights typically in the range 15–25 m (Figure 2). The tampers are dropped repeatedly at each impact point on a pre-determined pattern over the treatment area. Depending on soil conditions, two or more treatment

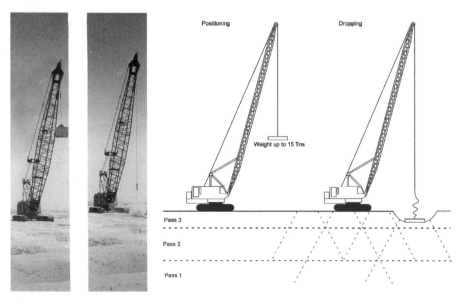

Figure 2. Dynamic compaction operation.

passes are applied on a progressively closer grid pattern until the required ground improvement is achieved. Further information on the technique is given in Greenwood and Kirsch (1983), Mayne et al (1984), Lukas (1986) and BRE Report 458 (2003).

There is a correlation between depth of compaction and the energy per blow for falling weight dynamic compaction. This has typically been expressed as:

$$Z_e = K(MH)^{0.5} \tag{1}$$

where Z_e is the depth of compacted soil in metres, M is the mass of the falling weight in tonnes, H is the height of fall in metres and K is a coefficient which depends on fill type. Mitchell (1981) and Mayne et al (1984) have presented summaries of data from many sites which demonstrate $0.3 < K < 0.8$. In the case of Municipal Solid Waste (MSW), Van Impe and Bouazza (1996) report that the K value can vary from $K = 0.35$ representing a lower limit for older landfills to about $K = 0.65$ representing an upper limit for younger landfills.

3 GEOTECHNICAL BEHAVIOUR AND TREATMENT OF LANDFILL WASTE

In the UK successful redevelopment of old landfill sites relies heavily on predictive modelling of engineering behaviour, notably long-term settlement characteristics. Field observations and measurements are particularly important for assisting the development of methods for estimating landfill settlements. Whilst it is not possible within the scope of this paper, to consider in detail the complex settlement characteristics of landfill wastes, useful commentary on these aspects is provided in Edil et al (1990), Powrie et al (1998), Watts and Charles (1999) and Sarsby (2000) among others.

3.1 *Landfill settlement*

In summary, the main causes of settlement in landfill waste deposits are typically defined as being due to:

- *Mechanical compression,* resulting from crushing, distortion, re-orientation of the waste particles as vertical stresses are increased, either during compaction or under self-weight as further material is deposited on top.
- *Degradation,* attributed to biological decomposition of organic wastes and physio-chemical change such as oxidation, corrosion and combustion.
- *Ravelling,* whereby finer particles gradually migrate into the larger voids and which can occur both during both mechanical compression and degradation.

Waste settlement is conventionally classified as either primary or secondary, depending on the timescale over which it occurs, Powrie et al (1998). Such settlement will be influenced by the waste composition, initial bulk density and moisture content, the level of leachate within the fill and also the degree of compaction achieved at the time of placement. Where former landfills have been redeveloped, ground treatment techniques have been used to improve geotechnical properties, with the principal aim/objective of reducing compressibility and post-construction settlements.

In the context of DC it is important to make a distinction between older landfills and more recent (younger) landfills, when considering ground response and long-term settlement after treatment with this technique. Slocombe (1993), reports that energy input for MSW can be some 50–100% higher than energy regimes typically applied to inert soils. Van Impe and Bouazza (1996) undertook a review of published data on DC of MSW landfills, which is presented in Figure 3. This clearly demonstrates that enforced settlement depends on the age of the landfill. Van Impe and Bouazza (1996) also recognise that depth of influence of treatment and compaction behaviour is also influenced by age.

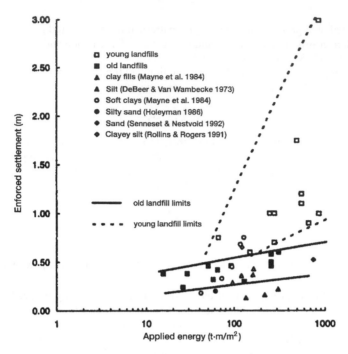

Figure 3. Enforced settlement vs. applied energy for landfill wastes and other types of soils (after Van Impe and Bouazza, 1996).

Table 1. Treatment of older and younger MSW landfills.

Site location in UK	Completion date of landfilling (year)	Age of landfill at time of treatment (year)	Depth of refuse (MSW) (metres)	Applied energy (t.m/m^2)	Induced settlement (metres)	Volume reduction (%)
East London[1]	1935	40	6.5	260	0.58	8
Redditch[1]	1960	15	6.0	260	0.50	10
Hertfordshire[1]	1960	15–20	8.0	260	0.50	10
North Shields[2]	1971–1981	15–25	5.0	92	0.46	10
M25, Bell Lane[3]	1986–1990	7–10	4.0–5.0	281	0.7–1.00	17.5–20

Sources: 1. Charles et al. (1981); 2. Mapplebeck and Fraser (1993); 3. Perelberg et al. (1987).

3.2 Old landfills (more than 15–20 years old)

In the UK, decomposition is generally cited as taking more than 25 years to occur. Hence for older landfills around or greater than this age, organic decomposition is likely (dependent upon site specific circumstances), to have largely taken place, leaving a landfill material usually consisting of a dark coloured soil matrix containing varying amounts of bottles, metal fragments, wood and debris. This tends to be relatively inert and inherently a better engineering material (generally having had a higher ash content when placed) compared to younger landfill. The material will also have densified to some extent due to self-weight settlement. The observed response to compaction is similar to inert soils, with enforced settlements in the landfill, typically in the range 5–10% (Table 1), of the treated depth. Densification as a result of treatment results in higher unit weight, with little long-term subsidence under load, by virtue of the lack of significant degradable materials.

3.3 Young landfills (less than 15 years old)

For more recent (younger) landfills, where organic decomposition is still underway, the main objective of the DC is to collapse the voids due to degradation (and the presence of, for example, drums and car bodies), and compact the remaining more inert constituents into a denser matrix, in the case of essentially granular soils, and close up void spaces between constituent lumps of fill, in the case of any cohesive soil strata, thereby reducing the magnitude of surface deformations.

Consequently, younger landfills show generally higher enforced settlements (Figure 3, Table 1), typically in the range 10–25% of the treated landfill thickness. It is important to recognise that the DC technique should not be considered as a method of overcoming long-term decay of degradable constituents. However, from an environmental standpoint, it is perhaps also important to recognise, that in landfills containing hazardous substances, e.g. chemicals, asbestos etc, the major advantage of DC over alternative methods is that greater control can be exercised, to avoid exposure of hazardous material to the atmosphere, while facilitating compaction of the soil at depth.

4 TREATMENT OF "OLD" LANDFILLS IN THE UK

Charles et al (1981), Charles (1991) and Watts and Charles (1999) provide an example of where DC was utilised to improve the bearing capacity characteristics of an old domestic refuse (MSW) tip associated with the development of a dual carriageway highway, with interchanges and slip roads at Redditch (UK). The domestic refuse was approximately 5–6 m deep and typically comprised black ash and clay with fragments of brick, woods, rags, plastic bottles and metal. Bulk densities were of the order of 1.8 Mg/m^3 and moisture contents around 30%. Standard penetration tests (SPT's) suggested the fill was of loose to medium density. Underlying natural strata comprised stiff Marl (Mercia Mudstone). The fill materials had been in place for about 15 years at the time of treatment in 1975.

Treatment was carried out using a 15 tonne weight dropped from a height of 20 m onto a granular working blanket producing 1 m deep craters. The average energy input was 260 t.m/m^2. Post treatment levelling indicated that treatment produced a 10% volume reduction in the 5–6 m deep refuse. Construction of a 3 m high slip road embankment produced immediate settlement of 20 mm in the treated refuse, whilst settlement of untreated refuse was 3 times greater under a similar load. Refuse settlements between 60 mm and 80 mm have been measured over an 18-year period following road completion, Figure 4. Although treatment did not eliminate long-term movements (creep settlements), it is understood that the performance has been satisfactory and within serviceability limits. Further case histories are cited in Table 1.

5 TREATMENT OF "RECENTLY-PLACED AND YOUNG" LANDFILLS

5.1 Waste compression

Compression of newly landfilled waste occurs both during placement (due to machine compaction and overburden effects), and in the long-term (as a result of processes such as degradation and ravelling). Within the UK, reliance is generally based on the use of steel-wheeled compactors and self weight settlement for compaction. It is important to recognise that modern compactors have a mass ranging from about 10 to 40 t and are able to achieve significantly higher densities than in the past. Waste fill has a density varying from about 0.12 to 0.13 t/m^3 when tipped. After compaction using steel-wheeled compactors, bulk densities typically range from a low of 0.4 t/m^3 to an upper limit of around 1.2 t/m^3, Sarsby (2000). It is further suggested by Sarsby (2000), that values in the range 0.65–1.0 t/m^3 may be assumed for analytical purposes and that a bulk density of about 0.8 t/m^3 seems to be the optimum for the biodegradation processes in mixed household waste.

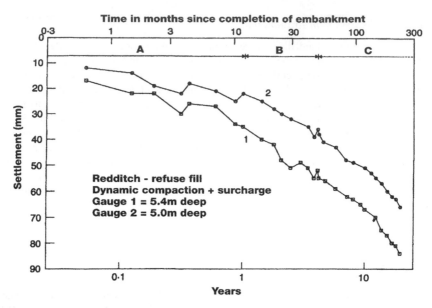

Figure 4. Settlement of surface of old refuse (MSW) following dynamic compaction and embankment construction (after Watts and Charles, 1999).

5.2 *Deep compaction of waste*

Internationally, very little data currently exists regarding the evaluation of the effectiveness of DC on recently placed MSW, including the provision of landfill space gains and improvement of the integrity of the final cover system. From a review of literature, Welsh (1983) describes the use of DC on a recently placed sanitary landfill in the USA. The treatment of this refuse produced enforced settlements of up to 2.5 m or 25% of the original fill thickness. Galante et al (1991) have also presented a case history of field trials, conducted to evaluate the effect of incorporating the DC technique into the normal daily operations for disposal of MSW, at a landfill facility in eastern Pennsylvania, USA. Intake of landfill during the trial comprised around 79% MSW, 16% dry municipal waste sludge and 5% miscellaneous non-hazardous waste, placed in around 2.5–3.0 m lifts. The field trials compared landfill performance for areas constructed with "normal" refuse placement and compaction procedures, utilising steel wheeled landfill compactors, to an area that had the DC technique applied in addition to the normal operation. In place unit weights of around 9.9–10.9 kN/m^3 were reported with steel-wheeled landfill compactors. With DC, in-place densities were increased to around 12.5 kN/m^3 for an optimum (saturation) energy input of around 300 t.m/m^2. In volumetric terms, this corresponded to an overall airspace gain of the order of 17–20%. It is stressed, however, that this is to some extent an upper bound or theoretical maximum which may not be achievable in practice on an active landfill due to operational constraints restricting application of optimum energy inputs. The average airspace gains generally produced by DC in the trials were reported to be in the order of 8%. In a production situation, for a suitable size and layout of cell, an improvement of around 1% is cited as being practical. Hence, notwithstanding the operational constraints of the DC equipment/plant, there has perhaps been little incentive for drop weight DC to be considered in the UK for newly placed MSW.

The usefulness of deep compaction methods such as DC, both as a means to improve the quality and rehabilitation of disposal areas and for extending the life of younger landfills by increasing the density of MSW, is being increasingly recognised, particularly in Continental Europe and the USA. Differential settlements are also reduced which is important for the integrity of the final cover system. Useful research/case histories have been published by Singleton (1996), Van

Impe and Bouazza (1996) among others. Singleton (1996) describes a novel approach to MSW landfill space gains, using a combination of insertion of depth vibrators (vibroflots) and DC. This combination produces a resultant improvement in geotechnical behaviour of the landfill materials, which is described as particularly beneficial for the final sealing measures/integrity of the final cover system. The use of vibroflots in this context is described as the "refuse depth compaction technique" and involves the insertion of a depth vibrator into the landfill (typically on a grid pattern ranging from 2 to 3 m). A displaced open shaft is formed which is immediately filled with introduced refuse material. The depth vibrator is then utilised to laterally vibrate and tamp the infill material into the surrounding landfill, using vibration and the impact force of the vibrator. This is repeated in stages up to the fill surface. The refuse depth compaction is typically carried out first to address the deeper, and also the shallower fill, to provide a more stable platform for the heavy crawler cranes associated with DC of the shallower upper 5.0 m or so of the MSW. It is indicated that large continuous areas of minimum size in the range 0.2 to 0.5 ha are required to make the ground treatment process practical and economic.

6 RISK ASSESSMENT

As with any foundation or ground treatment system, it is critically important that the likely hazards associated with the technique are identified at an early stage and are adequately addressed. Potential hazards which should be assessed prior to the design of DC include; existing services, vibration, flying debris and noise. Further guidance on this is given in BRE report 458 (2003), including guidance and references on requirements for safe, stable working platforms – falling weight DC rigs are generally cranes or large track-based vehicles. The working surface must be suitable to safely support the plant, which is subject to varying static and dynamic loads during compaction treatment. Since dynamic treatment is usually employed where there are poor ground conditions an appropriate working platform is often required. The working platform should be maintained at an appropriate standard throughout the works.

Dynamic compaction of both new and older landfills may cause the release of gas and may also generate contaminated ground water (leachate). After a small 1940's landfill in the UK was dynamically compacted, gas concentrations increased and preventive measures were required for the proposed structures (Institute of Wastes Management, 1990); compaction had caused a change in water level, saturating previously dry waste. At Surrey Docks in London (UK), Thomson and Aldridge (1983) found that the rate of emission of methane into installed perforated plastic borehole tubes generally increased after dynamic compaction. Therefore, following appropriate risk assessment, consideration should also be given to the installation of venting trenches around the boundary of the site or treatment area to prevent gases migrating laterally off the site. Foundation design for the development may also need to incorporate measures to prevent the ingress and build-up of gas. Consideration should also be given to the integrity of any existing landfill liner systems.

It is also important to ensure that all parties involved in a DC contract are aware of their particular responsibilities to ensure that the works are carried out in a safe manner and appropriate risk assessment carried out.

7 DYNAMIC COMPACTION PRIOR TO FORMING NEW LANDFILL CELL

In the 1990's landfill philosophy in the UK moved to the objective of total containment and isolation of wastes and there was a major upsurge in the development of engineered waste disposal by landfill. The current preferred method of landfilling is in individual cells because it encourages a philosophy of progressive filling and restoration. Downward shear stresses resulting from total and differential compression effects form part of the considerations required in the design of a landfill lining system. Structures such as the landfill liner therefore need to be designed to resist or accommodate

Figure 5. Dynamic compaction treatment on grid pattern in loose sandy soils.

the shear stresses imposed by settling waste. The geotechnical properties of existing soils below the base of the proposed landfill is therefore also an important consideration and may necessitate employment of ground treatment techniques prior to liner installation.

7.1 Case history

At a site in Wangford, Suffolk (UK), ground treatment utilising DC was required to improve the geotechnical properties of a loose sand fill deposit (Figure 5), prior to construction of a new landfill cell.

The site has been utilised for sand and gravel extraction and subsequently turned over to land-filling in phases. The area subjected to DC corresponded to the base of a pit (from which sand and gravel had been recovered), and was to serve as a repository (landfill cell) for disposal of household and limited commercial and industrial wastes (Figure 6).

Some regrading and surface compaction of the site was required to form a level working platform (+4.0 m AOD) for the DC crane. The treatment area for this particular phase of the works was approximately 14,900 m². The underlying soil profile typically comprised about 5.0 m of very loose to loose light orange-yellow-brown gravely slightly silty to silty sand fill (with some light grey silt-bound horizons). This is in turn underlain by natural medium dense becoming very dense orange-brown sandy gravel. A typical grading envelope for the sand fill deposit is given in Figure 7. Groundwater was typically at around 3.5–4.0 m below working platform level.

The DC treatment was required to achieve an average cone resistance (q_c/C_r) value over any 1 m within the upper 6 m of the soil profile (from the commencement level of +4.00 m AOD), of at least 6 MN/m², prior to composite liner construction comprising:

- 300 mm layer of bentonite enriched soil with a maximum permeability of 1×10^{-10} m/sec
- 2 mm thick HDPE lining geomembrane
- 300 mm thick 40 mm gravel drainage layer incorporating a "herringbone" pipe system

Provision of a leachate collection point and re-circulation system was also implemented. The applied DC energy regime is illustrated in Table 2. The layout of the applied compactive energy and the chronological sequence of its application are critical to the effectiveness of the DC treatment. Initial grid spacing was approximately equal to the anticipated treatment depth. As can be seen from

74

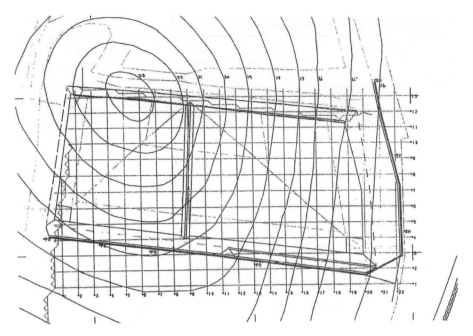

Figure 6.　Cell location within landfill site (MSW landfill contours in metres).

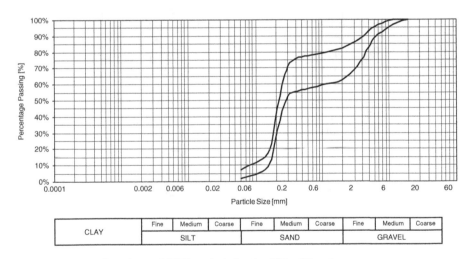

Figure 7.　Grading envelope for sand fill/deposits below landfill cell location.

Table 2, the compaction sequence comprises two types of passes, namely "grid" and "contiguous" (continuous) treatment passes. The grid passes are designed to compact the deeper soil layers, whereas the contiguous passes are primarily used to re-compact the shallow disturbed zone between grid passes. The ground is therefore effectively improved from depth to the surface (Figure 2).

Monitoring of the DC as it progressed formed an essential part of the works. This included level surveys on a 10 m grid pattern before and after each main treatment pass, to permit assessment of enforced settlement; heave tests and measurement of imprint depths. The compaction operation and ground response is continually monitored so that the applied energy, the location, and the sequence and timing of the drops can be adjusted to achieve the required results. Further detail on

Table 2. Typical applied DC energy regime at Suffolk landfill site.

Grid pass no.	Treatment pass (type)	Mass of weight/tamper (tonnes)	Drop height (metres)	No. of drops at each tamper position (no.)	Applied energy $(E = t.m/m^2)$
1	5 m × 5 m grid	10	15	9–10	54–60
2	5 m × 5 m offset grid	10	15	8	48
3	Contiguous	10	5	1	12.5
				Total energy	114.5–120.5 t.m/m²

Note: Upon completion of DC, the treated area was proof rolled utilising a minimum 6 passes of a 6 tonne smooth-wheeled vibrating roller.

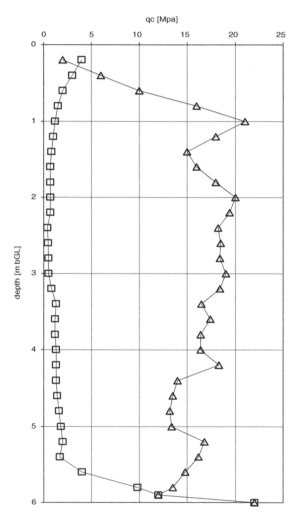

Figure 8. Typical pre (□) and post (△) treatment CPT plots for sand deposits.

quality control during dynamic compaction may be found in Serridge (2002). Recorded enforced settlements (void closure) were of the order of 6–7%, which is sufficient to change a loose sandy soil into a medium dense to dense state of compaction. Mean pre- and post-treatment cone penetration test (CPT), data are summarised in Figure 8, and from which the effectiveness of the treatment

Figure 9.   Rapid Impact Compaction (RIC) technique.

can be clearly judged. The results demonstrate that the specified performance requirements were satisfactorily achieved.

8   FUTURE TRENDS FOR LANDFILLING IN THE UK

Within the UK landfilling of waste has been an integral part of the approach to waste management. According to DEFRA (2004), "Some 400 million tonnes of waste are produced in the UK each year, a quarter of which is from households, commerce and industry. The remainder is made up of construction and demolition wastes, mining and agricultural wastes, sewage sludge and dredged spoils. Most waste currently ends up in landfill sites…". "Under the EU Landfill Directive the UK must dramatically reduce, over the next 20 years, the amount of biodegradable municipal waste sent to landfill. Municipal waste has been increasing at some 3–4% per annum and it is estimated that if it continues to increase at this rate it will have doubled from the 1995 level by 2020." It is also cited in the literature that current government strategy, prompted by the EU Directives, is to promote disposal practices that will achieve stabilisation of landfill sites within one generation.

Based upon experiences in continental Europe and the USA it is evident that techniques are being researched and implemented to extend the life of existing landfills (e.g. DC, vibroflot insertion, surcharging, landfill mining etc.). There are therefore potential research areas/opportunities in the UK of a practical and applied nature that lend themselves to field trials (on both active and older landfill sites) and large scale laboratory test programmes to investigate potential landfill gains, utilising the best of the European and global technologies (including, for example, the Rapid Impact Compaction (RIC) technique (Figure 9) and surcharging with monitoring) and hopefully permit more accurate prediction of what have historically been regarded as highly complex time related settlements and improve/enhance the integrity of the final cover systems (in addition to assisting in the achievement of these government targets for waste reduction). The RIC technique can be described as an intermediate technique between surface compaction methods such as vibratory or impact rolling, and deeper treatment compaction methods such as DC. The technique is reviewed more fully in Watts and Charles (1993), BRE Report 458 (2003).

9 CONCLUSIONS

The age and composition of municipal solid waste (MSW) landfill is important in evaluating suitability for development. Young landfills behave differently from old landfills. This phenomenon results from the initial degradable content and biodegradation of the waste. Each site must be assessed on its own merits and appropriate risk assessment carried out prior to ground treatment using DC, and subsequent site development. This should also include an assessment of the stability of the working surface for the DC rig and the impact of the technique on the integrity of existing structures, such as embankments, liner systems, and on leachate levels and landfill gas emissions.

Depth of influence of the treatment and enforced settlement are affected by the ageing process in MSW landfill. Within the UK, enforced settlements of around 10% have been reported for older landfills, increasing to about 20% for younger landfills. This is attributed to initial composition, but also to a transition to a more soil like matrix during the ageing process.

The DC technique has been used to good effect on older MSW landfill (> 15 years), prior to road embankment construction on a site in Redditch, UK. The principal aim was to increase density and decrease compressibility by reducing voids in the fill. Measurements showed that DC had substantially reduced the compressibility of the refuse fill although treatment did not eliminate long term settlements. Nevertheless it appears to have provided serviceable structures.

In Suffolk (UK) the geotechnical properties of loose sand fill deposits were significantly improved by the application of DC prior to construction of a new landfill cell. Close monitoring and validation testing during execution of the works contributed significantly to its successful application.

Current landfill sites in the UK are reaching their capacity and European Legislation is restricting development of new landfill sites. Ground treatment techniques such as DC and perhaps Rapid Impact Compaction (RIC), together with other emerging technologies, warrant further research and investigation in both recent and younger landfills based upon experiences gained internationally.

REFERENCES

Building Research Establishment (2003) *Specifying Dynamic Compaction,* BRE Report BR 458 CRC, Garston, UK.

Charles J A, Burford D and Watts K S (1981) *Field studies of the effectiveness of "dynamic consolidation",* Proc of the 10th Int. Conf on Soil Mechanics and Foundation Engineering, Stockholm, pp 617–622.

Charles J A (1991) *The causes, magnitude and control of ground movements in fills,* Ground Movements and Structures – Proc 4th Int. Conf. held in Cardiff, July, London, Pentech Press, pp 3–28.

DEFRA (2004) *UK-Environmental Protection,* Recycling and waste web page, www.defra.gov.uk/environment/waste/intro.htm

Department of the Environment (DOE) (1995) *Waste Management Paper 26B – Landfill design, construction and operational practice,* HMSO, London.

Edil T B, Ranguette V J and Wuellner W W (1990) *Settlement of municipal refuse,* Geotechnics of waste fills – theory and practice, ASTM, Special Technical Publication No. 1070, pp 225–239.

Frantzis I (1991) *Settlement in the landfill of Schisto,* Proc of 3rd International Sanitary Landfill Symposium, Cagliari, Italy, Vol. 1, pp 487–492.

Galante V N, Eith A W, Leonard M S M and Finn P S (1991) *An assessment of deep dynamic compaction as a means to increase refuse density for an operating municipal waste landfill,* The Planning and Engineering of Landfills, Midland Geotechnical Society, pp 165–169.

Greenwood D A and Kirsch K (1983) *Specialist ground treatment by vibratory and dynamic methods,* Piling and ground treatment, Thomas Telford Ltd, London, pp 17–45.

Institute of Wastes Management (1990) *Monitoring of landfill gas,* IWM, Northampton, UK.

Lewis P J and Langer J A (1994) *Dynamic compaction of landfill beneath an embankment,* Proc Settlement 94, ASCE, Geotechnical Special Publication No. 40, Vol. 1, pp 451–461.

Lukas R G (1986) *Dynamic compaction for highway construction,* Vol. 1, Design and Construction Guidelines. Federal Highway Administration Report No. RD-86/133, Washington DC, pp 204–219.

Mapplebeck N J and Fraser N A (1993) *Engineering landfill by dynamic compaction to support highways and buildings,* Proc Int Conf on Engineered Fills, Newcastle-Upon-Tyne, pp 492–504.

Mayne P W, Jones J S and Dumas J C (1984) *Ground response to dynamic compaction,* ASCE Journal of Geotechnical Engineering, Vol. 110, No. 6, June, pp 757–774.

Mitchell J K (1981) *Soil Improvement,* State of the art report, Proc 10th Int. Conf on Soil Mechanics and Foundation Engineering, Stockholm, Vol. 4, pp 509–565.

Perelberg S, Boyd P J H, Montague K N and Greenwood J R (1986) *M25 Bell lane pit: Ground Improvement by dynamic compaction,* Proc Conf Building on Marginal and Derelict Land, Glasgow, Thomas Telford, London, pp 267–280.

Powrie W, Richards D J and Bevan R P (1998) *Compression of waste and implications for practice,* Geotechnical Engineering of Landfills (eds N Dixon, E J Murray and D R V Jones), Thomas Telford, London, pp 3–18.

Sarsby R W (2000) *Environmental Geotechnics,* Thomas Telford, London.

Serridge C J (2002) *Dynamic compaction of loose sabkha deposits for airport runway and taxiways,* Proc 4th Int Conf on Ground Improvement Techniques, March, Kuala Lumpur, pp 649–656.

Singleton M (1996) *Extending the life of existing landfill sites through the use of deep compaction processes,* Proc 4th Int Conf on Polluted and Marginal Land, London, pp 301–305.

Slocombe B C (1993) *Dynamic Compaction. Ground improvement* (ed M P Moseley), Blackie, London, pp 20–39.

Thomson G H and Aldridge J A (1983) *London docklands – problems associated with their development with special reference to the Surrey Docks scheme,* Reclamation 83, Proc Int Land Reclamation Conf, Grays, Essex, Industrial Seminars, pp 124–132.

Van Impe W F and Bouazza A (1996) *Densification of domestic waste fills by dynamic compaction,* Can. Geotech. J. 33, pp 879–887.

Watts K S and Charles J A (1993) *Initial assessment of new rapid ground compactor. Engineered Fills* (eds B G Clarke, C J F P Jones and A I B Moffat), Proc Int Conf, Newcastle-Upon-Tyne, Thomas Telford, London, pp 399–412.

Watts K S and Charles J A (1999) *Settlement characteristics of landfill wastes,* Proc Institution of Civil Engineers, Geotechnical Engineering, Vol 137, October, pp 225–233.

Welsh J P (1983) *Dynamic deep compaction of sanitary landfill to support superhighway,* Proc 8th European Conference on Soil Mechanics and Foundation Engineering, Helsinki, Vol 1, pp 319–321.

Geotechnical and Environmental Aspects of Waste Disposal Sites – Sarsby & Felton (eds)
© 2007 Taylor & Francis Group, London, ISBN 978-0-415-42595-7

The Mock-up-CZ experiment

J. Svoboda

Centre of Experimental Geotechnics, Czech Technical University, Prague, Czech Republic

ABSTRACT: Because the time when a deep nuclear waste repository will be necessary is approaching (not only in the Czech Republic, but also in other countries in Europe and worldwide) most countries are performing extensive research toward developing a safe system of radioactive waste disposal. This consists of material search and testing (for example, bentonites as backfill or buffer, materials for canisters), finding suitable host environment (granites in the Czech Republic, salt formation, clays) and also experimental simulation tests (like the Mock-up-CZ experiment).

1   INTRODUCTION

Electricity is an important part of our life. It warms up our food in microwave ovens, allows us to work with computers, to watch TV, etc. Part of electric power is produced in nuclear power plants. Although producing electricity in nuclear power plants is a clean technology we end up with a certain amount of highly radioactive spent fuel.

There are several options for handling spent fuel and other highly active radioactive wastes, but the only one used today is to put it into deep repository (Figure 1). The only deep repository in operation today is an underground repository in the USA. Other countries are planning to build repositories in the not-too-distant-future.

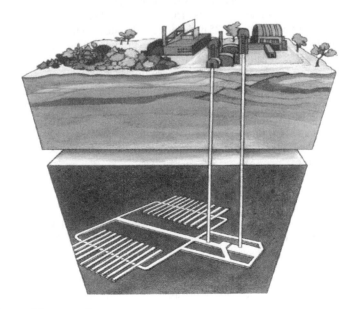

Figure 1.   Scheme of deep repository.

This type of waste storage is based on a concept which envisages several barriers acting together:

- Engineered barriers
  - Glass matrix and the waste container itself
    - This is the first barrier preventing radionuclides from getting into the environment
  - Geotechnical barrier which should give:
    - isolation of container from underground water
    - transfer of heat from container into surrounding environment
    - hindrance to radionuclides travelling from the container
    - structural and filling material
  - Structural barrier should fulfill spatial stability of the repository. Included in this category are lining, shaft reinforcements, etc.
- Geological barrier (the host geological formation) is the last barrier in the system. This must be stable and homogeneous. Depending on local conditions it can be a salt formation, clays or crystalline rocks (such as granites in the Czech Republic).

The Mock-up-CZ experiment simulates vertical placement of a container with radioactive waste according to the Swedish KBS-3V system. The model consists of a barrier of bentonite blocks and a heater. The heater, which simulates a container with radioactive waste, is placed inside the barrier. Water inflow through the barrier from its outer surface is possible (it simulates the impact of groundwater). The heater (like the container) loads the barrier with temperature. The whole experiment is enclosed in a cylindrical box, which is designed to withstand high pressure (due to the bentonite swelling). A number of sensors (monitoring the changes in temperature, pressure and moisture) are placed inside the bentonite barrier. A data collection and storage system allows continuous monitoring of experiments and all data coming from sensors is immediately available on-line.

2 APPARATUS

2.1 *Geometry*

The Mock-up-CZ experiment is placed inside a box-shaped test silo (with dimensions 3 m × 3 m × 3 m) and consists of several parts. The first part, which is visible from the outside, is an 8 mm thick cylindrical steel box, with 25 mm thick steel plates (on top and bottom) and 16 bolts. These elements hold the experimental structure together (Figure 2). Four small vertical tubes, used for injection of water, are attached inside the steel box, as well as two nets (acting as a filter) in order to circulate water all around the surface of the cylindrical box.

In the centre of the experiment is the heater which simulates a canister of spent fuel (Figure 3). The heater is placed on bentonite blocks and is surrounded by a small air gap (approximately 20 mm wide). The gap is enclosed by big blocks of highly-compacted bentonite mixture. These blocks are 70 mm high and pie-shaped (Figure 4) to surround the heater and the air gap. Smaller blocks are used below and above the heater. The last 50 mm between the blocks and the steel box is filled with a hand-compacted mixture.

2.2 *Hydration system*

Because one of the problems in the ground is water, we have to simulate the inflow of ground water. In our experiment we use synthetic granitic water, which is injected using a hydration system (Figure 5). The hydration system has two parts – the first outside the experiment (water tank, pressure generation system and piping) and the second part inside the experiment.

The second part consists of two concentric filters and four vertical perforated tubes. The filter closest to the steel box is coarse-meshed and the other one is fine-meshed having voids of less than 100 μm.

Figure 2.   The Mock-up-CZ experiment.

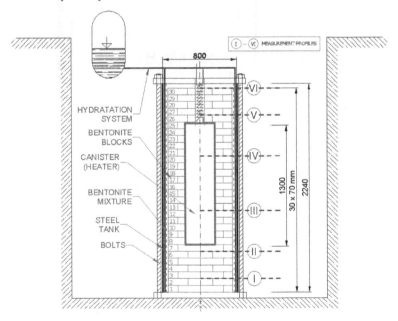

Figure 3.   Vertical cross-section through apparatus.

The water flows from a water tank through piping into tubes and then through the filter inside the bentonite. The two-layer filter ensures equal distribution of water on the outer bentonite surface.

2.3   *Heater*

The canister containing highly radioactive waste is simulated using a heater. The heater consists of a steel cylinder with an outside diameter of 320 mm and a height of 1300 mm, filled with oil.

83

Figure 4.   Pie-shaped bentonite block.

SCHEME

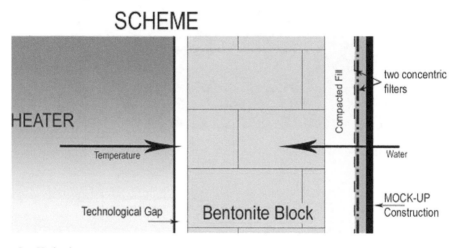

Figure 5.   Hydration system.

Into the oil medium are submerged two spiral heating elements. Each heating element can be switched on or off (to full or reduced power). This gives us an opportunity to select the maximum power used by the heater as 0.34, 1 or 2 kW.

The heater is equipped with two thermometers placed inside the oil medium (one above the heating spirals, the second below). These thermometers are used by a control unit of the heater to maintain selected temperatures (to an accuracy of 1°C) by switching heating on or off.

The outlets from the heating spirals and thermometers above the canister are spiral-shaped to allow the movement of the canister inside the barrier (due to swelling) and, at the same time, to avoid their failure.

3   BUFFER MATERIAL

Several types of bentonite have been tested in the Centre of Experimental Geotechnics (CEG) since 1999. Bentonites tested have included natural bentonites from 4 locations (Rokle, Stránce, Černý Vrch and Hroznětín) and industrial bentonite G (so-called geological).

Based on the findings from Czech bentonite research the most appropriate material for the buffer has been found to be a mixture of treated bentonite (from Rokle), silica sand and graphite. The basic mixture contains:

- 85% of bentonite
- 10% of silica sand (reduces swelling)
- 5% of graphite (increases thermal conductivity)

Figure 6. AD-SYS logger.

This mixture can be found in the Mock-up-CZ in two different forms:

- highly compacted blocks (with a dry density of $\rho_d = 1760\,\text{kg/m}^3$ and a swelling pressure of 3–5 MPa)
- hand-compacted mixture of the same composition as the blocks (dry density of $\rho_d = 788\,\text{kg/m}^3$ and swelling pressure of 370 kPa).

4   INSTRUMENTATION

The Mock-up-CZ experiment is monitored using over 150 sensors. The values from those sensors are gathered and stored for further analysis each 10 minutes.

Sensors and corrosion samples installed in the experiment can be divided into several parts:

- Thermometers (a total of 52)
- Pressure Cells (a total of 50)
- Moisture Content Indicators (a total of 13)
- Samples for Corrosion Tests (a total of 36)
- Tensometres (a total of 20 outlets).

These sensors are organized in so-called profiles. In addition to these sensors, heater power and total energy consumed by the heater is monitored.

The whole system of registration, evaluation and transfer of data uses 3 small portable AD-SYS data loggers (Figure 6) connected to a CEG server where they are stored in a database.

All the recorded data are immediately available on a web-site (using the apache web server at http://ceg.fsv.cvut.cz).

4.1   *Measurement profiles*

Six measurement profiles are mounted along the height of the Mock-up-CZ experiment (two measurement profiles are mounted below the heater, two within the heater's height and two above the heater). Each profile is fitted with a number of sensors as illustrated in Figure 7.

For each measurement profile the blocks used had to be modified to enable perfect mounting of the sensors. Some sensors are mounted in the layer of blocks situated below the measurement profile, others above them – the term "measurement profile", therefore, must be understood as including not only the juncture between the respective blocks, but also the layer of segments "above and below" the marked line.

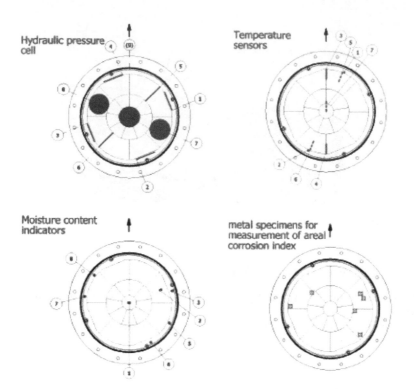

Figure 7. Instrumentation scheme.

4.2 *Core sampling*

The structure of the Mock-up-CZ experiment allows samples to be taken from inside the experiment by core drilling. This gives an opportunity to monitor directly water content changes and temperature distribution inside the experiment.

The Mock-up-CZ experiment is equipped with 6 entry ports with a diameter of 32 mm for taking samples and 11 entry ports with a diameter of 16 mm for taking samples. Those positions are organized in three horizontal layers as shown in Figure 8.

Sampling from a position is undertaken by making a succession of cuts. The length of each cut is highly dependent on the state of the material but on average it is 6 cm. After each cut the temperature in front of the drill is measured and the water distribution inside the sample is measured. When the last cut is performed, the excavated place is backfilled with a prepared plug of compacted bentonite mixture.

5 TEST PROCEDURES

The running of the Mock-up-CZ experiment can be divided into several parts:

- Phase 1 – the filter is kept dry and power is switched on to reach a maximum temperature of 90°C in the bentonite.
- Phase 2 – the power is maintained at a constant level and the filter is filled with synthetic granitic water.
- Phase 3 – saturation and temperature loading of the buffer by the heater is stopped. When the heater is switched off, the cooling phase commences, lasting approximately 2 weeks.

86

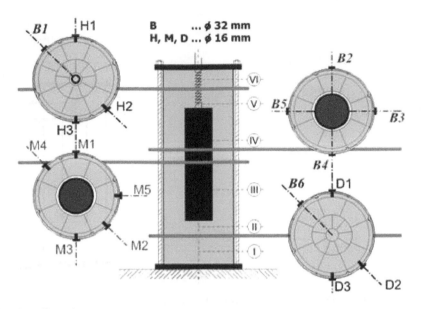

Figure 8. Sampling scheme.

A dismantling process is then conducted according to a detailed scheme which includes a scientific programme, a sampling plan and a scenario. This also includes testing of used equipment – sensors, container, etc. The testing is intended to reveal the changes in material properties due to long-term loading with moisture and temperature.

6 RESULTS

6.1 *Phase I*

The experiment Mock-up-CZ began at the beginning of May 2002 when the heater was switched on. During the first 3 months the temperature of the heater was increased in steps in order to safely reach maximum temperature 90°C in the buffer material. After that period thermal equilibrium was reached. However the process of water redistribution is much slower. The moisture travels from the inner part of the test specimen (test model) to the outer part. The moisture transfer is driven by the thermal gradient and tends to cause swelling in places where the moisture content increases (mainly in the backfill).

In order to directly monitor water content distribution inside the test specimen a decision was made to take samples at predetermined intervals by core drilling. The first samples (of diameter 32 mm) were taken before the end of Phase I at the very beginning of October 2002. These samples were taken from mid height of the model (Figure 9).

At the same time, the samples was used to determine permeability and swelling pressure – Figure 10. The results of these tests (performed by Geodevelopment a.s. Sweden) did not reveal significant changes of the material.

6.2 *Phase II*

Phase II began at the beginning of November 2002 when the water was introduced into the experiment. Initially about 80 l of synthetic granitic water was introduced into the hydration system. This water caused an immediate drop of the temperature in the buffer. After approximately two weeks a new thermal equilibrium was reached. At the same time as the water was let in the saturation level and swelling pressure started to increase significantly – Figure 11.

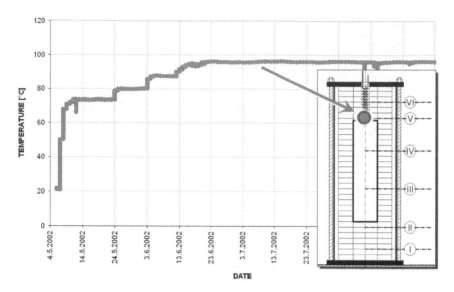

Figure 9. Temperature in the Mock-up-CZ experiment, Phase I.

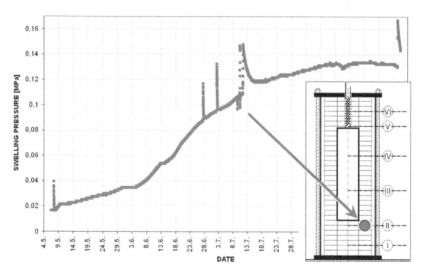

Figure 10. Swelling pressure, Phase I.

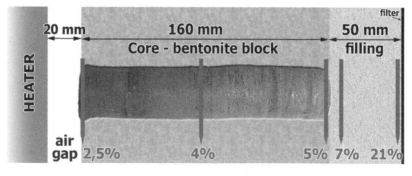

Figure 11. Water content distribution – core sample.

During the first month of the Phase II water was introduced several times (Figure 12). After one month (and after the control of model structure) continual saturation started.

The water was allowed to flow freely into the experiment. Up to June 2004, 386.5 l of water had been consumed and the monthly water consumption was 0.2 litres – Figure 13. The changes of swelling pressure (Figure 12) and changes of temperature distribution in the model clearly follow the water content distribution (Figure 14).

The results from core drills and the water intake show that the end of the Phase II (June 2004) full saturation of the model was approaching (Figure 14).

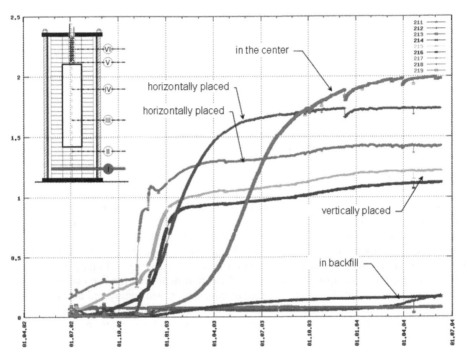

Figure 12.   Swelling pressure, Phase II.

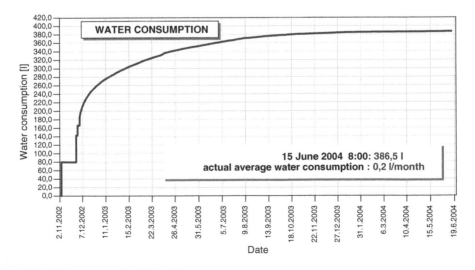

Figure 13.   Water consumption, Phase II.

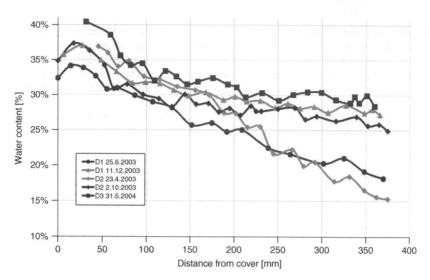

Figure 14. Water content distribution, Phase II.

7 CONCLUSIONS

The Mock-up-CZ experiment has functioned successfully for two years. Because the costs of its construction and operation are very much lower than analogue experiments running abroad the results obtained are considered to be of great success for the CEG.

From May 2002 (when the experiment was launched) to June 2004, the total data measured (due to a continuous data measurement technique) amounts to over 15 million items. The technical quality of the experimental apparatus will allow the experiment to be run for a very long time and will produce usable outputs for the safe design of bentonite-based engineered barriers.

Another positive fact is the decision to make all the data and information related to our research freely accessible to everyone interested, not only to a narrow circle of professionals. It is hoped that this approach will contribute to better dissemination of information relating to the safe handling of high-level radioactive waste.

ACKNOWLEDGEMENTS

The experiment would not have taken place without the support of the Czech Science Foundation (G103/02/0143).

We would like to thank to RAWRA, SKB (the Swedish Nuclear Fuel and Waste Management Co.), Geodevelopment AB Sweden and CTU in Prague for their help and co-operation.

REFERENCES

Anon (2002) *Mock-Up-CZ General Report*, CEG/SKB Report, CTU CEG in Prague, Czech Republic
Anon (2002) *Mock-Up-CZ Quarterly Report 1*, CEG/SKB Report, CTU CEG in Prague, Czech Republic
Anon (2002) *Mock-Up-CZ Quarterly Report 2*, CEG/SKB Report, CTU CEG in Prague, Czech Republic
Anon (2003) *Mock-Up-CZ Quarterly Report 3*, CEG/SKB Report, CTU CEG in Prague, Czech Republic
Anon (2003) *Mock-Up-CZ Quarterly Report 4*, CEG/SKB Report, CTU CEG in Prague, Czech Republic
Anon (2004) *Mock-Up-CZ Quarterly Report 5*, CEG/SKB Report, CTU CEG in Prague, Czech Republic
Pacovský J (2001) *Example of Utilization of Physical Modeling in the Design of an Engineered Barrier*, Environmental Geotechnics, Newcastle, Australia, pp. 347–352
Pacovský J (2002) *Experimental Research of Prefabricated Bentonite Barriers Used for High Radioactive Waste Isolation in Underground Disposal*, Proceedings of Workshop 2002, CTU in Prague, Prague, Czech Republic, pp. 956–957

Geotechnical and Environmental Aspects of Waste Disposal Sites – Sarsby & Felton (eds)
© 2007 Taylor & Francis Group, London, ISBN 978-0-415-42595-7

Landfilling and re-use of a calcium sulphate waste

D.M. Tonks
EDGE Consultants UK Ltd, Manchester, UK

A.P. Mott
MWH, New Zealand

D.A.C. Manning
School of Civil Engineering and Geosciences, University of Newcastle, UK

ABSTRACT: Various calcium by-products are produced from mining and chemical industries. Many are still regarded as wastes and are sent to landfill for disposal. However, they have useful properties which can be of value within the landfill or for wider uses. This paper focuses on one such case, a calcium sulphate, produced as dry powder in the anhydrite form which slowly hydrates in the presence of water to the gypsum form with the strength of a moderately weak to moderately strong rock, having low permeability. Geotechnical and mineralogical investigations are described in the paper. Field applications have demonstrated the suitability of the materials for use in various applications including low-cost, site roadways, pond lining and to cap slurries. Potential other applications include use in various lime/cement/PFA mixes ranging from treatment of very wet slurries through to formulation of high strength mixes.

1 INTRODUCTION

Anhydrous calcium sulphate ($CaSO_4$) arises as a by-product of various industries. The work described here concerns such materials produced from kilns as an anhydrite powder. This hardens with time and uptake of moisture as it converts to gypsum, having the strength of a moderately weak to moderately strong rock. It has useful properties for construction and other purposes, including plasters and lightweight screeds. It has been used for many years for various Civil Engineering purposes. The studies described here have been directed at better assessing the properties with a view to further assisting and enabling appropriate civil engineering application of these materials. Various tests have been undertaken to investigate the hardening/conversion process and to relate this to geotechnical parameters.

The material is produced as a fairly uniform, silt to sand sized white powder (Figure 1). It is generally off-white to buff in colour, light brown when less pure (which it tends to be). It is typically more than 98% $CaSO_4$ with trace CaF_2 and typically around 2% of added lime $Ca(OH)_2$, to give a slightly alkaline final condition.

James and Kirkpatrick (1980), Bell (1987) and Zanbak and Arthur (1986) give useful data on the physical, chemical and engineering properties of natural anhydrite and gypsum. Typical geotechnical properties are listed on Table 1.

James (1992) gives useful information on solubility and solution effects – he notes that the conversion processes depend on the specific conditions and on access to water. Under normal conditions anhydrite $CaSO_4$ converts slowly in the presence of water to gypsum $CaSO_4 \cdot 2H_2O$. This involves a decrease in specific gravity from 2.93 for anhydrite to 2.36 for gypsum, with an increase in molecular weight from 136 to 172. Theoretically this leads to about 63% increase in volume of the solids in an open system, i.e. water freely available. There may be significant swelling

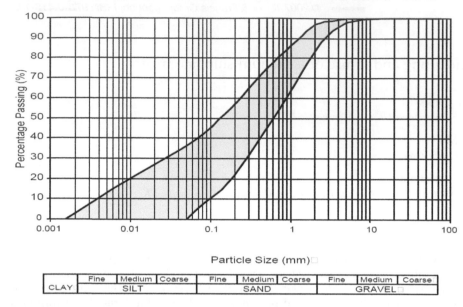

Figure 1.   Particle size distribution of calcium sulphate waste.

Table 1.   Typical geotechnical properties of natural anhydrite and gypsum rocks.

Property	Unit	Gypsum	Anhydrite
Specific gravity	Mg/m^3	2.36	2.93
Dry density	Mg/m^3	2.19	2.82
Porosity	%	4.6	2.9
Unconfined compressive strength	MPa	27.5	97.5
Point load strength	MPa	2.1	3.7
Young's modulus	$\times 10^3$ MPa	24.8	63.9
Permeability	$\times 10^{-10}$ m/s	6.2	0.3
Solubility (Cs) in pure water	kg/m^3 at 10°C	2.5	2.0
Solution rate constant (K) at 10° (flow velocity 0.05 m/s)	m/s $\times 10^5$	0.3	0.2
Limiting seepage velocity	m/s	1.4×10^{-6}	1.6×10^{-6}

Cs is dependent upon temperature and the presence of other dissolved salts.
K is dependent on temperature, flow velocity and other dissolved salts.

pressures. However, the gypsum may expand into pore space, where available, e.g. in less dense or less confined situations. From an engineering point of view this means that gypsum tends to seal water paths in contact with anhydrite (James, 1992). This gives the interesting situation that excess waters are absorbed and the site does not tend to produce leachate.

2   GEOTECHNICAL TESTING

A wide variety of laboratory tests has been carried out over many years. Most have been standard geotechnical tests to BS1377, with some concrete type testing procedures to BS1881. Particular test specifications have been developed in some cases to address the hydration and conversion-related characteristics of the material, discussed below. Typical results are summarised in Tables 2, 3 and 4.

Table 2. Permeability and strength of industrial anhydrite/gypsum.

Sample	Soaking (days)	Swell (mm)	Testing (days)	k^* (m/s)	Strength (N/mm^2)
1	7	0.73	21	9.3×10^{-9}	–
2	28	0.95	14	9.7×10^{-10}	–
3	28	1.56	14	5.7×10^{-10}	13
4	7	1.33	21	7.0×10^{-9}	15.5
5	28–50 kPa	−1.14	14	1.3×10^{-10}	–
6	28–50 kPa	−1.14	14	1.2×10^{-10}	14

* Cylinder 100×100 mm triaxial permeability (BS1377 Part 6, Method 6).

Description	k^{**} (m/s)	Comp. strength (N/mm^2) Typical	Comp. strength (N/mm^2) Range
Lab. tests on cylinders 28–42 days	7.4×10^{-9}	13	9.8–16.5
In-situ approx. 12 years	–	13	5–20.0
Cored field cylinders (90 days)	–	15	7–>40

** 3 day falling head test at omc.

Table 3. UCS testing.

Sample	Panel	Core no.	UCS	H/D
In specification	1	1	28.7	1.9
	2	2	11.5	2.0
	2	3	7.2	2.0
	2	4	15.5	1.5
	3	5	15.4	1.7
	3	6	18.1	2.0
	3	7	17.2	1.2
Excess lime	5	8	35.4	2.0
	5	9	39.7	2.0
Excess acid	6	10	10.8	1.3
	6	11	14.6	1.0
	6	11A	19.2	1.0
Excess fluoride	4	12	21.4	1.7

Table 4. Mineralogical testing.

Sample	Position	Time soaking (weeks)	Approx. anhydrite (%)	Gypsum (%)
Laboratory	Middle of core	6	57	42.1
Laboratory	Middle of core	4	56	43.9
Laboratory	Top of core	1, 3	94, 60	5.5, 40.0
Laboratory	Middle of core	1, 3	72, 50	27.2, 50.0
Field	Surface	Approx. 6	79	21.0
Field	Surface	Approx. 104	23	77.0

Methods used: Differential Scanning Calorimetry-Thermogravimetric Analysis (DSC-TGA), X-Ray Diffraction (XRD), bulk weight loss.

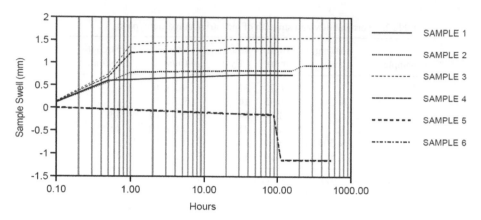

Figure 2. Swelling as a result of soaking.

2.1 *Test procedure*

Dry powder was taken directly from the production plant and stored in sealed plastic bags within airtight containers while not in use. As-produced, it is completely dry (w < 0.1%). Compaction tests (2.5 kg rammer) give optimum moisture content (omc) of 10–11% with maximum dry density (mdd) of 2.1–2.2 Mg/m^3. In the very dry state, conventional earthworks compaction in layers is not practical or effective. However, this is not necessary or desirable. In-situ tests show dry (and bulk) densities in the range 1.80 to 1.97 Mg/m^3 after placement, with about 33% porosity (voids ratio 0.5) in-situ. Samples were lightly tamped into 100 mm nominal diameter moulds to model the nominal site compaction at around 1.9 Mg/m^3. After 7–28 days soaking, the bulk densities increased to 2.24 and 2.33 Mg/m^3, consistent with around 50% conversion to gypsum. Samples for permeability and strength testing were then soaked from the top and base maintaining 50 mm head of distilled water, for 7–28 days. Some samples underwent further saturation in a triaxial permeability cell, giving a total soak time of between 28 and 42 days. Samples expanded by up to 1.6% under free swell during this soaking, mostly in the first hour, with only very minor swell thereafter (Figure 2). Overall, these results indicate quite limited shrink/swell characteristics under these conditions, taken to be related to conversion "tightening" of the initially open texture.

2.2 Laboratory test results

Permeability testing by the falling head method gave $k \approx 10^{-9}$ m/s. Triaxial permeability test results (BS1377 Part 6, Method 6, with slight modifications to allow for the soaking/conversion process) are summarised in Table 2. Permeability reduced from around 6×10^{-9} m/s after 7 days soaking to around 6×10^{-10} m/s after 28 days soaking. It further reduced to about 1×10^{-10} m/s under 50 kPa vertical stress. The permeability can be expected to further reduce with time as further conversion occurs. Chemical testing of eluate gave around 1400 mg/l SO_4, consistent with the solubility of calcium sulphate under normal conditions.

Unconfined Compressive Strength testing to BS1881 Part 116 gave strengths of between 13.0 and 15.5 N/mm^2 (equivalent to 13,000 kPa and 15,500 kPa respectively) after 28 days soaking – Table 2. A sample under 50 kPa vertical applied stress gave similar strength of 14 MPa.

3   FIELD TRIALS

Six trial panels were formed and tested to assess field behaviour of the range of materials. The powders were classified visually and by reference to chemical testing at source. Powders were delivered to site dry (m ≈ 0.1%) and compacted by "D4" size bulldozer to around 1.8 Mg/m^3

94

(approximately 80% mdd). In earthworks 1.5 m layer thicknesses are commonly adopted – 0.3 and 0.5 m layer thicknesses were attempted to investigate options to compact the material. However, it was evident that in the dry state the powders were not amenable to much compaction, even placed in layers. Subsequent layers simply mixed with the dry powders below. Various methods of rolling were investigated and found to give no benefit. Vibratory compaction was particularly impractical, as may be expected. Adjustment of moisture content could assist compaction, but makes for considerable working difficulties for the plant. More importantly, compaction to greater in-situ densities would actually be undesirable, as it could lead to significant swelling potential on conversion.

Standard compaction tests showed optimum moisture contents (omc) similar for all 6 samples at around 10% and maximum dry densities at 2.1–2.2 Mg/m³, similar to the laboratory data. Tests on cylinders showed 28 day compressive strengths of around 10–16.5 MPa.

The panels hardened in the typical fashion observed over many years. No significant differences could be perceived visually in the various source powders. The test panels were cored after 90 days. This showed the strengthening to extend consistently throughout the depth. Unconfined compression tests on 100 mm diameter cores gave strengths in the range 7–40 MPa (Table 3). The normal "in specification" powders gave some variations in strength between and within panels. Panels classified as having slight excesses of acid, fluoride or lime, actually gave similar or better strengths, with the "excess lime" panels reaching 35–40 MPa.

Site trials were also carried out in connection with possible use of the anhydrite and some calcium slurries with geomembrane liners. Panels were placed and left to harden over the geomembrane, worked by heavy plant and eventually fully broken out. The hardened material had a relatively rough/abrasive surface and could damage an adjacent geomembrane. Inclusion of a conventional, thick geotextile protector or a layer of calcium slurry (produced from a filter press), was found to give adequate protection.

There have also been various boreholes taken through the materials. These gave strengths of moderately weak to moderately strong rock, in materials typically up to about 12 years age, with an average strength of approximately 13 MPa and a range of 5–20 MPa.

In essence these studies confirmed the properties and hardening. It was therefore decided to investigate the mineralogy and its influence on the behaviour of the material.

4 MINERALOGICAL TESTING

Mineralogical testing was carried out by Mineral Solutions Ltd in collaboration with the University of Manchester, to investigate the calcium sulphate conversion and the relative proportions of anhydrite to gypsum:

- *X-Ray Diffraction (XRD)* was carried out to identify the minerals present in each sample, using a Phillips PW1730 generator and goniometer system and Siemens DIFFRAC-AT interpretive software.
- *Differential Scanning Calorimetry-Thermogravimetric Analysis (DSC-TGA)* was carried out using a Stanton Redcroft PTA-STA 1500 system to quantitatively determine proportions of anhydrite, bassanite (if present) and gypsum. Samples (0.02 g) were heated under argon at 10°C/minute. The percentage of each mineral phase is determined by the weight loss at characteristic dehydration temperatures.
- *Scanning Electron Microscopy-Back-Scattered Electron Imaging (SEM-BSE)* was carried out using a JEOL JSM-6400 electron microscope fitted with a Link Systems energy dispersive analytical detector and interpretive software. This technique was used to provide compositional petrographic and textural information, and to determine the distribution of voids.

Results are summarised in Table 4. The powder was confirmed to be virtually 100% anhydrite. After 7 days soaking, the samples had undergone about 25% conversion to gypsum, increasing to 40% after 2 weeks. A sample soaked for 7 days, then placed in a triaxial cell for 3 weeks gave a gypsum content of 44%. Another sample was soaked for 28 days and then in the permeability cell

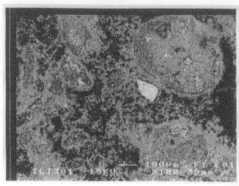

Field Sample after about 6 weeks.
Voids are black; darker grey is
gypsum, lighter grey is anhydrite
and white is iron oxide (or
similar) impurity.

Figure 3.   Scanning Electron Microscopy-Back-Scattered Electron images.

for 14 days, and showed 42% gypsum. By contrast, samples under field conditions showed 21% gypsum at 6 weeks and 77% gypsum at 2 years. SEM-BSE images show the heterogeneous nature of the material. The differing presence of anhydrite and gypsum and corresponding difference in structure and void space in the 6-week sample and the 2-year sample is shown:

- SEM-BSE images of the 6-week old sample (Figure 3) clearly show an open-textured nodular material, with some limited crystal growth of gypsum in fractures and cavities. Much of the sample is void space, indicated in the image as black.
- SEM-BSE images of the 2-year old sample (Figure 4) show an interlocking network of small crystals largely consisting of gypsum, with much less void space than for the 6-week sample.

As the calcium sulphate hydrates, it expands into the available void space, thus reducing permeability. Similarly the increased particle contact appears consistent with increasing hardening and strengthening.

5   ENGINEERING APPLICATIONS

Experience over many years has shown that industrial anhydrite can be a particularly useful engineering material. It has some unusual properties which can be beneficially used for a variety of applications. These need to be suitably understood and controlled, through appropriate testing, specification and Construction Quality Assurance (CQA).

Anhydrite has proved practical and effective for a variety of site applications such as unsurfaced roads, containment bunds and ditch lining. Figure 5 gives a general view, with the material forming a road, drainage ditch and capping of slurries. Figure 6 shows the typical final surface texture

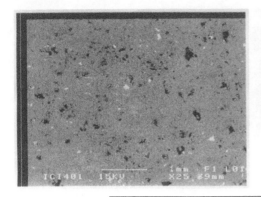

Field Sample after about 2 years. Voids are black; the bulk of the
sample is gypsum, with some residual anhydrite (lighter grey).

Figure 4.   Scanning Electron Microscopy-Back-Scattered Electron images.

Figure 5.   General view of the site.

achieved on an extensive working area. Once hardened, the material cannot readily be reprofiled.
It forms quite rough abrasive surfaces, with hardened profiles/ridges reflecting the placement
(Figure 7). It is difficult to work to a reasonably smooth surface and not amenable to smooth wheel
rolling. However, these rough surfaces have proved durable and very suitable for site roadways,
akin to a low grade concrete road, withstanding several years of regular use, loaded by several
(typically 8 to10) 20–40 t trucks per day.

The dry powder can be readily conveyed by tanker/covered wagon, tipped and handled by tracked
bulldozers. In this state it does not compact well with a tendency to "flow" under placement by
earthworks plant. However, it "self compacts" as it hydrates, expanding to reduce the void space as
it increases in strength and decreases in permeability with time. Confining loads also lead to lower
permeability.

Testing indicates that with controlled conversion, swelling need not present problems for many
practical conditions. This is consistent with experience for applications to date. However, there
could be cases where swelling is of significance and this aspect must be kept in mind in design.

Figure 6.   Finished surface of placed material.

Figure 7.   Rough surface skim.

A particularly useful application has been for capping areas of wet slurry (Figure 6). The material takes up the moisture as it mixes with the slurries to form a capping capable of supporting earth-moving plant. This has proved a very practical and effective means of placing initial cover/capping to old lagoon areas.

Figure 8 shows the material placed to form the lower part of the liner system for a surface water storage pond, with a 1 mm thick polypropylene geomembrane. It could be formed to the required geometries without undue difficulty. A 400 g/m^2 geotextile was used here for the geomembrane protection. The geomembrane was then quite easily placed, with anchor trenches around the perimeter and suitable weighting for operational use.

Figure 8.   Use in pond lining system.

The material can also have useful geochemical properties. Clearly wetting fronts would not penetrate far into it and any moisture would be taken up in the conversion process and self sealing effects. The calcium sulphate matrix has the capacity to absorb and effectively immobilise contaminants due to the alkalinity and ability to "fix" potential contaminants in stable insoluble compounds.

6   USE IN STABILISATION TREATMENTS

Anhydrite and gypsum are known to have beneficial applications in various stabilization treatment applications, including for wet clays and slurries. There is some experience of including gypsum, sometimes with PFA, in mixes with soils containing excess sulphate or sulphide, to control risk of heave – further information can be found in Kennedy (1997), Tonks et al (1998, 2000, 2003).

There is also considerable scope for inclusion of the anhydrite in various stabilised mixes with soils, PFA and cement for engineering applications. Consideration was given to use of PFA stabilised with anhydrite alone, i.e. without cement. However, published work by Hyde (1977) had indicated little beneficial effect on PFA strength unless more than 60% anhydrite was used. Limited testing of the effect of 10% and 20% anhydrite was carried out at this site in case this might be sufficient for some purposes. Results showed an increase in unconfined compressive strength from about 0.1 to 0.3 MPa at 7 days. Such a mix might be usefully employed on occasions.

Use of the anhydrite with PFA-cement mixes has also been examined. Hyde (1977) found that relatively small quantities of anhydrite significantly increased the strength of such mixes. For example, 10% anhydrite reduced the cement requirement for a 28-day strength of 5 MPa from about 10% to 5%. Testing was therefore carried out on a mix of PFA:OPC:anhydrite in the proportions 84:6:10, giving a 7-day compressive strength of 3.3 MPa – similar to the strength obtained using 6% cement without anhydrite. This work has opened the possibility of using such mixes as appropriate. A variety of "mixes" can be developed to meet specified engineering requirements. However, for this site no particular benefits were identified which could not be satisfactorily obtained from use of the anhydrite powders alone. These thus remain the preferred option here. However, for other sites or applications, such mixes could well be of value.

7 CONCLUSIONS

The geotechnical properties of industrial anhydrite as a dry powder and its conversion to gypsum, as a moderately weak rock, are amenable to consistent evaluation, by practical conventional testing, suitably specified and controlled to address the hardening process. The results have been shown to be consistent with logical expectation in the light of the chemical/mineralogical processes taking place. Field behaviour can be rationally evaluated and assessed.

The dry powder takes up water to convert to gypsum, strengthening to approach 15 MPa after 28 days soaking and saturation, under laboratory conditions. Permeability values of less than 10^{-9} m/s were achieved (unconfined) and as low as 1×10^{-10} m/s under confining stress after 28 days, further decreasing with time. Mineralogical testing confirmed the expected conversion process, with the samples after 28 days laboratory soaking giving about 40–50% conversion to gypsum.

The material use has been managed in the field by application of standard UK Highways Agency type specifications, adjusted to suit its specific properties of the material. It can be placed and lightly compacted in-situ, to give dry density around 1.9 Mg/m³. The conversion process leads to increase in volume (decreased specific weight) of the calcium solids. However, under these conditions, this gives useful self-compaction as the solids take up the available void space, to give a strong low permeability material. Laboratory testing simulating this gave quite limited swelling, ranging from less than 2% for vertical free swell to a decrease of about 1% under 50 kPa confining pressure.

Test results have been reasonably consistent on various occasions although carried out for different purposes and with some differences of detail in specification. This suggests reasonable consistency in relevant properties over a number of years. With increased knowledge and experience of test data and field applications, there are significant opportunities to make productive commercial use of a material once seen merely as a by-product, but now increasingly appreciated as having unique applications.

REFERENCES

Bell F G (1987) *Ground Engineers Reference Book,* Butterworths
BS1377 (1990) *Soils For Civil Engineering Purposes,* British Standards Institution
BS1881 (1983) *Testing Concrete,* British Standards Institution
Hyde A F L (1977) *The Properties of PFA/Anhydrite/Cement mixes*
James A N and Kirkpatrick I M (1980) *Design of Foundations of Dams containing soluble rocks and soils,* QJEG, 13, pp 189–198
James A N (1992) *Soluble Materials in Civil Engineering,* Ellis Horwood
Kennedy J (1997) *European Standards and Research – development of applications in soil improvement and pavements,* British Lime Association
Tonks D M, Hillier R P and Beeden H J (1998) *Lime/Cement Treatment to improve marginal materials*, AGS Symposium on Value of Geotechnics, ICE, London
Tonks D M and Gallagher W (2000) *Some environmental applications of Lime/Cement Treatment*, Proc Green3 symposium, Thomas Telford Ltd, London
Tonks D M, Nettleton I M and Needham A D (2003) *Sustainability, Maintenance & Re-Use of Existing Infrastructure – Some Geotechnical Aspects,* XIII Euro Conf. SMGE, Prague
Tonks D M (2003) *Geotechnical and Geoenvironmental Uses of Very Wet Materials, Dredgings and Slurries,* XIII Euro Conf. SMGE, Prague
Zanbak C and Arthur R C (1986) *Geochemical and engineering aspects of anhydrite/gypsum phase transitions,* Bull Assoc Eng. Geol, 23 (4), pp 419–433

Geotechnical and Environmental Aspects of Waste Disposal Sites – Sarsby & Felton (eds)
© 2007 Taylor & Francis Group, London, ISBN 978-0-415-42595-7

Habitat creation on old landfill sites

I.C. Trueman & E.V.J. Cohn
University of Wolverhampton, UK

P. Millett
Wolverhampton City Council, UK

ABSTRACT: The authors have been involved in creating examples of semi-natural grassland and woodland on old landfill sites in Wolverhampton (in the West Midlands) since 1984. This paper describes the methodology used in creating species-rich grasslands on such landfill sites. The results obtained suggest that it has been possible to produce attractive and species-rich vegetation with significant nature conservation and amenity value, and that the methodology described could be used to create a positive end-use for landfill.

1 INTRODUCTION

Habitat creation is a familiar activity but is difficult to define (Gilbert and Anderson, 1998). The authors have adopted 'The reconstruction or restoration of natural or semi-natural communities or assemblages of organisms' as a working definition. 'Semi-natural' is used to describe communities which are based on the organisms naturally occurring in an area but which have been affected by management by humans such as the felling of trees and the imposition of grazing to produce pasture. Typically the organisms concerned in habitat creation are the more obvious non-mobile constituents of the community, i.e. the plants. Generally the models have some antiquity, are species-rich and have significant nature conservation value.

2 REQUIREMENTS FOR CREATING SPECIES-RICH GRASSLAND

Natural and semi-natural woodlands and grasslands have both been used as models in habitat creation on landfill in Wolverhampton, but the present paper focuses on the re-creation of old species-rich neutral grasslands. These have become increasingly rare in the countryside due to agricultural intensification (Blackstock et al, 1999) and the best examples are now mostly nationally protected. They are particularly appropriate for habitat creation partly because they include many attractive and uncommon plant species in pleasing combinations. In addition they seem to be largely the product of a consistent and non-intensive system of management over a wide range of well-drained, low nutrient fertility soil types not marked by extremely high or low pH (Rodwell, 1992). Therefore potentially they can be created on a wide range of soils. Good examples are often managed by being cropped for hay in mid-summer, followed by 'aftermath' grazing through the autumn and sometimes into early spring and this should also be borne in mind when considering long-term site management.

Taking the foregoing characteristics into account, the requirements for creating species-rich neutral grassland include:

- The provision of a well-drained soil of neutral pH and low fertility.
- A reliable source of the plant species which make up the grassland vegetation.

- An achievable long-term management plan replicating the features of hay-making and aftermath grazing.

Diversity is rapidly overwhelmed by small numbers of competitive plant species such as Ryegrass (Lolium perenne) and Cock's-foot (Dactylis glomerata) on nutrient rich soils. Soil fertility criteria for grassland habitat creation have been widely based on levels of available nitrogen and phosphorus. Ash et al (1992) defined 'infertile' soil as containing less than $2\,\mu g/g$ mineralisable nitrogen and less than $20\,\mu g/g$ extractable phosphorus (as measured using the techniques in Allen, 1974) and considered such soils suitable for creating species rich grassland. They defined soil with more than $20\,\mu g/g$ mineralisable nitrogen and more than $80\,\mu g/g$ extractable phosphorus as 'fertile' and inappropriate for habitat creating. Intermediate values have an intermediate likelihood of success.

3   VEGETATION COVER ON LANDFILLS

In Wolverhampton landfill has typically been capped with layers of sandy subsoil which has many of the right requirements for habitat creation since subsoils are generally less rich than topsoils and light soils lose fertility relatively rapidly. However, tests for levels of available phosphorus, at least, are advisable. It has been found that an appropriate upper limit for extractable phosphorus seems to be circa $70\,\mu g/g$ (Truog's extraction) – McCrea et al (2004).

Seed is the most convenient source for the plant species. Commercial seed mixtures are available but may include inappropriate genotypes or even species, and there is no efficient way to test them for germination potential. In Wolverhampton, the Dutch method of taking a hay cut on an existing species-rich meadow and transporting the freshly cut green hay to the receptor site, has been adopted. A seed mixture similar in diversity and proportion to the source site will be generated as the hay dries on the new site.

It is difficult to replicate hay meadow management on urban landfill sites. Drying hay can be set on fire and grazing stock can escape. In Wolverhampton it has generally been found that a single cut and remove in the first week in August is adequate to maintain much of the diversity. A single cut in autumn has proved more difficult to implement due to deteriorating weather and has also encouraged increases in species not associated with hay meadows. At sites where it has been possible to graze the aftermath there have been quantifiable positive effects on diversity.

4   THE HAY STREWING METHOD

Details of the method are given in Trueman and Millett (2003). The process is summarised in the following sections:

Characterisation of the receiver site
- If already vegetated the prospective site should be subjected to a detailed ecological evaluation with a predisposition to reject habitat creation if any nature conservation value is identified.
- Consider the necessity for long-term management including access and funding.
- Undertake a chemical analysis of the actual or prospective soil and reject if it falls outside the guidelines in the literature.

Identification of a possible source site
- A suitable source site should be seriously species rich and most will be designated as Sites of Special Scientific Interest. This means that it will be necessary to negotiate with English Nature or its equivalent, and possibly the county Wildlife Trust as well as the site owners and/or managers well in advance.
- The site will probably be a hay meadow and will conform to the National Vegetation Classification type MG5 (Rodwell, 1992).

- One can expect to spread the hay quite thinly – twice the area of the source site will be adequate and the material will not need to be removed later.

Preparation of the receiver site
- Remove any existing vegetation at the receiver site, and particularly perennial weeds, using glyphosate.
- On a light, open-textured soil it does not seem to be necessary to raise a tilth.

Extraction and strewing of the hay
- The ideal time to extract the hay is when the owner would normally make hay on the site.
- Bale and remove the hay from the source site immediately after cutting.
- Spread the hay the same day to prevent heating up in the bale.
- Hay can be hand spread using volunteers or a muck-spreader could be used.

Site management
- Keep a close eye on the developing vegetation.
- It is not recommended to cut the developing sward until normal hay meadow management starts the following summer.
- Make hay on the site in July, or, if this is impossible because of potential vandalism, cut and remove the vegetation when most of the species are in seed (circa early August).
- Continuation of the annual cut each year indefinitely may be sufficient to maintain diversity indefinitely on poor soils.
- If possible graze or gang-mow throughout the autumn, otherwise a second cut in late autumn or very early in spring (not after the beginning of April) may be beneficial.

5   EXPERIMENTAL MEADOW CREATION IN WOLVERHAMPTON

The fist successful meadow was created at Bushbury Hill landfill in Wolverhampton in 1984 using hay from Pennerley Meadows SSSI in Shropshire. Freshly cut hay was strewn by hand onto areas of pre-existing amenity grassland which had existed on the landfill since it was completed in the early 1970s. Different areas were prepared by cutting short, killing with glyphosate and rotavating.

Table 1 shows quadrat sample data from 1987, three years after strewing. The data suggest that all the strewn areas have increased diversity measured in number of species per 1 m^2 sample compared with the unstrewn area and that establishment is encouraged by destroying the existing vegetation before strewing. There is however no obvious advantage of rotavation over glyphosating – it does not seem to be necessary to raise a tilth on the light sandy soil used at Bushbury Hill.

Table 1.   Species transfer at Bushbury Hill landfill site (three years after strewing with Pennerley Meadow hay).

		Bushbury Hill Meadow				
	Pennerley Meadow	Unstrewn plot	Strewn, rotavated	Strewn, glyphosated, rotavated	Strewn, glyphosated	Strewn onto original vegetation
Number of 1 m^2 samples	111	12	21	25	25	23
Total number of species recorded	48	17	30	32	30	24
Total number of species recorded both at Pennerley and at Bushbury Hill		12	20	24	22	17
Total number of species recorded at Bushbury Hill not recorded at Pennerley		5	10	8	8	7
Mean number of species per 1 m^2 sample	18.6	7.9	13.7	15.6	14.5	10.6

Figure 1.   Created meadow at Bushbury Hill landfill site 17 years after strewing (showing Ox-eye Daisy, Knapweed and Cat's-ear).

Table 2.   Success of species transfer at Kitchen lane landfill site.

	Eades Meadow	Kitchen Lane Meadow	
Year sampled	1993	1995	2001
Total number of 1 m^2 samples	116	91	117
Total number of species recorded in the samples	69	47	56
Total number of species recorded at Eades Meadow also found at Kitchen Lane		37	46
Total number of species recorded only at Kitchen Lane		10	10
Mean number of species per 1 m^2 sample	25.6	9.4	17.7

Although it has only been possible to cut and remove the vegetation once a year at Bushbury Hill much of the induced diversity has persisted – Figure 1 shows an area in 2001, 17 years after hay strewing.

Another source meadow, Eades Meadow in Worcestershire, was used at the Wolverhampton landfill site at Kitchen Lane in 1994. This site continues to actively vent carbon dioxide and methane. Table 2 compares the source and receptor sites one year and seven year after strewing. Initially the transfer seemed successful in terms of numbers of species only, but after seven years the transferred species had spread well. Figure 2 shows the Green-winged Orchis at the receptor site, which has successfully transferred from the source site.

6   CONCLUSIONS

Habitat creation is potentially a beneficial and economical use of landfill, both for public amenity and to promote nature conservation. It should however be understood that it is not impossible

Figure 2. Green-winged Orchis and Cowslip at the Kitchen Lane sandfill site (both successfully transferred from Eades Meadow).

for valuable and attractive vegetation to develop spontaneously on landfill. Therefore, before undertaking any habitat creation, it is an absolute prerequisite to obtain a specialist survey of the existing vegetation to make sure that nothing of value will be lost. Furthermore species introductions into the 'wild' are confusing and can sometimes be deleterious, so it is advisable to consult and collaborate with local conservation groups such as English Nature or the county Wildlife Trust.

ACKNOWLEDGEMENTS

The authors wish to thank Wolverhampton City Council, who have continued to support and aid meadow creation in Wolverhampton for many years.

REFERENCES

Allen S E (1974) *Chemical Analysis of Biological Materials,* Blackwell Publishers.
Ash H J, Bennett R and Scott R (1992) *Flowers in the Grass,* Publisher English Nature, Peterborough.
Blackstock T H, Rimes C A, Stevens D P, Jefferson R G, Robertson H J, Mackintosh J and Hopkins J J (1999) *The extent of semi-natural grassland communities in lowland England and Wales: a review of conservation surveys 1978–96,* Grass and Forage Science, 54, pp 1–18.
Gilbert O L and Anderson P (1998) *Habitat Creation and Repair,* Oxford University Press, Oxford.
McCrea A R, Trueman I C and Fullen M A (2004) *Factors relating to soil fertility and species diversity in both semi-natural and created meadows in the West Midlands of England,* European Journal of Soil Science, 55, pp 335–348.
Rodwell J S (1992) *British Plant Communities Volume 3: Grasslands and Montane Communities,* Cambridge University Press, Cambridge.
Trueman I C and Millett P (2003) *Creating wild-flower meadows by strewing green hay,* British Wildlife, 15(1), pp 37–44.

Geotechnical and Environmental Aspects of Waste Disposal Sites – Sarsby & Felton (eds)
© 2007 Taylor & Francis Group, London, ISBN 978-0-415-42595-7

Slope stabilisation measures at Hiriya waste site, Israel

A. Klein
Geotechnical Consultant, Israel

D.M. Tonks & M.V. Nguyen
EDGE Consultants UK Ltd., Manchester, UK

ABSTRACT: The Hiriya landfill site was for many years the main disposal site for municipal and other solid waste for the Greater Tel-Aviv area in Israel. It is a well-known national feature, due to its substantial size and proximity to Ben-Gurion airport. It covers an area of about 40 ha and reaches a height of up to 60 m with slopes of 45° or more in places. The Ayalon and Shappirim rivers run along the northern and southern borders respectively. The site is unlined and until recently was also without any cover. In the winter of 1997 a major slip occurred on the northern face, blocking the Ayalon river, which was later re-routed further north. The dump was finally closed in 1999, principally owing to the danger to planes landing at the nearby airport, from birds congregating at and above the site, during placing of MSW. The authors were retained in the year 2000 to report on the overall stability of the slopes. This paper details the design assumptions and the measures recommended, with a review of the measures taken to date to improve the overall stability of the slopes.

1 INTRODUCTION

Hiriya Landfill Site is a major disposal site for municipal solid waste (MSW). It covers an area of about 40 ha, is up to 60 m in height and has slopes over 45° in places. The tip is mostly surrounded by two significant rivers, the Ayalon to the north and Shappirim to the south. These pose risks of erosion which could undermine the tip. Leachate may be affecting the rivers and groundwater in the area. Failure of the tip would cause major environmental damage. It has suffered significant slips and there are concerns over stability and environment risks.

A preliminary assessment in 2000 gave grounds for concern, with serious risks of sudden and catastrophic failures which could;

- block adjacent river courses, also creating new risks from flood water upstream,
- cause major damage to areas affected,
- cause injury/death to anyone in the area at the time,
- cause pollution and long term damage of areas immediately affected and,
- affect substantial further areas in the longer term.

Recommendations have been given for some immediate actions, with a strategy to bring the site under control. This is a major task, which will take several years and involve substantial investigations and studies to find the most suitable solutions.

2 SITE INFORMATION

2.1 *Tip characteristics*

Figure 1 is a Schematic Plan of Hiriya tip and Figure 2 is a view of the original tip.

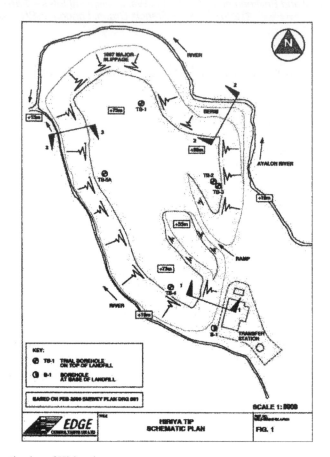

Figure 1.   Schematic plan of Hiriya tip.

Figure 2.   Hiriya landfill site, Tel Aviv, Israel.

The tip contains about 14 Mm^3 of material. It comprises a wide variety of waste materials from Greater Tel-Aviv domestic and industrial facilities over many years. The materials are quite variable with location and time and in varying states of composition. The slopes are far higher and steeper than is normal for landfills. The surfaces are mainly bare/very poorly vegetated, making them very prone to erosion and surface slippage (as shown in the photographs). This can undermine the slopes and trigger major instability. There are various tension cracks developing in places, indicating failure is imminent.

2.2   *Waste properties*

Five boreholes (TB1-5A) drilled from the crest showed mainly MSW (Municipal Solid Waste) which proved to over 30 m deep and quite variable in composition and properties. The waste heap is actually over 60 m high in places. The reported bulk densities were very low – 400 kg/m^3 in the upper MSW increasing with depth to more typically 600–800 kg/m^3. The densities indicate quite limited settlement and self-compaction to date. There will be ongoing settlements for many years (i.e. the tip will be subject to significant ongoing internal movements) which can lead to instability, as the wastes decompose. Moisture contents were around 20% in the upper 20 m, increasing to 30–40% below. The lower MSW is 'saturated with gas and is damp to wet'. There are thin bands of cover layers in places, typically 0.2–0.4 m of sandy clay (CL-CH). These are unlikely to be continuous, as the waste will have settled considerably and variably. There are probably complex leachate conditions associated with these.

At the toe near the transfer station, waste extends to 3.5 m depth. It is quite decomposed and weak, with SPT N values around 3–4. The basal area is particularly weak, with much leachate. Underlying the waste here is a dark brown high plasticity clay. This is stiff to hard with N values of 20 and above. It is much stronger than the MSW.

3   STABILITY

3.1   *Analysis*

Figure 3 gives a cross-section of one of the more significant areas. Three basis patterns of stability risk have been identified as indicated in Figure 3:

a. *Major overall instability.*

   This is the most serious involving a large quantity of material, possibly sudden, catastrophic and capable of travelling along distance across the river and beyond.

b. *Local instability – toe area.*

   There are extensive areas where the toe of the slope is significantly more vulnerable than else-where. Such slips pose risks of undermining the slope leading to overall failure, which could follow rapidly. The interim condition would be very dangerous and difficult to stabilise.

c. *Local instability – crest area.*

   Some upper slopes are significantly steeper than their lower slopes. Local failures here will have considerable impact. At the steep angles local erosion and surface instability can also be expected, especially during and following heavy rainfall. This is of much less consequence.

Certain areas and conditions were identified as more vulnerable from geotechnical engineering slope stability analysis. Stability has been analysed by computer program STABLE. Factors of Safety less than 1.0 indicate the slope is unstable. Factors of Safety greater than 1.4 would normally be required for slopes of this significance. Factors of safety vary with conditions. In particular they will decrease with;

- weakening of the MSW with time,
- rising leachate/groundwater levels.

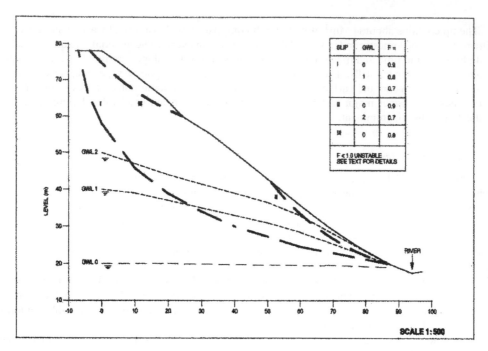

Figure 3. Section through a part of the tip which is of particular concern.

3.2 Parameters

Strength parameters of $c' = 0$, $\phi' = 25°$ were initially used for the MSW, based on the limited information available. This is reasonably typical and consistent with published knowledge and current experience (Eid et al, 2000; Dixon and Jones, 2003). However, MSW is very variable. There is clearly additional strength to explain the current stability. Back analyses indicate a range from $\phi' \geq 45°$ (assuming $c' = 0$) to $c' = 40\,kPa$ for the lower bound $\phi' = 25°$.

Stability was assessed for various groundwater/leachate levels:

- gw-0 for "dry" conditions, i.e. leachate below toe level
- gw-1 represents a moderate leachate level, emerging a little above the toe
- gw-2 is for a fairly high leachate level, which appears consistent with some current observations (but locally varying around the site) and/or credible in wet conditions.

3.3 Assessment

The south-east face (Figure 3) poses a significant threat to the transfer station. Slopes are steep at around 35° overall with signs of shallow slips on the lower part of this face. Overall stability (case a) is unacceptably low, even if dry. There is significant leachate here. The flatter lower slopes probably result from seepages and may be steadily creeping forward. Major failure is likely with any increase in leachate level/pressure. The present profile appears to act as a catch-ditch which may somewhat protect the transfer station.

The upper slopes (case c) were particularly steep at more than 40°. They are fairly dry but could fail with rainfall infiltration in the wet season. These have been pulled back in the south-eastern corner to about 10°, with the formation of a number of berms. The remaining length of the eastern slope to the north remains potentially unstable, with slopes of 50° and more in the north-eastern corner, above the Ayalon River.

The north-east corner is the area of highest tip (+80 m crest level) with steep slopes. A toe berm has improved overall stability to the river. The most important further action is to ensure leachate

110

is drained away and re-profile/seal to minimise infiltration. There are various local areas which are over-steep (case b). The long-term solution here involves diverting this section of river further away from the toe of slopes.

In the winter of 1997, a major slip occurred on the northern face which blocked the Ayalon River. This spread to a fairly flat profile at the toe, around 15° overall, indicating low strength of the waste, with much leachate ponded in this area. Leachate pressures were probably the main cause of the failure.

In the western face the overall slopes (case a) are again very steep and high, although less so than the northern and eastern area. They are also very close to the river. The toe area and lower slopes (case b) are particularly vulnerable, due to seepage. There is very little space for any toe berm. The long-term solution here will again probably involve diverting the river further away from the toe of slopes. Pending this, the most practical emergency actions are profiling and sealing the crest area to minimise infiltration and pulling back the crest, with priority given to the steepest areas and those with existing tension cracks.

4 LEACHATE

There are significant quantities of leachate arising from decomposition of the waste combined with rainfall infiltration. Local effects will be extremely variable and very significant to stability. Leachate levels would be expected to rise with time and could eventually lead to slope failures. However, leachate quantities and balance in arid climates can be far less than in temperate climate (Blight and Fourie, 1999) to the extent that they pose less environmental and stability risk. This aspect needs to be further assessed at this site.

The leachate is to be brought under control, primarily by drainage and cover to reduce infiltration. This includes profiling the crest area and creating channels to enable run-off. These measures will decrease the total amounts of leachate (more surface water run-off), reducing the environmental impact and will also improve stability.

5 LANDFILL GAS

Considerable landfill gas has been noted in some boreholes. This poses serious risks of explosion. It should be investigated and controlled as soon as possible, probably by venting. Gas pressures can cause slope instability, the pressures should be relieved by venting.

The transfer station appears particularly at risk from gas and should be monitored immediately and suitable safety precautions put in place. The possibility of landfill gas elsewhere and migration off-site should be assessed.

6 IMMEDIATE ACTIONS

It was concluded that Hiriya tip posed significant risks of instability. Remedial actions proposed, in approximate order of importance/likely cost-effectiveness included:

- Make the site secure from public access, with suitable warning signs and no access for personnel, except those working in the area, duly warned of the risks.
- No actions are to be taken which could worsen the present situation. In particular, there should be no excavation near the toe or in the lower slopes.
- No further waste should be placed on the crest.
- Decrease leachate levels – enable leachate to drain freely, lead it away to treatment. Infiltration at the crest is to be reduced as much as possible. Re-profile to shed rainfall where safe to do so. Profile the crest area inwards to shed rainfall back down channels next to the access ramps and then safely away to the river.

- Crest areas should be clay covered to seal the best profiles practical.
- Tension cracks should be exposed and sealed as far as practical.
- Surface water should be directed away.
- The river banks must be made secure and bank erosion prevented as far as possible. Further investigations and assessments should be made to stabilise the tip. Priorities should be established. Limited 'short-term-fixes' may substantially reduce risks.
- More detailed investigations – initially about 6 boreholes around the crest with careful logging descriptions of the waste by a geoenvironmental/waste specialist. Assess leachate strikes and equilibrium levels.
- Levels of leachate should be regularly monitored. Changes in leachate levels between dry and wet season and responses to heavy rainfall should be assessed.
- The leachate quality should be tested as soon as possible, not least to allow changes with time and landfill processes to be assessed.
- Pollution of the rivers and groundwater should be investigated. If the tip poses unacceptably large risks to the groundwater, other control measures may be needed.

7 DEVELOPMENT AND REMEDIAL WORKS

Toe berms are the most practical stabilisation measures for many areas. However, they must not block existing leachate drainage. Only granular materials should be placed immediately against the toes of slopes, in the form of free draining blankets. They will need to be quite extensive to have significant effect. For many of the slopes, this will require diversion of the rivers. Procedures to enable this are being pursued.

The excavated material should be used to best advantage on the site. The existing rivers will need to be infilled. Some free draining granular material is probably best, perhaps providing some leachate drainage and gas venting/barrier.

Berms could be formed of select engineered wastes, as dense, inert and granular as possible. With suitable design, this approach might create significant future capacity.

Pulling back the crest should be done only after securing of the toe areas and only in the dry season. Also create benches at around mid-height, suitable for access and monitoring and possibly also drainage, leachate and gas control. Reprofiling could allow processing of the wastes, with some recovery of materials (waste re-mining, re-use). Some treated wastes might be used in new tip areas.

Further study is needed of the landfill performance and processes at this site to better assess the extent of works really needed to ensure this landfill is sustainable (Tonks et al, 2003).

8 CURRENT SITUATION

There have been no major incidents to date, with only moderate rainfalls through recent winters. Plans are being developed for River diversion and options for reprofiling landfill for stability, settlement and future use (leisure/park).

A detailed survey was carried out in early 2001 and a system of monitoring was introduced. Figure 4 gives a view from the 3-D survey model showing patterns of leachate outbreaks and movements. About 50 leachate seepage areas have been identified. Most are near toe level, a few up to mid slope. Photographic records have been made and linked to the 3-D survey model. Flows reduce significantly from end of winter to summer, from fairly high reducing to slight. However, they increased to reach a peak in January 2002, with some ongoing movements and a landslip in one area (Figure 5). Five tension crack areas were monitored up to early to mid-2002. These cracks have now been covered, with the top of the dump being covered and re-profiled with a thin layer of local sandy clay fill.

Leachate recirculation to accelerate decomposition of the waste and methane generation is being assessed, subject to stability controls. More leachate collection may be designed, possibly including

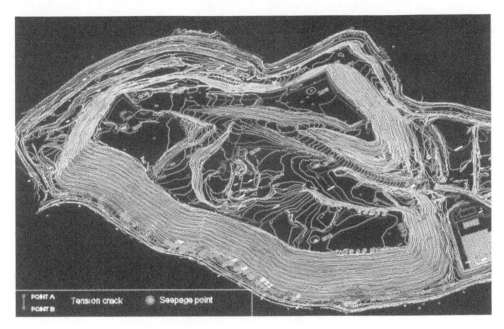

Figure 4. View of 3-D model which shows primary features.

Figure 5. Landslippage (January 2002).

horizontal drains beneath the base of landfills. Plans are in hand to investigate, monitor and relieve gas pressures. A number of boreholes have been drilled and gas collection pipes placed around the top of the dump, with the idea that the methane gas would be used to produce electricity and heat for local industries. These suggest that the leachate level reaches some 20 m below the top, i.e. some 50–60 m above the base and surrounding ground.

Figure 6 shows the north-eastern corner of the dump, along the Ayalon River. Note the steep slope – the waste was removed here following the slip in 1997 that blocked the river just to the west (around the corner). The river flows very close to the toe here.

113

Figure 6.   North bank.

Figure 7.   South bank and Shappirim river.

Figure 7 shows the similarly steep slopes of the southern bank adjacent to the Shappirim River. Figure 8 is a general view along the eastern slope, from north to south. The distance from the foot of the slope to the reactors is no more than 15 m. It is planned to build a berm along the southern section of the eastern slope, west of the waste transfer station. The slope of the lower (untouched) section of the eastern slope is approximately 35 degrees. In the upper part, the waste has been moved back to create a slope of about 10 degrees.

Figure 9 shows the crest areas, viewed from east to west. Note the new earth cover and landscaping measures. Also note the lined surface water run-off collection pond in the centre of the picture, to prevent ponding on the top of the dump.

Figure 8.   Eastern slope.

Figure 9.   Reprofiling of crest area 2004.

The Dan Area Municipal Cooperative that owns and runs the site, want to turn the Hiriya waste dump and the open farmland to the west and south-west into a park – the Ayalon Park. This involves some complicated statutory regulations, involving national, local and regional town planning committees and various interests. The process of expropriating land and obtaining permits to move the Ayalon River 100 m to the north is gradually moving forward, although final realisation of this may take some years. The owners of the site are developing a team to oversee the rehabilitation of the dump and the creation of the park. The team have begun to tackle implementation of parts of the overall concept for the creation of the park and stabilisation of the dump, including 'clawing back' waste from the upper section of the south-eastern slope, above the transfer station, and covering the top with clay and sandy clay fill, to reduce water infiltration, and to re-slope the area to prevent wholesale erosion of the fill (Figure 9). A lined pond has been built to collect water and leachate run-off during the winter months.

Figure 10 is a general view of reactors/digestors (these are digesting selected MSW) and the northern part of the transfer station, at the foot of the eastern slope. A quarter of the MSW generated

Figure 10.   View from crest showing treatment works at eastern toe.

in Israel passes through the Hiriya waste transfer station. The Hiriya team envisage turning this site into the largest waste recycling and waste-to-energy site in the country. A pilot digestor plant was built just north of the transfer station, using solid waste that had been screened and separated automatically at another part of the site. Plans are in motion for more buildings for recycling of waste below the north-eastern slope of the dump. All of this means that reinforced berms, possibly using crushed building waste, will have to be built along the foot of the eastern slope, even before the Ayalon River is moved to the north. These berms will help to stabilise the eastern slope leading to a reduction in the overall angle of the slope.

9   CONCLUSIONS

Hiriya tip has been standing over 60 m high at very steep angles, some in excess of 45°, for more than 10 years. Failures to date have been quite limited, but significant. There was a major failure in 1997 which blocked the Ayalon River. Similar conditions are present elsewhere the perimeter. The tip poses serious stability risks. The most immediate risks arise from heavy rainfall. Backanalyses shows substantially higher strength than normally relied on for MSW – most likely due to the climatic conditions which lead to relatively little leachate and very slow degradation. Conditions are likely to worsen as landfill processes progress.

Immediate measures have been introduced on a priority basis and have given limited improvements. A programme of monitoring has been introduced to assess ongoing risks using an observational approach. Assessments and priorities are reviewed as more information becomes available. Longer term plans are being developed for sustainable future use of the site. For stability, the rivers will need to be diverted well back from the toe, to allow substantial reprofiling, with suitable controls. Leachate and landfill gas are subject to ongoing investigations and remedial actions. The long term scheme will consider best options, timetables and phases to return the site to most suitable future use.

REFERENCES

Blight G E and Fourie A B (1999) *Leachate generation in landfills in semi-arid climates*, Proc ICE, Geot Eng, Oct, pp 181–188.

Dixon N and Jones D R V (2003) *Stability of landfill lining systems: Report No.1 Literature Review*, R&D Technical Report Pl-385/TR1, Environment Agency, UK.

Dixon N and Jones D R V (2003) *Stability of landfill lining systems: Report No.2 Guidance*, R&D Technical Report PI-385/TR2, Environment Agency, UK.

Eid H T, Stock T D, Evans D S and Sherrie P E (2000) *Municipal solid waste slope failure. Waste and foundation soil properties*, ASCE, Geotech and Geoenv Eng, May, pp 397–419.

Tonks D M, Nettleton I M and Needham A D (2003) *Sustainability, Maintenance & Re-Use Of Existing Infrastructure – Some Geotechnical Aspects*, XIII Euro Conf SMGE, August, Prague.

Landfill capping with microbiological treatment of landfill gases – materials and performance in test cells

M.M. Leppänen
Tampere University of Technology, Finland

R.H. Kettunen & H.M. Martikkala
Tritonet Ltd, Tampere, Finland

ABSTRACT: In Finland there is a need to find economical and robust methods, suitable also for small and remote municipal landfills, to collect and treat landfill gases. One possibility is the microbiological oxidation of landfill gases in landfill cover layers.

To evaluate the effectiveness of microbiological treatment of landfill gases in capping layers in full scale, four different test cells were built summer 2002. The main idea was to collect the landfill gases with collection layer, transfer them through the sealing liner and distribute them again to the bottom of top layer, which was built from biologically active material mixture. The follow-up of the test cells showed that landfill emissions can be controlled with such an ecological cover and methane oxidation around the year is possible, also in Finnish winter conditions. In this paper, the experiences gained during construction and monitoring about material properties and behaviour are discussed.

1 MOTIVATION AND BACKGROUND

Due to new landfill regulations, hundreds of small landfills were closed in Finland at the end of year 2001. Most of them are waiting for final capping structures. According to the European, and thereupon also the Finnish, landfill regulations, landfill gases should be collected and preferably utilised, or disposed by burning. Most old Finnish municipal solid landfills are small and have a relatively thin waste fill layer. In general, landfills are situated far from urban areas. Typically the gas production is low and uneven in different areas of a landfill. Therefore, possibilities for gas usage are limited and the cost of effective collecting and disposal of landfill gases is too high considering the environmental impact. Other solutions have been searched for, and one considered possibility is the microbiological treatment of landfill gases.

The work was based on research work made at University of Jyväskylä, in which several laboratory studies have been made to find out the methane oxidation activity of different capping materials and effect of conditions, e.g. effect of low temperature and water content.

2 TEST CELL MATERIALS

Four test cells, $200\,m^2$ each, were built on two municipal landfills, in Parkano and in Ylöjärvi in Southern Finland (Figure 1). Both landfills are relatively small, about 5 hectares, they started accepting waste at the late sixties and they both finished receiving waste at the end of year 2001.

The location of test cells was chosen on the basis of gas emission measurements. Especially in Ylöjärvi, which has received a lot of industrial waste, there are some filling areas with very low gas production. On both landfills, the test cells were built on top of recent waste fill in order to get constant and sufficient gas production. The temporary cover was removed.

The test cells were designed according to Finnish practice, and local, economical materials or suitable waste materials are used. The cover structure consisted of gas collection layer, impermeable

Figure 1. Location of Parkano and Ylöjärvi landfills.

Table 1. Layer thicknesses and materials used in test cells (from bottom to top).

	Ylöjärvi				Parkano			
	Test cell A		Test cell B		Test cell A		Test cell B	
Functional layer	Thickness mm	Material	Thickness mm	Material	Thickness mm	Material	Thickness mm	Material
Gas collection layer	300	Crushed concrete	300	Crushed concrete	300	Gravely sand	300	Gravely sand
Sealing layer	–	–	500	Fiberclay	10	Bentonite mat	400	Fiberclay
Drainage layer	400	Gravel + moraine	300	Gravel	300	Gravely sand	300	Gravely sand
Oxidation layer	400	Mixture of composted sewage sludge, sand and wood chips	300	Mixture of composted sewage sludge, sand and chemical sludge	350	Mixture of composted sewage sludge, sand and peat	450	Mixture of composted sewage sludge, sand and fiberclay

mineral liner, drainage layer and vegetation layer, as the EU directive on landfills requires. One aim of the study was to develop conceptual design criteria for a landfill cover suitable for controlling methane emissions via biological methane oxidation. Therefore the layer thicknesses and materials were varied to evaluate their effect on performance. The thicknesses of drainage layers and vegetation layer, i.e. oxidation layer, were less than directive requires in all cells, because the directive does not fully apply to landfills closed before year 2002 (Table 1). The motive was to save

Figure 2.   Openings for gas migration in bentonite mat structure.

material and further work expenses. One structure was built without sealing liner to investigate its effect on the structure performance.

The requirement for the gas collection layer and drainage layer material was simply good water and gas permeability: i.e. coarse and nonhumous material, like gravel or crushed aggregate. In Parkano, the city had a gravel pit next to the landfill hence the use of natural pure sand was economical. In Ylöjärvi landfill, a crushed concrete was used to evaluate its suitability and clogging.

The permeability requirement for the liner was $k \leq 10^{-8}$ m/s. Several options were available: Trisoplast®, bentonite mat, fiberclay (i.e. de-inking or other sludge containing clay) from paper and pulp industry, bentonite soil mixture or natural clay. The fiberclay was the most cost-effective alternative on both landfills and it will remain so also in future, and it has been widely used on old municipal landfills in Southern Finland. Hence, it was used in both landfills in one cell. The fiberclay material was, however, coming from two different producers, the other from Finncao® (Mänttä) and the other from Georgian Pacific Finland Ltd (Nokia). The other cell in Parkano was built with bentonite mat to evaluate the effect of openings (Figure 2) and thinner structure. And in Ylöjärvi, the other cell was built completely without the liner to investigate its effect on the structure performance, as mentioned above.

The most demanding and major task was to choose the materials for the oxidation layer. Important variables were porosity, water content, amount of organic matter and nutrients and water holding capacity, since the aim was to provide optimised conditions for methane oxidation. The goal was to achieve a mixture containing 10–20 percentage of weight of organic matter to provide nutrients and possibly micro-organisms, and 40–50 percentage of weight non-organic inert aggregate to maintain the porosity. In addition, small amount, 5 percentage of weight at the maximum, fine fraction can be added to improve the water retention capacity. Finnish Association of Landscape Industries has set goal values for, e.g., amount of organic matter, pH, conductivity and nutrients, and those values were used as guidelines when choosing the mixture portions.

The source of organic matter can be decomposed peat, humus, composted sewage sludge, composted bio waste or composted sludge from pulp and paper industry or other composting product. They all have an excellent water retention capacity. If waste materials are used, the quality of leaching water should be considered.

In test cells, a mixture of sand and organic matter was used. One organic component was always composted sewage sludge, while another part was either the de-inking waste (fiberclay), wood chips, decomposed peat or chemical sludge from flotation of food board factory effluents.

3   TEST CELL DESIGN

The cells were designed to enable the follow up of the hydrological balance of the structure. The test cells were surrounded by low permeable dike. The surface of the test cell was inclined 5% and the surface runoff and water running in the drainage layer on top of the sealing layer were collected separately to measuring wells. The details of the edge cross section is illustrated in Figure 3. The wells were equipped with a triangular measuring dam and a pressure sensor measuring the water level (Figure 4).

The gas redistribution system was a bit different in each test cell. In the bentonite mat cell, the gas distribution was done by 1metre-wide openings (see Figure 2). In both fiberclay structures, the landfill gas collected by the gas collection layer was guided through the sealing liner with wells and then redistributed to the top of the drainage layer through perforated pipe system. The layout of pipe system was radial in Parkano and a rake-like teeth system in Ylöjärvi (Figure 5).

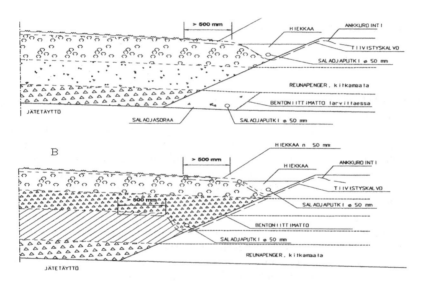

Figure 3.   Cross-section of dyke in Ylöjärvi test cells.

Figure 4.   Collection well with triangular measuring weir and pressure sensor.

122

4   TEST CELL CONSTRUCTION

Material properties were investigated in the laboratory in advance. During the construction, normal quality control measurements were performed.

Densities of the layers were measured by radiometric Troxler device. For the sealing layers material specific requirements for the wet density were set. The Troxler has some problems with materials like fiberclay, which contain both free and bound moisture, and often some affecting chemicals, too. Therefore the water content and onwards the dry density values given by the Troxler device should be corrected or use a material specific calibration. We chose to follow the wet density, hence that is nearly correct, and to determine the water content from the samples in the laboratory to obtain the correct dry density values. For example, for Finncao® fiberclay the limit wet density was 1.3 t/m³, which was achieved (Figure 6). The fiberclay was compacted by caterpillar excavator.

Figure 5.   Construction of radial gas distribution system from perforated pipes.

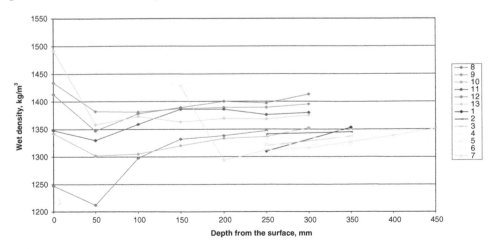

Figure 6.   Results of Troxler measurements from liner built from fiberclay (in Parkano).

123

Figure 7.   The screener-crusher bucket used for mixing.

The oxidation layer should not be compacted to maintain the high porosity. The Troxler was used also to follow the density and air void content of the oxidation layer. The target value for the air void content was 25% of volume and according to the Troxler measurements it was achieved in average in all test cells; 23–43%. The wet density was $1.2–1.5\,t/m^3$.

The oxidation layer material was mixed on the volumetric bases and it is not possible to give very small-scale amounts. The mixing was done with screener-crusher bucket (Figure 7).

Samples were taken from each layer during construction for water content and grain size distribution determination, if relevant. Especially, the quality of oxidation layer materials was investigated.

In oxidation layer materials, the amount of total solids varied from 70–80% of wet mass except in the layer where chemical sludge was used were it was under 65%. In spite of different materials and their varying humus content, the amount of organic matter in oxidation layers varied from 7–12% of TS, while the guideline values were 6–12%.

5   MONITORING

Test cells were monitored for one year. Gas emission measures were made regularly. For monitoring, a climate station, temperature and moisture gauges, settlement plates and pore gas pipes were installed to the test cells. Location of the gauges is illustrated in Figure 8. The performance of structures was also monitored. The amount of surface runoff and water collected by the drainage layer were measured with collection wells.

Temperature gauges were pretty reliable, but with moisture gauges and pressure sensors installed to the wells there were some problems every now and then, probably due to the aggressive chemical environment. There were problems with collection wells in the beginning, because the factory did not understand that the measuring dams should be watertight and it was not an easy thing to repair once they were already installed and surrounded with waste.

Wells were supposed to be emptied regularly but they were no longer personnel on the landfill and there were some periods when no measurements were done. Automatic pumps could have prevented these human errors.

The magnitude of total settlements were 100–250 mm, depending on the weight of the layers and the waste underneath. Predictably, the waste fill under the test cell structures was responsible

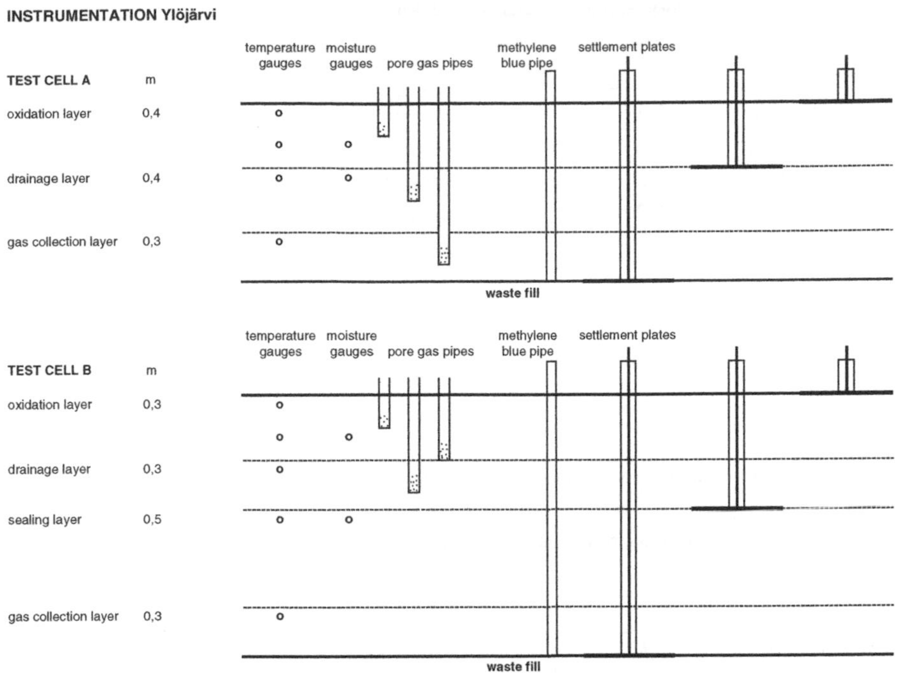

Figure 8. Instrumentation plan for Ylöjärvi test cells.

for the largest portion of the settlement. Surprisingly, the fiberclay layers seemed to expand rather than compress.

6 WATER BALANCE AND WATER QUALITY

The precipitation and evaporation were measured. In theory, the total amount of infiltrating water (precipitation minus evaporation) during the follow up period calculated from daily average was in Parkano 266 mm (7.6.2002–30.9.2003) and in Ylöjärvi 171 mm (19.6.2002–24.6.2003), which is 60–65% of the precipitation. In Parkano, the rainfall to the test cell area was $84 \, m^3$ (420 mm) and in Ylöjärvi $68 \, m^3$ (339 mm).

Evaporation was in general larger than precipitation. Only 3% of the infiltrating water was running through the drainage layer and 97% was temporarily adsorbed by the oxidation layer and later evaporated. Surface runoff was a bit bigger in Parkano, where the vegetation covered the surface. Even there the surface runoff was only 3% of rainfall. The openings in bentonite mat test cell did not increase the infiltration measurably.

Quality of the water was determined five times during the follow up period (pH, alkalinity, conductivity, TOC, COD_{Mn}, COD_{Cr}, total nitrogen, ammonium nitrogen, total phosphorus). The amounts of heavy metals were studied once. The water running in the drainage layer on top of the sealing layer contained COD_{Cr} over the limit for municipal wastewater in two test cells. The first was the oxidation layer containing fiberclay and the other the layer containing chemical sludge. Also total nitrogen values were high, but phosphorus was not a problem.

The changes in the relative moisture of the oxidation layer in the test cells are presented in Figures 9–10 with rainfall and air temperature, for one cell from each landfill. These are not absolute values, because moisture gauges were not calibrated. The years 2002 and 2003 had exceptionally light rainfall. Therefore, the oxidation layers were dry at the end of the summer 2003.

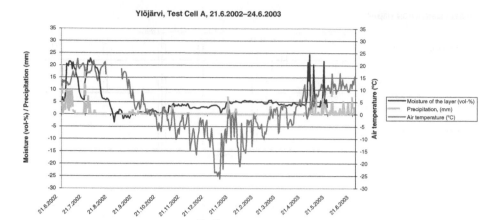

Figure 9.   Relative moisture of the oxidation layer in test cell A at Ylöjärvi.

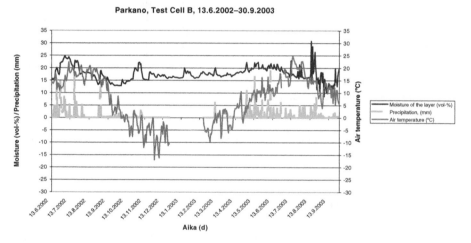

Figure 10.   Relative moisture of the oxidation layer in test cell B at Parkano.

7   TEMPERATURE AND FROST

The temperature profiles of the test cells are presented in Figures 11–14. The approximate installation depth of the temperature gauges is presented in Figure 8. According to the temperature measurement in Parkano all layers in both test cells remained below zero, even during the severe frost period in winter 2002–2003. In Ylöjärvi temperature fluctuations were wider. In both test cells there were periods below zero in oxidation layer, lasting two weeks in wood chip cell (A) and one month in the other test cell (B). In wood chip cell the temperature raised very high, over 40°C, during the first summer, probably due to the decomposition of the wood chips. The high temperature might have retarded the activity of methane oxidation bacteria, because their optimum temperature area is 25–40°C.

The thickness of snow cover and frost depth were measured once a month. The thickness of snow cover varied 0–300 mm and frost depth was less than 100 mm, as also temperature measurements indicate.

When choosing materials for oxidation layer special attention should be paid to the thermal conductivity of materials so that methane oxidation is possible during winter, too.

126

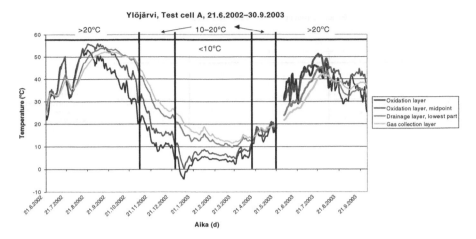

Figure 11.   Temperature profile from Ylöjärvi test cell A.

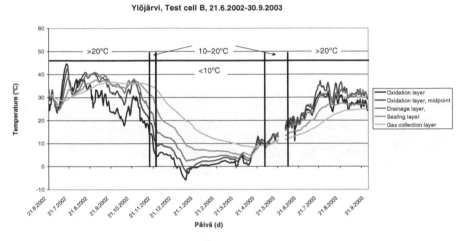

Figure 12.   Temperature profile from Ylöjärvi test cell B.

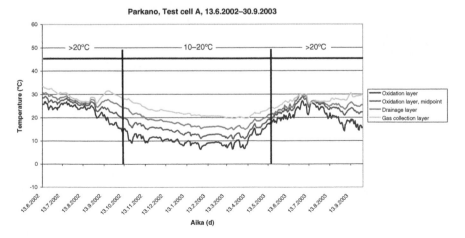

Figure 13.   Temperature profile from Parkano test cell A.

127

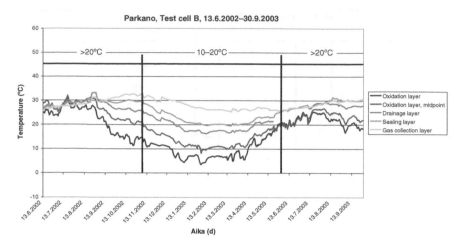

Figure 14.   Temperature profile from Parkano test cell B.

Figure 15.   View over Parkano test cells.

8   VEGETATION

Turf was planted on each test cell. In Parkano the turf was flourishing, thanks to the watering during the rooting. Only minor damage caused by the landfill gases were noticed (Figure 15). In wood chip test cell in Ylöjärvi the turf had problems to root due to the poor substratum and landfill gases discharging through the permeable cover. In other test cell in Ylöjärvi turf coped better, but there were noticeable damage on top of the gas distribution system. During summer weeds spreaded out, while the turf was suffering from the drought (Figure 16).

128

Figure 16.   View over Ylöjärvi test cell A.

9   CONCLUSIONS

Both bentonite mat and fiberclay are suitable for landfill cappings with methane oxidation. Methane can be redistributed either by well and pipe systems or with openings. The frequency of openings and pipes should be sufficient to collect gas. According to gas emission measurements, gas migrates about 2 m sideways from the pinpoint of the perforated pipe or from the edge of an opening.

The recommended minimum thickness of an oxidation layer is 40–50 cm considering the depth where the oxygen migrates, the possible drying of the oxidation layer during dry periods and the decrease of temperature of the layer during winter. It is recommended to plant shallow rooted vegetation on top of the oxidation layer to prevent erosion, to maintain the porosity through root and worm activities, and to level temperature differences. While the requirements for a good oxidation layer are pretty much equal to the requirements for a good substrate for plants, a functional oxidation layer is also landscaping the landfill.

The materials used in capping layers contain nutrients and contaminants, which might leach out from the structure. This should be taken into account in material selection and when estimating the environmental impacts of the landfill. Local and suitable waste materials can and should be used to obtain ecological and economical capping. Each layer is essential for the proper function of the capping, but the thicknesses can be varied depending on local conditions, chosen materials and the functional targets of the capping.

ACKNOWLEDGEMENTS

This paper is based on reports composed by Riitta Kettunen and Heidi Martikkala. On the basis of this study, recommendations for design of landfill capping structure to oxidate the methane were written and published at Landfill seminar organised by Finnish Environment Centre.

REFERENCES

Kettunen R. (2004). Kaatopaikan pintakerroksen suunnittelu ja metaanin biologinen hapettuminen. (*Design of landfill capping and biological oxidation of methane*). Landfill seminar, 23–24.3.2004 Finnish Environment Centre, Helsinki. In Finnish. Available from web-site of FEI http://www.ymparisto.fi/default.asp?contentid=35111&lan=fi.

Leppänen M. (2003). Metsäkylän kaatopaikka, koerakenteet. Rakentamisen laadunvalvonta. (*Metsäkylä landfill, test cells. Quality control during construction*) 10.11.2003. Tampere University of Technology. In Finnish. Unpublished.

Leppänen M. (2003). Kangaslammin Kaatopaikka, Parkano, koerakenteet. Rakentamisen laadunvalvonta. (*Kangaslammi landfill, Parkano, test cells. Quality control during construction*) 29.10.2003. Tampere University of Technology. In Finnish. Unpublished.

Martikkala H. (2004). Metaanin biologinen hapettuminen kaatopaikan pintarakenteessa. (*Biological oxidation of methane in landfill capping*). MSc thesis, Tampere University of Technology, Environmental Engineering and Biotechnology, Tampere.

Martikkala H. & Kettunen R. (2003). Tutkimus kaatopaikkojen metaanipäästöistä ja niiden hallinnasta Parkanon ja Ylöjärven kaatopaikoilla. (*Investigation on landfill methane emissions and their control in Parkano and Ylöjärvi landfills*). Final report. 62 pp. In Finnish, extended English abstract. Available from web-site of Finnish Solid Waste Association http://www.jatelaitosyhdistys.fi/.

Contaminated land

Geotechnical and Environmental Aspects of Waste Disposal Sites – Sarsby & Felton (eds)
© *2007 Taylor & Francis Group, London, ISBN 978-0-415-42595-7*

Effect of rates of biological transformation of additives on the efficiency of injection in soil under hydraulic and electric gradients

A.N. Alshawabkeh, X. Wu & D. Gent
Northeastern University, Boston, USA

J. Davis
US Army Engineer Research and Development Center, Vicksburg, USA

ABSTRACT: This study evaluates the relationship between degradation rates and transport rates of lactate, a biodegradable additive. Transport and transformation rates are evaluated in clay, and the effect on injection is assessed using a theoretical model. The results show that hydraulic injection of lactate into clay is limited due to the relatively faster rate of lactate transformation compared to the slow hydraulic transport. Injection by electric fields required application of a relatively high current density (of the order of $5\,A/m^2$) in order to be able to effectively deliver lactate into the soil. The modelling results showed that lactate could be injected at a rate of 10 cm/day in low permeability soil, if no transformation occurs. However, the biological transformation limits the transport efficiency to around 1 to 3 cm/day.

1 BACKGROUND

In-situ bioremediation is one of the most promising and cost effective technologies to clean up contaminated soil. The process requires indigenous active microbes, electron donour, macronutrients (e.g., nitrogen and phosphates), micronutrients, trace nutrients and electron acceptor. For highly chlorinated aliphatic hydrocarbons (CAHs), such as tetrachloroethylene (PCE), carbon tetrachloride (CT), and 1,1,1-trichloroethane (TCA), reductive dechlorination is more pronounced and electron donours are required (McCarty, 1997).

Recent studies have shown that hydrogen gas could serve as a key electron donour for biological transformation of highly chlorinated aliphatic hydrocarbons, such as PCE. Many organic acids can be additives as a source of hydrogen gas (Freedman and Gossett, 1989; Ballapragada et al, 1997), but the effective delivery of these additives is usually a constraint to the technique. Traditional injection methods have been generally based on hydraulic flow systems (Erickson et al, 1994; Boyle et al, 1999). Such systems are very effective in high permeability soils (k greater than 10^{-3} cm/s) such as sandy aquifers, but few systems have been developed to address soils with lower permeability and heterogeneous deposits.

Studies have shown that electric fields can injections into low permeability soils (k $< 10^{-7}$ cm/s) at efficient and uniform rates of up to a few centimeters per day (Acar and Alshawabkeh, 1993; Acar et al, 1996 and 1997). The major transport mechanisms of fluids and additives in clay under electric fields include electro-osmosis (EO) and electro-migration (EM) or ion migration (Mitchell, 1993). Acar et al. (1997) demonstrated uniform transport rates (up to 10 cm/day) of inorganic additives in sand and kaolinite (ammonium and sulfate) by electrokinetics. Thevanayagam and Rishindran (1998) showed that nitrate was transported by ion migration in kaolinite. Rabbi et al. (2000) evaluated transport of benzoic acid to enhance TCE transformation in soil while circulating electrolytes. Reddy et al. (2003a) and Suer and Lifvergren (2003) showed that electrokinetic injection of iodide could enhance the removal of mercury as an iodide complex from glacial till. Reddy et al. (2003b) evaluated electrokinetic delivery of ammonium, acetate and phosphate for biostimulation of Cr(VI) reduction in kaolinite.

Most studies have focused on inorganic additives, but for organic additives, it is necessary to assess the feasibility and efficiency of transport because of limitations due to adsorption and biological transformation of the additives. Rabbi (2000) injected benzoic acid to enhance the dechlorination of TCE, but TCE transformation was only limited at the boundaries, not in the middle sections. The study concluded that injection was not effective due to the fast rate of consumption of benzoic additives by the soil microorganisms. Reddy et al. (2003b) observed the same phenomenon when trying to deliver ammonium, acetate and phosphate for reduction of Cr(VI) in kaolinite. The study showed that only 2% acetate, as compared to 90% of phosphate, was injected into soil. Microorganisms may affect the concentration of acetate in the soil. It is necessary to assess the transformation rates of these additives compared to the transport rates by electrokinetics.

The study reported herein evaluates the relationship between degradation rates and transport rates of a biodegradable additive. Electrokinetic injection of lactate, a commonly used electron donour for anaerobic biodegradation (Carr and Hughes, 1998; Fennell et al, 1997; Song et al, 2002), in clay was evaluated. A numerical model is proposed and the delivery of lactate was predicted in soils. Measured injection rate and the prediction rate were compared, and the effect of degradation rate was analyzed.

2 THEORY OF ELECTROKINETIC INJECTION

Since the principle of electrokinetic injection was introduced by Mitchell (1993) and Acar and Alshawabkeh (1993) a lot of models had been proposed to analyze or solve different problems. Yu and Neretnieks (1996) developed a one-dimensional model to simulate the removal of copper from sand, a process with cathode rinsing, and a process with ion exchange membrane. Alshawabkeh and Acar (1996) presented a mathematic model for multi-component species. Kim et al. (2003) formulated a numerical model to simulate Cadmium transport in Kaolinite. Alshawabkeh and McGrath (2000) comprehensively described the theoretical basis of electrokinetic remediation. Generally, under an electric field, contaminant transport is through hydraulic flow, electro-osmosis flow and ion migration. Electrolysis reactions at electrodes and physico-chemical soil-contaminant interaction make the system extremely complex. This is coupled with microbial activities and the soil system under electric field has not been fully understood.

The total mass flux under a hydraulic gradient is given by,

$$J = -D_L \frac{\partial c}{\partial x} + u^* c i_e - k_e c i_e \pm k_h c i_h \qquad (1)$$

where D_L is coefficient of hydrodynamic dispersion, k_h is hydraulic permeability, i_h is hydraulic gradient, i_e is electrical gradient, k_e is electro-osmotic permeability, u^* is effective ionic mobility and c the additive concentration.

Hydrodynamic dispersion includes molecular diffusion and mechanical dispersion, which equals $(D^* + a_L V)$. D^* is effective diffusion coefficient, a_L is the dispersivity and V is the flow velocity (Fetter, 2001). The direction of hydraulic flow may enhance or retard ion delivery. The dispersivity in the laboratory ranges from 3.96×10^{-3} to 7.07×10^{-2} m (Klotz et al, 1980). Lactate ions are negatively charged and therefore will migrate from the cathode towards the anode. However, electro-osmosis generally occurs from anode to cathode and will retard the ionic migration of lactate. In low permeability soil, it was demonstrated that ion migration was the dominant mechanism of injection (Acar et al, 1997; Thevanayagam and Rishindran, 1998).

Electrolysis reaction, adsorption/desorption, precipitation/dissolution, and biodegradation were the most important reactions in the soil system. Electrolysis reaction at electrodes may result in system pH change, zeta potential change and other physicochemical reactions (Yeung et al, 1997). For bioremediation a neutral condition is the optimum state. Feeding acid and base in the cathode and anode, respectively, is one of the options (Acar et al, 1997). Recycling electrolyte is another easy option (Rabbi et al, 2000; Lee and Yang, 2000). The first method was used in this paper.

Adsorption of additives onto soil decreases the delivery rate, limits the access of microbes. It is reflected by retardation factor R (Reddi and Inyang, 2000). If a single component linear isotherm is assumed the retardation factor is given by,

$$R = 1 + \frac{\rho_b K_d}{n} \tag{2}$$

where ρ_b is the bulk density, K_d is the adsorption equilibrium constant, and n is the soil porosity.

Significant bioactivity may exist in a soil system where certain conditions apply, this is the basis of natural attenuation. Biodegradation generally follows the Monod equation;

$$\frac{dC}{dt} = -\frac{\mu_m x C}{(K_s + C)} \tag{3}$$

where C is the lactate or substrate concentration (mg/l), μ_m is the maximum specific growth rate (or maximum rate of substrate [lactate in this case] utilization per unit mass of microorganisms, expressed as mg lactate/l per mg biomass/l per day) multiplied by the microbe biomass concentration (mg biomass/l), and K_s is the half-saturation constant (mg/l) – (Rittman and McCarty, 2001).

The equation of conservation of mass in a unit volume of the soil can be derived for the above analysis. Certain assumptions were made to make the equation simple. Firstly, it was assumed that the soil and water volumes are not changing. Secondly, k_e and i_e were assumed to be constant and equal to the average value. Thirdly, the dynamic dispersivity was constant at 1×10^{-2} m. First order transformation models were assumed for lactate, and no soil-contaminant interactions other than adsorption was considered.

The flow rate for calculating dispersion coefficient would be,

$$v = \frac{(k_e i_e + k_h i_h)}{n} \tag{4}$$

The total ion migration rate would be given by,

$$V = \frac{(u^* i_e - k_e i_e \pm k_h i_h)}{n} \tag{5}$$

The partial differential equation would be,

$$R\frac{\partial C}{\partial t} = \frac{(D^* + a_L v)}{n}\frac{\partial^2 C}{\partial x^2} - V\frac{\partial C}{\partial x} - \mu C \tag{6}$$

with boundary conditions $C(x,0)=0$, $C(0,t)=C_0$, $\frac{\partial C}{\partial x}(L,t)=0$.

An approximate solution was given by van Genuchten and Alves (1982) as below (equations 7 and 8) and Matlab was used to calculate the solution:

$$u = V\left(1 + \frac{4\mu D}{V^2}\right)^{\frac{1}{2}}$$

$$
\begin{aligned}
B_3(x,t) = &\frac{1}{2}\exp\left[\frac{(V-u)x}{2D}\right]\text{erfc}\left[\frac{Rx - ut}{2(DRt)^{\frac{1}{2}}}\right] + \frac{1}{2}\exp\left[\frac{(V+u)x}{2D}\right]\text{erfc}\left[\frac{Rx + ut}{2(DRt)^{\frac{1}{2}}}\right] \\
&+ \frac{(u-V)}{2(u+V)}\exp\left[\frac{(V+u)x - 2uL}{2D}\right]\text{erfc}\left[\frac{R(2L - x) - ut}{2(DRt)^{\frac{1}{2}}}\right] \\
&+ \frac{(u+V)}{2(u-V)}\exp\left[\frac{(V-u)x + 2uL}{2D}\right]\text{erfc}\left[\frac{R(2L - x) + ut}{2(DRt)^{\frac{1}{2}}}\right] \\
&- \frac{V^2}{2\mu D}\exp\left[\frac{VL}{D} - \frac{\mu t}{R}\right]\text{erfc}\left[\frac{R(2L - x) + Vt}{2(DRt)^{\frac{1}{2}}}\right]
\end{aligned} \tag{7}
$$

$$B_4(x,t) = 1 + \left(\frac{u - V}{u + V}\right) \exp(-uL/D),$$

$$B(x,t) = B_3(x,t)/B_4(x),$$

$$C(x,t) = C_0 B(x,t)$$

(8)

3 METHODOLOGY

3.1 *Materials*

Lactate was selected as the electron donour additive. Lactate is introduced as an acid in order to provide the lactate required for remediation and balance the alkalinity of the catholyte to maintain neutral pH conditions. Bromide was used as the tracer as it is non-degradable and not adsorbed in soil.

The soil was a finely-graded, low plasticity silty clay (CL), named Gessie. The soil was mixed with tap water at 23% water content and was compacted in 6 layers with 8 blows per layer using a 2.49 kg (5.5 lb) hammer and 30.48 cm (12 in.) drop to give a final soil porosity of 0.44.

3.2 *Test procedure*

Adsorption tests were conducted to assess the role of retardation due to lactate adsorption onto the soil organic content. Concentrations of 40 ml lactate solutions were mixed with 10 g of soil in a 50 ml polyetheyle centrifuge vial. Concentrations of lactate in blank samples (without soil) were assumed as the initial value. A sodium azide concentration of 200 mg/l was used to prevent microbial activity and biodegradation. All vials were continuously mixed for one day and centrifuged at 2000 rpm for 20 minutes to separate the liquid from the soil. Lactate concentration in the supernatant liquid was measured and solid phase concentration was calculated based on the mass balance principle.

Degradation of injected lactate by indigenous microbes in soil is expected to further retard lactate transport and injection efficiency. Anaerobic biodegradation tests were conducted using the non-sterile clay without innoculation. Anaerobic conditions were selected because lactate is injected at the cathode and it will ultimately be used for anaerobic dehalogination of chlorinated solvents. Samples of 10 g soil were mixed with 50 ml lactate solution in serum bottles, and continuously mixed by shaker. Supernatant lactate concentration was measured to detect the biodegradation kinetics.

Electrokinetic injection experiments were conducted using a typical electrokinetic transport setup. Figure 1 shows a schematic arrangement of the setup, which is made of acrylic and consists of three compartments for anode, cathode and soil cell. Both electrode compartments are similar and can hold up to 4200 ml effective volume of electrolyte. A liquid tank with a volume of 1000 ml was connected to the cathode, to provide the ability to add liquid in the cell tank and pH measurement without electrical interference. The soil cell compartment had a rectangular cross section and was 5 cm wide, 15 cm deep and 40 cm long. Along the cell, voltage probes and sampling ports were placed every 8 cm at two vertical levels (total three vertical level). Soils in the cell were separated from the reservoir solution by a hyper synthetic non-reactive permeable membrane. Graphite electrodes were placed in the reservoirs. Cathode electrolytes were circulated and pH-control pumps were used to provide lactic acid to neutralize it in the cathode tanks. Anode electrolytes were neutralized manually everyday by adding base sodium hydroxide. DC power supply was used to give constant direct current at 9 mA or 40 mA, which would generate current density of 1.2 A/m^2 or 5.3 A/m^2 along the soil cross-area. Overflow at the cathode compartment and inflow at the anode compartment was measured by volume cylinder each day. Pore water samples were taken and analyzed using Ion Chromatograph (Dionex 120). The column used was AS15, with mobile eluent 40 mM sodium hydroxide and flow rate at 1.2 ml/min. The standard curve was updated every three months, and it was checked by two levels of standard solution each time analyzing.

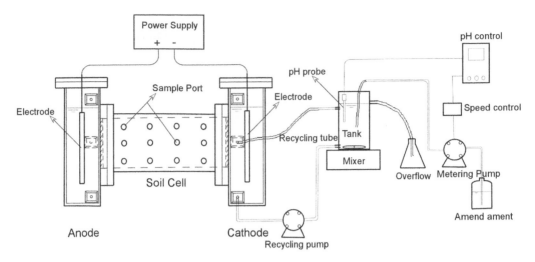

Figure 1.   Injection experiment set-up.

Table 1.   Injection experiments description and variables.

Test label	Current density (A/m²)	Average electrical gradient (V/cm)	Duration (h)	Electrode condition	
				Anode	Cathode
Clay L1	1.2	0.2	669	Manually adjust	Control pH 8
Clay L2				pH 7 daily	10,000 mg/l
Clay H1	5.3	0.8	1606	Manually adjust	Control pH 8
Clay H2				pH 7 daily	10,000 mg/l
Clay 21	5.3	0.5	1841	Manually adjust	Control pH 8
Clay 22				pH 7 daily, 10,000 mg/l	10,000 mg/l
Clay B1	5.3	0.48	288	Manually adjust	Control pH 8
Clay B2			310	pH 7 daily	Bromide 1000 mg/l
Clay C1	Hydraulic		2184	No control	10,000 mg/l
Clay C2	grad. = 1				

A series of injection experiments were conducted. Duplicate tests, control tests with unit hydraulic gradient and non-reactive control tests with bromide were conducted to ensure quality data were obtained. Table 1 lists the experimental tests and the variables identified for testing.

4   RESULTS AND DISCUSSION

4.1   *Lactate adsorption and biodegradation in clay*

The adsorption tests were conducted under sterile conditions. It was found that the equilibrium concentrations of lactate were always a little higher, from 3.5 mg/l to 15 mg/l with an average of 10 mg/l, than the initial concentration (blank concentration). The results demonstrate that there was no adsorption of lactate onto the clay, but some lactate (approximately 10 mg/l) existed in the original soil and was released into the solution. Under such conditions, the soil would not exhibit any retardation of lactate injection due to adsorption, and the retardation factor R equals 1.

The biodegradation tests were conducted non-sterile with indigenous microorganism cultures. The Monod equation (equation 3) was used to simulate the biodegradation. No growth of microbe mass was assumed during the anaerobic degradation experiments and similarly for the electrokinetic bioremediation. The concentration was monitored for more than 10 days till stable. The results

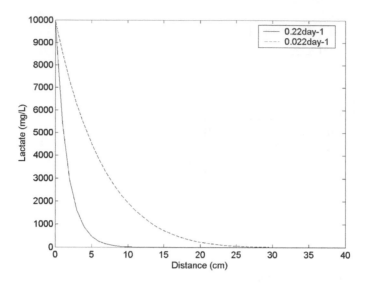

Figure 2.   Lactate transport under unit hydraulic gradient with different degradation rates.

show that first order biodegradation was followed, and lag time was less than half day. Average degradation rates resulted in a first order transformation rate ($\mu = \mu_m/K_s$) of 0.22/day. The results clearly show that lactate is easy biodegradable under the conditions of this study. Lactate is a safe additive for remediation that will not develop as a secondary contaminant. But the degradation rate can be a limiting factor of effective injection and should be assessed together with transport rates.

4.2   Transport control experiments

Two types of transport control experiments were conducted; hydraulic injection control and non-reactive tracer injection control. The hydraulic injection control experiments were conducted to assess the transport of lactate in soil under a unit hydraulic gradient. After two months of testing, limited lactate concentration was detected in the ports. In fact, the concentration did not exceed 100 mg/l in the first port (8 cm distance). This is expected due to the low hydraulic conductivity of the soil (3.6×10^{-7} cm/s). At the same time, lactate degradation further limits the transportation and results in an efficient hydraulic injection. From the theoretic model calculation the same conclusion could be drawn. With degradation rate as high as $\mu = 0.22$/day, 100 mg/l of lactate can never penetrate 8 cm into clay under unit hydraulic gradient. As the degradation rate decreases injection could be more effective. But even at $\mu = 0.022$/day, which is ten times slower than experiment result, the 100 mg/l lactate front can only penetrate 24 cm (Figure 2). The results demonstrated that lactate could not be successfully injected in clay under unit hydraulic conductivity, and the high gradation rate of lactate could be the key limitation factor for effective delivery of additives.

Bromide was selected as the non-reactive ionic tracer because of its negative charge, and its negligible initial concentration in soil. Experiments were conducted under 5.3 A/m^2. The average voltage gradient in this experiment was about 0.48 V/cm and the net ion migration rate was calculated as of the order of 5.7 cm/day. The average electro-osmosis permeability was 2.0×10^{-5} cm^2/Vs. Using the parameters given above, the theoretically-predicted concentration would be much less than the measured concentration at the beginning (data not shown). This may result from using the average electro-osmosis and electric gradient. During the first 50 hours, the voltage gradient was much higher than average (about 0.86 V/cm) and the electro-osmosis permeability was as low as 1.2×10^{-5} cm^2/Vs. Figure 3 shows the bromide concentration measured in the soil and predicted using the model at 26 hrs and 42 hrs. The measured results and the predicted results agree very

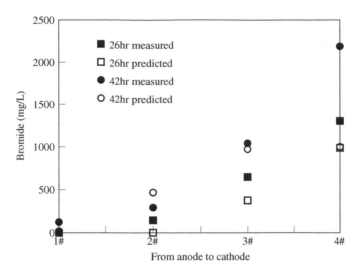

Figure 3. Measured and predicted bromide concentrations in clay under $5.3\,A/m^2$.

well. The trace experiments demonstrated that the potential of electrokinetic delivery rate could reach as high as 13.5 cm/day with electro-osmosis control under electrical gradient 1 V/cm.

4.3 *Lactate injection in clay under DC*

Low current density $1.2\,A/m^2$ was first tried to inject lactate into clay and which resulted in a potential gradient of 0.2 V/cm and no significant electro-osmosis flow. The lactate concentration in the ports was not consistent and fluctuated over a range of a few hundreds of mg/l. There was no clear increasing trend or significant accumulation of lactate concentration in the clay even after four weeks. It was concluded that the current density was not sufficient to produce efficient injection of lactate into the soil and that the biological transformation rates significantly influenced injection rates. Theoretical model prediction showed that the injection front of 1000 mg/l could not penetrate more than 25 cm in the soil column. The front of injection decreased sharply with higher degradation rate as shown in Figure 4. The unsuccessful injection of lactate in clay under low current density may result from possible higher bioactivity and exist electro-osmosis flow which cannot be measured in the experiment because of evaporation and soil swelling.

Experiments using high current density $5.3\,A/m^2$ produced a voltage gradient of 0.8 V/cm and high electro-osmosis with an average k_e $5.4 \times 10^{-5}\,cm^2/Vs$. Lactate was only introduced at the catholyte. These experiments showed interesting results. Within the first few days, significant concentration of lactate was detected in port 3 and port 4, and it reached higher than 1000 ppm at 400 hrs. But there was some retardation for advance beyond port 2 towards the anode. Figure 5 shows that it took another 800 hrs for lactate to reach port 2. Non-uniform electro-osmosis flow across the electrodes may limit the transport towards the anode. The soil close to the anode resulted in an EO flow that is more than 5 times the EO produced in soil close to cathode. This behavior and its effect on the mechanical properties of soil was reported by Alshawabkeh et al (2004). As a result, higher flows occur near the anode and decreased flows occur towards the cathode. This is clearly the reason why lactate migrates up to middle sections of the soil and does not move further. The delayed transport by EO combined with the microbial activity of the soil limits the transport of lactate across the clay.

139

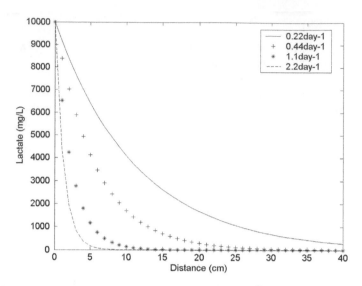

Figure 4.   Effect of degradation rate on lactate delivery under 1.2 A/m^2.

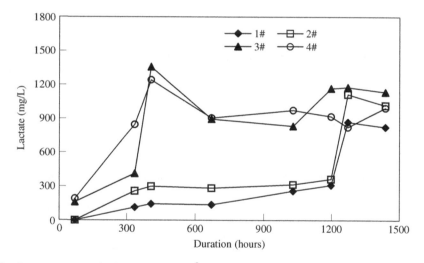

Figure 5.   Lactate transport in clay under 5.3 A/m^2 with high concentration at cathode.

To increase the extent of lactate injection and transport into the soil, it would be beneficial to reduce the electro-osmotic flow rate, particularly near the anode. As

$$k_e = \frac{\varepsilon \xi}{\eta} n \qquad (9)$$

where n is soil porosity, ε is permittivity of the medium, ξ is zeta potential and η viscosity, it is mainly dependent on zeta potential (Thevanayagam and Rishindran, 1998). Zeta potential is a complex function of both the pore water and the solid phases (Hunter, 1981). It is sensitive to pH and electrolyte concentration (Eykholt and Daniel, 1994). The effect of electrolyte chemistry was represented by Kruyt (1952) as,

$$\xi = A - B \log C \qquad (10)$$

where A and B are both constants.

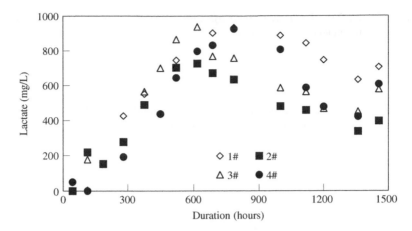

Figure 6.   Lactate transport in clay with high concentrations at both electrodes.

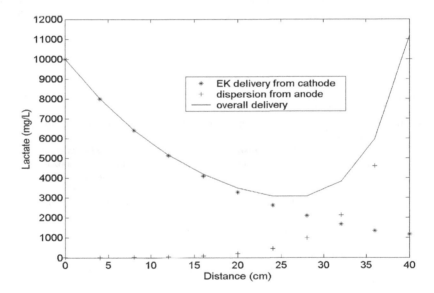

Figure 7.   Dispersion effect of high concentration in anode.

Increasing the ion concentration in the pore fluid could be a good option to control the electro-osmosis. Experiments with initial concentration of lactate 10,000 ppm added in both anode and cathode tanks were conducted under current density of 5.3 A/m². In these experiments, average effective potential gradient was around 0.5 V/cm, and the electro-osmotic flow was decreased to an average of 2.5×10^{-5} cm²/Vs, which is half of experiments without control. Although substantial electro-osmotic flow occurred, the results showed that ion migration from cathode to anode was dominant and effective injection of lactate was achieved. Lactate appeared in the soil after 100 hours of testing. After 520 hours (618 hours in the duplicate), a sustainable and relatively uniform lactate concentration of about 800 mg/l was achieved in the soil across the cell (Figure 6).

Net effective injection rate of 3.7 cm/day (3.1 cm/day in the duplicate) was achieved in soil, which is more than 120 (100 in duplicate) times faster than injection rate under a unit hydraulic gradient (the hydraulic advection rate in the clay was 0.031 cm/day). The uniform concentration achieved in the soil may result from the migration from cathode and the diffusion/dispersion from anode (Figure 7). With superposition of dispersion from anode added the concentration across the soil was more uniform. Hydrodynamic dispersion could affect the contaminant transportation significant due to high electro-osmosis flow.

5 CONCLUSIONS

Delivery of lactate in clay hydraulic gradient and electrical fields was evaluated in this paper. Even though there was no adsorption of lactate onto soil, the efficiency of lactate delivery could be limited due to high degradation rate. Hydraulic injection of lactate may not be successful due to the relatively faster rate of lactate transformation compared to the slow hydraulic transport.

Injection by electric fields required application of a relatively high current density (of the order of 5 A/m^2) in order to be able to effectively deliver lactate into the soil. Electro-osmosis flow control helped to improve the efficiency of electrokinetic injection a lot. The lactate can be injected at rates that may reach 10 cm/day in a low permeability soil, if no transformation occurs. However, the biological transformation limits the transport efficiency to around 1 to 3 cm/day. It is necessary to evaluate lactate consumption by indigenous microorganisms and its effect on injection rate. Transport rate should higher than degradation rate in order to achieve effective transport. Otherwise, lactate will always be consumed in the first few centimeters of the soil, causing excessive biological growth (biofouling) and ineffective injection.

REFERENCES

Acar Y B and Alshawabkeh A N (1993) *Principles of electrokinetic remediation*, Environmental Science and Technology, 27(13), pp2638–2647

Acar Y B, Ozsu E, Alshawabkeh A N, Rabbi F M and Gale R (1996) *Enhanced soil bioremediation with electric fields*, CHEMTECH, 26(4), pp40–44

Acar Y B, Rabbi M F and Ozsu E (1997) *Electrokinetic injection of ammonium and sulfate ions into sand and kaolinite beds*, Journal of Geotechnical and Geoenvironmental Engineering, 123(3), pp239–249

Alshawabkeh A N and McGrath C J (2000) *Theoretical basis for simulation of electrokinetic remediation*, Remediation Engineering of Contaminated Soils, Marcel Dekker, Inc, New York, pp155–171

Alshawabkeh A N, Sheahan T C and Wu X (2004) *Coupling of electrochemical and mechanical processes in soils under DC fields*, Mechanics of Materials, 36, pp453–465

Ballapragada B S, Stensel D H, Puhakka J A and Ferguson J F (1997) *Effect of hydrogen on reductive dechlorination of chlorinated ethenes*, Environmental Science and Technology, 31(6), pp1728–1734

Boyle S L, Dick V B and Ramsdell M N (1999) *Enhanced in-situ bioremediation of a chlorinated VOC site using injectable HRC*[TM], Hazardous and Industrial Wastes-Proceedings of the Mid-Atlantic Industrial Waste Conference, pp47–56

Erickson L E, Banks M K, Davis L C, Schwab A P, Muralidharan N, Reilley K and Tracy J C (1994) *Using vegetation to enhance in-situ bioremediation*, Environmental Progress, 13(4), pp226–231

Eykholt G R and Daniel D E (1994) *Impact of system chemistry on electro-osmosis in contaminated soil*, Journal of Geotechnical Engineering, Vol. 120, No. 5, pp797–815

Freedman D L and Gossett J M (1989) *Biological reductive dechlorination of tetrachloroethylene and Trichloroethylene to ethylene under methanogenic conditions*, Applied and Environmental Microbiology, 25(9), pp2144–2151

Hunter R J (1981) *Zeta Potential in Colloid Science*, Academic Press, London, UK

Kim S S and Han S J (2003) *Application of an enhanced electrokinetic ion injection system to bioremediation*, Water, Air, and Soil Pollution, Vol. 146, pp365–377

Klotz D, Seiler K P, Moser H and Neumaier F (1980) *Dispersivity and velocity relationship from laboratory and field experiments*, Journal of Hydrology, Vol. 45, pp169–184

Kruyt H R (1952) *Colloid Science*, Volume 1, American Elsevier Publishing Company, Inc., New York, NY

Lee H and Yang J (2000) *A new method to control electrolytes pH by circulation system in electrokinetic soil remediation*, Journal of Hazardous Materials, B77, pp227–240

McCarty L P (1997) *Biotic and abiotic transformations of chlorinated solvents in ground water*, Proceedings of the Symposium on Natural Attenuation of Chlorinated Organics in Ground Water, pp7–11

Mitchell J K (1993) *Fundamentals of Soil Behavior*, 2nd ed, John Wiley and Sons Inc, New York, NY

Rittman B E and McCarty P L (2001) *Environmental Biotechnology: Principles and Applications*, McGraw-Hill Inc, Boston, MA

Rabbi M F, Clark B, Gale R J, Ozsu-Acar E, Pardue J and Jackson A (2000) *In-situ bioremediation study using electrokinetic cometabolite injection*, Waste management, 20, pp279–286

Reddy K R, Chaparro C and Saichek R E (2003a) *Iodide-enhanced electrokinetic remediation of mercury-contaminated soil*, Journal of Environmental Engineering, 129(12), pp1137–1148

Reddy K R, Chinthamreddy S, Saicheck R E, Cutright T J and Teresa J (2003b) *Nutrient amendment for the bioremediation of a chromium-contaminated soil by electrokinetics*, Energy Source, 25(9), pp931–943

Reddi L N and Inyang H I (2000) *Geoenvironmental Engineering: Principles and Applications*, Marcel Dekker Inc, New York, NY

Suer P and Lifvergren T (2003) *Mercury-contaminated soil remediation by iodide and electroreclamation*, Journal of Environmental Engineering, 129(5), pp441–446

Thevanayagam S and Rishindran T (1998) *Injection of nutrients and TEAs in clayey soils using electrokinetics*, Journal of Geotechnical and Geoenvironmental Engineering, 124(4), pp330–338

Van Genuchten M T and Alves W J (1982) *Analytical Solution of the One-Dimensional convective-Dispersive Solution Transport Equation*, US Department of Agriculture, Technical Bulletin Number pp1661

Yeung A T, Hsu C and Menon R M (1997) *Physicochemical soil-contaminant interactions during electrokinetic extraction*, Journal of Hazardous Materials, Vol 55, pp221–237

Yu J and Neretnieks I (1996) *Modeling of transport and reaction processes in a porous medium in an electrical field*, Chemical Engineering Science, Vol 51, No. 19, pp4355–4368

Geotechnical and Environmental Aspects of Waste Disposal Sites – Sarsby & Felton (eds)
© 2007 Taylor & Francis Group, London, ISBN 978-0-415-42595-7

Nickel attenuation dynamics through clayey soils

S. Ghosh, S.N. Mukherjee & R. Ray
Department of Civil Engineering, Jadavpur University, Kolkata, India

ABSTRACT: A laboratory scale study has been conducted to explore the possibility of nickel ion attenuation by clayey soil adjacent to an ash disposal pond of a Thermal Power Station, situated in Kolaghat, West Bengal, India.

1 INTRODUCTION

The leaching out of toxic metals from waste disposal ponds and subsequent migration through subsurface soil-drainage system is a complex phenomenon in the geoenvironmental engineering discipline. Amongst various other heavy metals, nickel is one of such leachable constituents in ash pond containment structures that leave out and impart the ground water pollution during the course of its movement through soil and sub-surface strata.

Several phytotoxic and health effects from nickel due to its presence as $+2$ oxidation state in soil and subsequent dissolution in ground-water in this process are reported in the literature. However, the mobility of Ni^{2+} ions depends on dynamics and hydraulic system of subsequent clay and nature of ground water gradient.

2 INVESTIGATIONS

In this research investigation, an attempt has been made to evaluate the nickel uptake capacity by clayey soil through vertical column and horizontal adsorption test setups as shown in Figures 1 and 2 respectively.

The vertical column was fabricated with 90 mm diameter 620 mm height steel cylinder as shown in Figure 1. The proximate homogeneous soil brought from the field was placed inside the cylinder and compacted in three layers at the rate of 200 mm depth interval, achieving nearly 85% of relative compaction. The density of the soil in the column was tested before and end of the experiment.

The ash pond water quality was characterised through several sampling programme and found that nickel was present more than the prescribed limit besides the other prevalent metallic ions. The adjacent soil from ash-pond sites and beneath the pond was also collected for following physico-chemical analysis of soil; field moisture content, bulk unit weight, Atterberg limits, permeability and compaction. Necessary relevant Indian Codes of standard specification were followed in determining the above properties of soil. The water quality was also evaluated by following standard methods (APHA, WPCF-1987). The heavy metal tests were carried only by using Atomic Adsorption Spectrophotometer (AAS). The pH of the soil and water samples were carried out with the help of a digital pH meter (ORION, USA Make).

The synthetic feed solution was prepared by dissolving SA salts of $NiSO_4$ (A R Grade, E-Merck make) in deionized water. The initial concentration was 5 mg/L and the pH value of the solution was 6.5. The solution was allowed to percolate through the soil column and migrated samples were collected from the sampling ports provided at the middle and bottom end of the column. In the horizontal migration test, the nickel spiked solution was allowed to travel horizontally from a particular feeding point. The tank was initially filled up with soil upto 400 mm depth with relative compaction 85%. When the soil was fully saturated, the feed solution containing nickel concentration as 5 mg/L was added in an injection well located at one end of the tank. The solution could travel by gravity through the soil and samples were collected at different time intervals from the observation wells. All the liquid samples collected from different sampling points were tested for residual nickel concentration after digesting the sample with concentrated HNO_3 in a micro-wave and subsequently the residual Ni concentration was tested by Atomic Adsorption

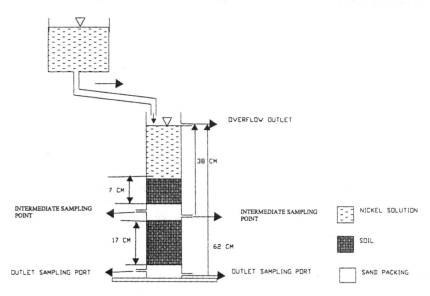

Figure 1.   Experimental setup (leaching column test).

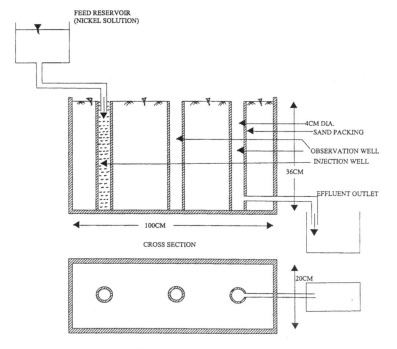

Figure 2.   Experimental setup (horizontal migration test).

Spectrophotometer (AAS) (Perkin-Elemer, USA make). In a separate study, pH effects on kinetics of removal was investigated by varying pH value in the range of 4–9.

The experimental results were tested for fitting to one-dimensional mathematical model.

3   TEST RESULTS

The physical test of the field soil reveals that the soil is clayey in nature with good insulation capacity for transmission of leachate from the contaminant source. It was further observed that the permeability of the soil was affected by pH of the soil. The permeability was reduced by 48% at a pH 6.2 of the soil near the ash-pond as compared to the soil sample collected at a distance from the site for which the pH was found to be 7.5. The permeability was also found to be decreased to the extent of 90% due to variation of pH. Since the net charge on the surface of the clay particle is negative, the Nickel ions in the solutions were expected to accumulate near the surface of the particles because of its affinity to form complex anions present in soil sample. The interaction between a negatively charged soil surface and the cations in the soil-water generates an electric-double layer in which the diffuse swarm of ions form a diffuse double layer. Due to the increase of pH value, the thickness of diffuse double layer increases perhaps due to light precipitation which leads to the decrease of permeability as shown in Table 1.

Table 1.   Values of coefficient of permeability.

Model Test No.	pH	Coefficient of permeability (cm/s)
1	4	8.44×10^{-8}
2	6.2	3.85×10^{-8}
3	7	2.29×10^{-8}
4	7.5	2.01×10^{-8}
5	9	0.94×10^{-8}

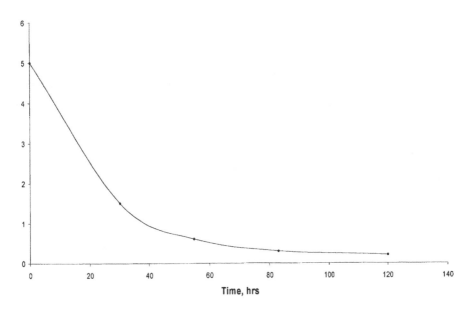

Figure 3.   Variation of Ni concentration with time in leaching column test (end effluent point) for pH 9.0 – a typical curve.

Figure 3 presents the variation of nickel concentration against time during the course of study in a vertical column test for different pH values of the solution. The concentration of nickel reaches an equilibrium concentration which was nearly zero after traveling through a distance of 240 mm as evident from the above figure. The nickel attenuation rate was found to be very rapid during the initial migration. However, the reduction of nickel from a value of 5 to 0.5 mg/L was influenced by pH values and the adsorption rate was decreased due to increase in pH value.

The test results indicated that due to more available sites as a surface phenomena more nickel ions were trapped within the soil pores and it was confirmed that after a period of 80 hours the soil column acted as adsorption layer of nickel in the steady state conditions.

REFERENCES

Anandarajah A (2003) *Mechanism controlling permeability change in clays due to changes in pore fluid,* ASCE Journal of Geotech and Geoenv Eng., vol. 129, no. 2, pp. 163–172.

Kawabata A and Nakamura I (1997) *Evaluation of contaminant transport characteristics through in-situ tracer test and column tests,* Proceedings Conference organized by the British Geotechnical Society and Cardiff School of Engineering, University of Wales, Cardiff.

Leij and Dane (1990) *Analytical solutions of the one-dimensional advection equation and two or three dimensional dispersion equation,* Water Resources Research, vol. 26, no. 7, pp. 1475–1482.

Munro R, Macquarrie J, Valsangkar A and Kan J (1997) *Migration of landfill leachate into a shallow clayey till in southern New Brunswick: a field and modeling investigation,* Canadian Geotechnical Journal, vol. 34, pp. 204–219.

Rowe K R and Badv I (1996) *Contaminant transport through a soil liner underlain by an unsaturated stone collection layer*, Canadian Geotechnical Journal, 33, pp. 416–430.

Yong R N and Phadungcgewit A (1993) *pH influence on selectivity and retention of heavy metals in some clay soils,* Canadian Geotechnical Journal, 30, pp. 821–833.

Geotechnical and Environmental Aspects of Waste Disposal Sites – Sarsby & Felton (eds)
© 2007 Taylor & Francis Group, London, ISBN 978-0-415-42595-7

Covering the mass graves at the Belzec Death Camp, Poland; geotechnical perspectives

A. Klein
Independent Geotechnical Consultant, Haifa, Israel

ABSTRACT: The Belzec Death Camp in south-east Poland was used by the Nazis for about ten months of 1942, to murder between 500,000 to 600,000 people, mostly Jews, together with some thousands of Gypsies. The bodies of the victims were interred in mass graves within the camp site. In late 1942/early 1943 the Nazis returned to the camp, dug up the bodies, burned and then crushed them in a large bone-crushing machine before re-burying the bodies in existing and new mass graves. Approximately 50% of the area of the camp is covered with mass graves. In 2003 to 2004 the American Jewish Committee together with the Polish government constructed a new Memorial to the Jewish Victims of the Nazis at Belzec, including a cover layer of blast-furnace slag, so that no plants will grow across the site. Subsequently it became apparent that the proposed solution for covering the site was not adequate. Among the problems encountered was the fact that ash and human bone fragments were moving across the surface of the site, as a result of soil erosion and rainwater washing across the site. The paper briefly describes the new Memorial, and provides a detailed description of the geotechnical aspects of the site, and the solution finally constructed.

1   INTRODUCTION

Belzec is a small village in south-east Poland, near the border with Ukraine and about 130 km south-east of Lublin (Figure 1). The Belzec death camp was established by the Nazis in February 1942, for the extermination of the Jewish population in the surrounding towns and villages.

From March to November 1942, between 500,000 and 600,000 Jews, together with some thousands of Gypsies and some Polish citizens from the Lwow area who had been caught helping the Jews, were brought to the camp by train and gassed within hours of their arrival (Kola, 2000; Solyga et al, 2003). The bodies of the victims were placed in mass graves dug principally in the northern and western sections of the site (with some mass graves in the eastern part of the camp) – Figure 2.

The camp was closed in November 1942 after all the Jews in the surrounding area had been murdered. All the buildings in the camp were dismantled and removed. In December 1942, the Nazis returned to the site, exhumed the bodies and over the following months, up to and including the spring of 1943, they burned the bodies of the victims in open pyres, and then proceeded to crush the remaining bones in a large bone crusher they had brought to the site. The ashes and bones were returned to the existing mass graves (Kola, 2000). Finally the Nazis planted trees over the whole site, in an effort to hide their past activities.

In the 1960's a memorial was set up at the camp site to commemorate the mass murders that took place there. In the years 1997 to 1999, an archaeological survey was made of the site under the direction of the University of Torun. The survey included traditional archaeological excavations that uncovered the foundations of the camp buildings. In addition a series of hand-drilled probes, on a 5 m by 5 m grid, were executed across the site to depths of up to about 8 m. These hand-drilled boreholes established the approximate diameters of the mass graves (Figure 3) and showed that they covered about 50% of the roughly 6 hectare site. The actual size of the memorial site is smaller, only about 3 to 4 hectares – the discrepancy is due to the fact that the exact dimensions of the camp

Figure 1.   Location of Belzec Death camp in Poland (Kola, 2000).

Figure 2.   Plan of the camp showing approximate positions of mass graves (Kola, 2000).

Figure 3.   Mass grave (outlined by the tape) within the camp.

are hard to determine, given the fact that the Nazis tried hard to completely eradicate all traces of the camp, including planting trees across the site.

In the period 2003–2004, a memorial and museum were constructed on the site, with the official opening ceremony taking place in June 2004. The memorial consists of a narrow concrete trench/walkway up to 10 m deep, which passes through the central part of the site (from south-west to north-east), a museum along the southern border of the camp, a concrete walkway around the whole camp with the names of the towns and villages from where the victims were brought in 1942, and a layer of clinker, or blast furnace slag over the whole site 'with local distinction to the shape of the mass graves' (Solyga et al, 2003). In the original design, a layer of thin perforated LDPE was to be placed over the whole site, after the removal of the trees and the leveling of the area, with the layer of slag being placed over this sheet of LDPE. However, when it was realised that rainwater run-off on the LDPE layer would cause the slag to move downhill towards the museum and would also create the possibility of large quantities of rainwater flooding the museum, the geotechnical work reported herein was undertaken.

2   TOPOGRAPHY AND SOIL CONDITIONS

The Belzec death camp is situated on a slope that descends from north-east to south-west at an angle of between 5° to 10°. The site was formerly covered with trees planted for the most part in the period 1943 to 1944, after the camp was closed. Almost all the trees were removed and their roots killed as part of the building of the new memorial site in 2003/04.

The soil profile across most of the site consists of a thick layer of yellow, fine to medium, sand. According to information supplied by the contractor's project manager, a layer of loam or light clay was found at the southern corner of the site, next to the museum. This layer of clay was removed from the site in the framework of the works for the new memorial, and a layer of medium hard yellowish chalk, at least 3 m thick, was found underneath. Groundwater was not found on site within the depth of the slurry walls excavated to construct the central concrete trench structure, i.e. it was at least 20 m below ground level.

3 ARCHAEOLOGICAL SURVEY

An archaeological survey of the death camp was carried out between 1997 and 1999, initially to define the location and dimensions of the mass graves. The survey was executed using small bore hand probing drills, 65 mm in diameter, on a 5 m by 5 m grid across the site, to depths of up to 8 m. The survey revealed the approximate dimensions and thicknesses of the mass graves, and gave a good indication of the composition of the graves. The thickness of the graves range from at least 2/3 m up to approximately 7.5 m, with plan dimensions of up to about 18 m by 24 m (Grave no. 10 – Kola, 2000) and 10 m by 37 m (Grave no. 14).

The upper layers of the mass graves consist for the most part of interchanging layers of ash, sand and human bone fragments (Figure 4). The lower layers of some of the mass graves consist of a spongy, semi-saturated layer of 'wax-fat', or homogeneous human remains, up to 2 m in thickness, that has collected at the bottom of the graves. The archaeologists reported finding ground water in the mass graves (Kola, 2000), at depths of around 4 m to 5 m below ground level. In view of the fact that no ground water was found when excavating up to 20 m into the sand layers in order to construct the concrete trench walls, it is clear that the ground water found in the mass graves is rainwater that has collected within and above the layer of 'wax-fat' at the bottom of some of the mass graves – this has created a type of 'perched' water table.

From a careful review of the historical sources and the evidence on site it seems that when the Nazis removed the human bodies in 1942/1943, some of the human remains were left in the graves. These remains became the layer of 'wax-fat' which was found at the bottom of some of the mass graves. Lime was placed above these human remains in at least some of the mass graves (Kola, 2000). The other bodies and human remains were burnt in huge pyres in late 1942/early 1943, and the bones were crushed in a special machine brought to the death camp site. The ash and bone fragments were then returned to the mass graves, sometimes "placed in layers with sand" (Kola, 2000), above the layers of the 'wax-fat' and the lime.

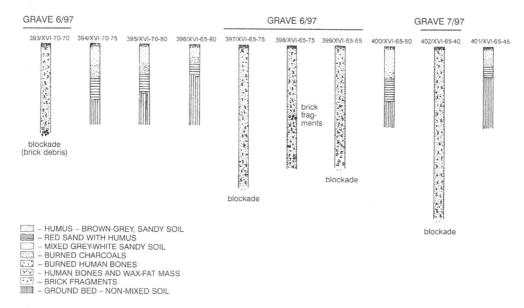

Figure 4. Typical results from the aracaeological probing drills within the camp (Kola, 2000).

4 CONSTRUCTION OF THE MEMORIAL AND COVERING THE MASS GRAVES

4.1 *The initial design*

The artistic design of the memorial at Belzec called for;

- removal of most of the trees planted across the site,
- levelling of the site with a layer of sand (some 20 cm in thickness) to the upper level of the tree stumps,
- covering of the whole area of the camp within the new concrete walkway with a layer of perforated LDPE (black PVC material)
- placement of a layer of sand over the perforated PVC,
- covering the whole site with a layer of blast furnace slag of sufficient thickness to prevent plant growth.

A slightly different colour of slag was to be placed over the mass graves, so as to delineate them for the visitors to the memorial site.

After the construction work had begun on site it was realised that there was a problem with the drainage across the site, and that in times of heavy rainfall the mass of sand and slag sitting on the perforated PVC might move down the slope towards the cemetery. In addition, the heavy rainfall could cause flooding in the museum area. The Client also became concerned that with the removal of most of the tree cover, human bone fragments and ash were working their way up through the sandy soil, out of the mass graves and moving across the site. Clearly the initial design had to be modified and the difficulty was how to provide a reasonable engineering solution whilst keeping the inevitable cost overrun to a minimum. A modification to the design was implemented at the end of 2003 (Klein, 2003).

4.2 *Revised construction*

Firstly a 10 to 20 cm thick levelling layer of sand was placed over the whole area of the camp, including above the mass graves. Then the mass graves were covered by one layer of high strength woven geotextile, placed above the first levelling layer of sand. The geotextile chosen was STABILENKA 120/120, with an breaking strength of 120 kN/m in each direction, and a working strength of 60 kN/m in each direction at 5% strain (Klein, 2003). The primary reason for this choice was its immediate availability from the suppliers, owing to the tight time constraints on the project with the approach of the harsh Polish winter, where after the first snowfalls of the winter it would be impossible to place the geotextile over the mass graves.

The geotextile (Figure 5) was placed above the approximate positions of the mass graves as detailed in the archaeological survey of the University of Torun (Kola, 2000) and for 10 m beyond each grave in all directions. This meant that large areas of the camp were covered by the woven geotextile, since some 50% of the area of the camp had been shown to be mass graves. The decision to choose a high strength woven geotextile was governed by two considerations;

- firstly the need to cover the mass graves so as to lessen possible settlements in the future,
- secondly, but no less important, to prevent the movement of human bone fragments and ash out of the mass graves and across the camp site.

Another layer of sand some 10 cm thick was placed above the geotextile to provide both stability to the woven geotextile and a base for the upper layer of blast furnace slag (Figure 6), which was placed over the whole cemetery to a thickness of about 30 cm.

A further problem was the possibility of water runoff on the woven geotextile towards the museum constructed at the lower, southern end of the camp site. The chosen geotextile would in fact allow a certain amount of water to flow through itself and into the ground below but it was clear that during periods of heavy rainfall, such as occur in this area of Poland, much of the water would not penetrate the geotextile and would run along the surface of the geotextile and down towards the

Figure 5.   Geotextile partially covered with sand drainage layer.

Figure 6.   Spreading of blast furnace slag cover over the camp site.

museum. The solution developed was to install a series of surface drains buried in the upper layer of sand, directly below the slag covering. The drains was designed by a local drainage engineer, and were placed across the site in a 'herring-bone' fashion, with collector drains leading the water run-off towards the entrance to the memorial site, and from there to the south-east.

5 CONCLUDING REMARKS

The site reported in this paper was the location of a former death camp (Belzec) in south-east Poland. Within the ground were mass graves (occupying about 50% of the total area of the site) containing the remains (after burning and crushing of the bodies) of up to 600,000 people. Consequently any construction work on this site needed to be carried sensitively and sympathetically with respect to the victims.

In order to create a new memorial at the camp site it was proposed to place a layer of thin perforated LDPE was to placed over the whole site, to protect the mass graves, and to cover the LDPE with a layer of blast furnace slag. During construction of the new memorial it was found that there was a problem with water run-off on top of the perforated LDPE and this water was moving towards the museum at the site. In addition, after the removal of the trees covering the site it was noticed that human bone fragments and ash were working their way out of the mass graves and beginning to move across the site. As a consequence the geotechnical works were significantly amended to include a layer of sand over the whole site and a high strength woven geotextile over the mass graves. A further layer of sand was placed over the geotextile and a 'herring-bone' drainage system was installed to collect the rainwater that did not penetrate the woven geotextile. A covering of blast furnace slag was then placed across the whole site as originally planned.

ACKNOWLEDGEMENTS

The Author is grateful to the Council for the Protection of Memory of Combat and Martyrdom and The American Jewish Committee for allowing this work to be published and to the Chief Rabbi of Warsaw (Michael Shudrich) for his personal support.

REFERENCES

Klein A (2003) *Cemetery/Memorial to the Jewish Victims of the Nazi Death Camp in Belzec, Poland,* unpublished geotechnical consultancy report.

Kola A (2000) *Belzec, the Nazi camp for Jews in the light of archaeological sources – Excavations 1997 – 1999,* The Council for the Protection of Memory of Combat and Martyrdom and the United States Holocaust Memorial Museum, Warsaw – Washington.

Solyga A, Pidek Z and Roszczyk M (2003) *Belzec – Nazi death camp 1942,* The Council for the Protection of Memory of Combat and Martyrdom and The American Jewish Committee.

Geotechnical and Environmental Aspects of Waste Disposal Sites – Sarsby & Felton (eds)
© 2007 Taylor & Francis Group, London, ISBN 978-0-415-42595-7

The effects of persistent anthropogenic contamination on the geotechnical properties of soil

E. Korzeniowska-Rejmer
Krakow University of Technology, Institute of Geotechnics, Poland

ABSTRACT: Nowadays water and soil contamination by chemical compounds is one of the greatest problems of our time. It is observed more and more often in our country, especially in industrial areas. It was necessary to systemize a concept of hazard to soil environment. Introduction of maximum allowable concentrations of toxic substances for groundwater and soil can stop the process of change of original geotechnical characteristics. In the article the following test results are referred to: changes of particle size of soils exposed to long-term anthropogenic contamination, changes of compressibility of contaminated soils in the context of their applicability for building purposes (oedometer tests, additional settlement of soils).

1 INTRODUCTION

Waste storage with related leachates, as well as technological wastewaters, fertilizers and pesticides washed out and infiltrating deep into the ground are the principal factors contributing to the degradation of the geological environment. The quantities of toxic substances contained therein have risen systematically. The intensive industrial activity most often results in significant contamination of the ground-water environment by chemical substances.

The effect of pollutants on the ground and water medium should be considered in a comprehensive manner because of the multidirectional nature of changes caused thereby. The infiltration of chemical substances into the substrate makes the latter a toxic medium. The danger is further enhanced by the persistence of these substances in the ground medium, particularly with respect to oil-related contaminants. On the other hand, once introduced into the ground medium, chemical contaminants affect significantly its properties both as a construction sub-base as well as a material for earthworks and road construction. The contaminants retained in the soil for longer times, chiefly oil-related ones, affect the rheological properties of the soil, changing its original characteristics. The issue is of essential importance to the design options applied in engineering structures built on contaminated areas and to the implementation of suitable measures to secure the existing structures erected on contaminated areas.

2 CHANGES OF PROPERTY IDENTIFIED DURING THE STUDY

Many years of study of construction sub-bases in industrialized regions (sites of chemical plants, smelters and oil refineries), carried out by the Institute of Geotechnics) have shown significant levels of contamination with chemical substances throughout the whole soil profile, reaching below the level at which foundations of engineering structures are laid. The infiltration of industrial waste water, surface run-off (when the area around structures is not protected and/or drainage systems are lacking) and extended underground infrastructure, have all resulted in increased moisture levels of the soils in the subsoil. In the cases of the Oświęcim Chemical Works and the Nowa Huta

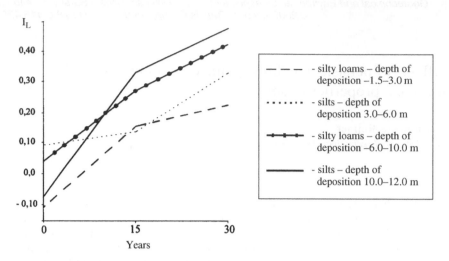

Figure 1. Variation of liquidity index (I_L) in soils subject to persistent chemical contamination in the Oświęcim Chemical Plant.

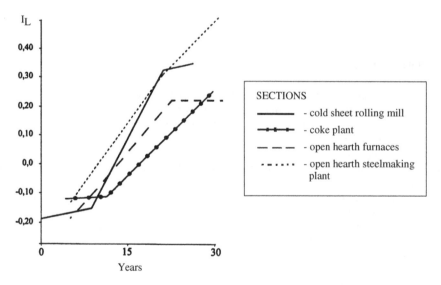

Figure 2. Variation of liquidity index (I_L) in the silts (loess cover) at the Tadeusz Sendzimir Steelworks – depths up to 6.0 m.

Steelworks, this increase in the moisture content amounted to more than 30% over the 1950–1980 period covered by the studies, which pertained to the construction, extension, and modernization phases in both plants. The higher moisture content caused an increase in the soil liquidity index (I_L). The increase in the moisture content in subsoil over time was found in all departments of the Nowa Huta Steelworks and Chemical Plants in Oświęcim. The nature of this process is illustrated in the graphs showing empirical regression lines for changing values of liquidity index over time (Figures 1, 2, 3 and 4).

The marked increases in parameters were related to the nature of operations of the facilities involved as well as the geological conditions in the subsoil. The soils which provide the construction

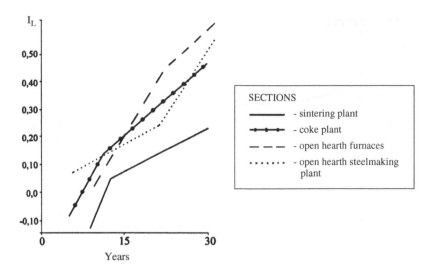

Figure 3. Variation of liquidity index (I_L) in the silt (loess cover) at the Tadeusz Sendzimir Steelworks – depths from 6.0 to 10.0 m.

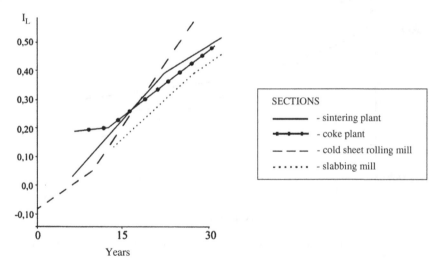

Figure 4. Variation of liquidity index (I_L) in the loams and silty loams in the Tadeusz Sendzimir Steelworks – depths from 10 to 15 m.

sub-base for the studied facilities, constitute a thickness of more than 10 m of loess soils, mainly silts and silty loams.

In the study period, the chemistry of waters collected from inspection boreholes showed some changes. In three decades of the operational life of the facilities, the levels of sulphates, aggressive carbon dioxide, chlorides, ammonia, heavy metals and, around dumpsites, phenols, cyanides and iron, have multiplied. The waste water infiltrating through cracks in tiled floors and industrial pipelines often contributed to significant additional soil settlement. In the 1990's when the open pits of foundations were made, some hollow spaces were found beneath tiled floors, which evidently pointed at long-term chemical aggression as a reason for the deformation of the ground medium.

Chemical substances in high concentrations which fill the pores of soil can affect the mineral composition and grain-size distribution of the soil, destroying structural bonds (and then the

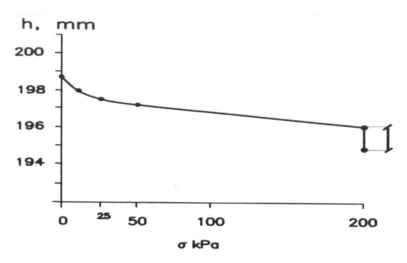

Figure 5.   Effect of diesel oil on settlement of silt sample (Nowa Huta area).

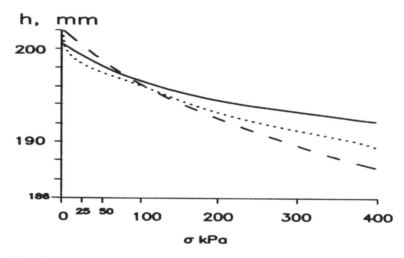

Figure 6.   The effect of diesel oil (DO) on the oedometer settlement of silt sample (———— – DO 0% ········ –
DO 4% - - - - – DO 8%).

disintegration of skeletal particles displays some features of weathering). In soils less resistant to chemical contamination, with fine size particles, significant alterations of physical and mechanical properties occur.

Tests of soil samples collected from test pits near foundations of engineering structures undergoing long-term effects of oil-related contaminants (oil refineries in Jedlicze and Trzebinia) with significant levels of structural damage, showed large decrease in physical and mechanical properties. These were caused by significant levels of moisture in the subsoil caused by waste water and oils penetrating into the soils. Examination of soil samples using the oedometer revealed significant soil settlement, as well as changes in the settlement process, over time. The deformations of the ground medium occur even at minor contamination levels. A significant proportion of soil movement is of the additional settlement nature (creep), which is characteristic of soils of less stable structure ($i_m > 0,02$) – Figures 5 and 6. For the soil in Figure 5 the additional settlement of a previously consolidated sample after treatment with diesel oil was 10%.

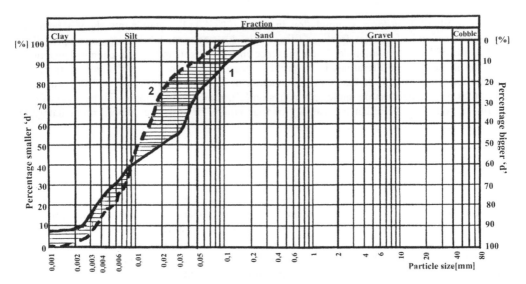

Figure 7. Changes of grading of silt subject to persistent oil-related contamination (——— – 1 outside the contaminated zone, – – – – 2 from open pit of foundation (soil with oil-related substances >25%), ▭ – range of changes in grain-size).

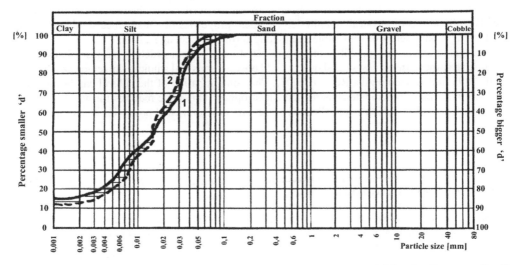

Figure 8. Changes of grading of silt loam in the Nowa Huta area (——— – 1 "clean" soil, – – – – 2 soil contaminated by diesel oil (DO-10%) duration of cantamination-5months. ▭ – range of changes in grain-size).

Major alterations were also found in the grain size distribution curves in soils subject to persistent effect of oil-related contamination. The results of particle size analyses of soil samples taken from the foundations of fuel storage facilities showed major shifts in the proportions of particular fractions of the soil. The analysis revealed a decrease in sand fraction content and a significant increase in silt fraction as a result of breaking aggregates of the sand fraction. Also there were clearly noticeable trends in the flocculation of particles in clay fraction, which are the most liable to the pressure of contamination. In the case of silt soils, the scope of changes in particle size for particular fractions can exceed 30% – Figures 7 and 8.

As a result of long-term accumulation in soils, oil-related contaminants show a tendency to "concentration" of heavier fractions (diesel oils, heating oils, lubricants), accompanied by the release of hydrogen sulphide. Their presence in the subsoil intensifies chemical reactions, resulting in permanent alterations in its particle size distribution. These changes in grain-size distribution (increase of silt fraction and reduction of sand fraction) are of vital importance to the values of internal friction angle (ϕ'), cohesion (c'), modulus of compressibility (m_v) of contaminated soils – these affect the stability of foundations of structures placed immediately on contaminated subsoils.

3 CONCLUSIONS

Because of its practical importance, the elucidation of long-term effects of various chemical contaminants on various types of soils, considered both in terms of chemical and mechanical processes becomes necessary. It is required because of the need to protect structures erected in highly industrialised areas from excessive settlement. It is also important to identify both contaminants as well as changes in the subsoil resulting from chemical aggression when undertaking any reclamation measures or selecting suitable methods for reinforcing contaminated soils.

The studies carried out in the field indicate that chemical contaminants infiltrating into the subsoil (leached from dump sites for solid and liquid waste or caused by deposition of industrial particulate emissions, waste waters, or gaseous pollutants) bring about irreversible alterations in the grain-size distribution of soils, quantities and types of clayey minerals, the moisture level, concentration, vulnerability to weathering and erosion, as well as the chemistry of porous water etc. The alterations of physical and physico-chemical properties of soils reduce significantly the load strength parameters of soils subjected to long-term chemical aggression.

The evaluation of the suitability of contaminated soils for construction purposes is essential because in some extreme cases (as shown in field observations and studies; Korzeniowska-Rejmer et al, 1995 and Korzeniowska-Rejmer, 2001). The contamination of the subsoil can bring about accident hazards or precipitate construction catastrophes. This issue requires further studies and analysis, the need for which is also dictated by general environmental protection requirements.

REFERENCES

Herzig J (2001) *Wpływ zanieczyszczeń organicznych na wybrane parametry fizyczne gruntów spoistych*, Inżynieria Morska i Geotechnika, nr 3/2001.
Korzeniowska-Rejmer E (2001) *Wpływ zanieczyszczeń ropopochodnych na charakterystykę geotechniczną gruntów, stanowiących podłoże budowlane*, Inżynieria Morska i Geotechnika. nr 2/2001.
Korzeniowska-Rejmer E, Motak E and Rawicki Z (1995) *Wpływ zanieczyszczeń olejowych na stan techniczny podłoża gruntowego i budynku magazynowania paliw płynnych*, Przegląd Budowlany, nr VIII.
Korzeniowska-Rejmer E (2003) Unpublished research within Research Project Ś-2/504/BW/2003.

Improvements in long-term performance of permeable reactive barriers

K.E. Roehl
Department of Applied Geology, Karlsruhe University, Karlsruhe, Germany

F.-G. Simon & T. Meggyes
Federal Institute for Materials Research and Testing (BAM), Berlin, Germany

K. Czurda
Department of Applied Geology, Karlsruhe University, Karlsruhe, Germany

ABSTRACT: Permeable reactive barriers (PRB) are used for passive in-situ groundwater remediation. In the European research project PEREBAR the long-term behaviour of PRBs has been studied with special respect to reactive material properties and physico-chemical processes. Laboratory and field studies showed that the most efficient materials for the removal of uranium from contaminated groundwater are elemental iron and hydroxyapatite. Geochemical studies of alteration processes were conducted on these materials and on activated carbon used for the sorption of hydrocarbons, to analyse their long-term behaviour in PRB systems. Innovative materials for the selective sorption of uranium were developed. The concept of electrokinetic fences to support the performance of PRB systems was evaluated and proved to be the most promising approach in this respect. With its empirical work the project contributed to the knowledge and expertise in the field of passive groundwater remediation using permeable reactive barriers.

1 INTRODUCTION

The technology of permeable reactive barriers (PRBs) is a novel groundwater remediation method which enables physical, chemical or biological in-situ treatment of contaminated groundwater by means of reactive materials. The reactive materials are placed in underground trenches or reactors downstream of the contamination plume forcing it to flow through them. By doing so, the contaminants are removed from the groundwater without soil or water excavation. The two main types of PRBs are continuous reactive barriers enabling a flow through its full cross-section, and 'funnel-and-gate' systems in which only special 'gates' are permeable for the contaminated groundwater. Generally, this cost-effective clean-up technology impairs the environment much less than other methods. General reviews of the development of research and application of the PRB technology are given by Gavaskar et al. (1998), EPA (1998, 1999), Simon and Meggyes (2000), Vidic (2001) and Birke et al. (2003).

Feasibility of PRBs largely depends on the life expectancy of the reactive materials which may be impaired by the remediation processes themselves (e.g. oxidation) or the reaction products (e.g. precipitates), or exhaustion of sorption capacity. Operational life is expected to be 10 to 20 years, experience over such periods is however not yet available. There are no reliable methods yet available to predict the long-term behaviour of PRB systems, i.e. allowing statements on how long the reactive materials will maintain their full function. Nor do the present approaches to design and installation encourage solutions where the active medium could be more regularly recharged or replaced.

2 OBJECTIVES AND SCOPE OF THE PEREBAR PROJECT

2.1 Research programme

The integrated research programme 'Long-term Performance of Permeable Reactive Barriers used for the Remediation of Contaminated Groundwater' (PEREBAR) supported by the European Union under the contract $N°$ EVK1-CT-1999-00035 was carried out from 2000 to 2003 with the participation of research groups from Austria, Germany, Greece, Hungary and the United Kingdom. The overall objective of the project was to evaluate and enhance the long-term performance of permeable reactive barrier (PRB) systems with the main emphasis on sorption and precipitation of heavy metals plus sorption and decomposition of organic compounds. Processes impairing barrier performance and technologies enhancing their long-term efficacy have been studied.

2.2 Test conditions

Two test sites formed the central points of the project. One of them was the former uranium ore mining and processing area in Pécs, Southern Hungary. The second site is located in Brunn am Gebirge, Austria, where an activated carbon PRB system is installed. Especially due to the conditions at the first site the project had its focus on uranium contamination. Reactive materials for PRBs and contaminant attenuation processes in these materials, such as sorption and degradation mechanisms, and geochemical processes in the barrier material that are governing the efficiency of PRB systems on the long-term, especially the influence of groundwater constituents, have been studied. To investigate the long-term performance of PRBs, the properties of the reactive material itself and the geochemical processes taking place inside the porous matrix of the reactive material have been looked into. Therefore, the project has been primarily dealing with the reactive material and not touching in a wider sense hydro-geological aspects such as hydraulics, catchment zones etc. Details on the work plan have been given by Roehl and Czurda (2002). Further information on the project is also available at its website at http://www.perebar.bam.de/.

3 SCIENTIFIC WORK

3.1 Materials

Reactive materials potentially suitable for use in PRBs have been considered and characterised, with a special view on groundwater contaminated with uranium. Those materials tested included zeolites, hydroxyapatite (HAP), activated carbon, hydrated lime and elemental iron. Contaminant attenuation and other geochemical processes taking place in the reactive materials were investigated in a number of batch, column and container experiments. HAP and elemental iron proved the most efficient materials in removing uranium from aqueous solutions (Simon et al. 2003, Krestou et al. 2004). The most likely uranium attenuation mechanisms are bulk precipitation when HAP is used and a combination of reductive precipitation and adsorption for elemental iron. Uranium attenuation by elemental iron and HAP was not affected by sulphates. Dissolved carbonate impaired the uranium attenuation with elemental iron, while it did not exhibit detrimental effects on HAP.

Accelerated column and container experiments were designed and performed to study ageing mechanisms caused by alteration of the reactive material, change in permeability, formation of precipitates, coatings etc. Ageing mechanisms of elemental iron and HAP were different. The experiments on elemental iron showed a reduction in hydraulic permeability which may be attributed to precipitation and formation of secondary minerals such as $CaCO_3$, $MgCO_3$ and $FeCO_3$. HAP showed no such effects, here the geochemical behaviour was dominated by a gradual consumption of HAP due to its solubility in groundwater.

3.2 Developments

An innovative approach adopted by the project to improve long-term performance of PRB systems was the development of selective contaminant-binding chemical compounds. A material named

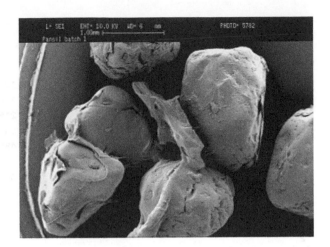

Figure 1.   Electron micrograph of PANSIL.

PANSIL, a silica coated with modified polyacryloamidoxime (PAN), was developed which efficiently removes uranium from aqueous solutions (Figure 1). Both the support matrix (sand) and the coating are durable, the sorption is most effective when the solution pH is between 4 and 8, specificity for uranium is fairly high, and there is no risk of precipitation of by-products that can block porous barriers (Bryant et al. 2003).

The potential of electrokinetic methods to enhance the long-term efficiency of PRBs was studied. An electrokinetic fence may be installed upstream of the barrier to reduce the concentration of groundwater constituents that might impair barrier function, by preventing them from being transported by the groundwater. Laboratory experiments and modelling showed that charged species can indeed be hindered from moving with the groundwater flow while uncharged species will still be transported by the hydraulic flow (Czurda et al. 2002). Groundwater constituents that might otherwise migrate towards the PRB and precipitate in the reactive material due to strong redox and pH changes in the elemental iron barrier, are precipitated around the electrodes, i.e. removed from the groundwater flow.

4   LABORATORY EXPERIMENTS

4.1   *Radiotracers*

Knowledge of the uranium behaviour in contact with reactive materials is needed if PRBs are to be designed, and their operation life estimated. Therefore, ^{237}U (a short-lived uranium isotope with a half-life of 6.75 days) was used as a radioindicator in column experiments to track the movement of uranium through the column without disturbing the system by taking samples or dismantling the apparatus. Soon after the detection of radioactivity, radioindicators (also called radiotracers) have been utilised for the investigation and analysis of processes and behaviour of material components in various fields of application, e.g. medicine, chemistry, physics, material science etc. (Gardner and Ely 1967, Gardner et al. 1997, Schulze et al. 1993). Results from the experiments with such radiotracers will be used to gain a better understanding of the uranium uptake capacity and thus of the long-term performance of permeable reactive barriers using elemental iron or hydroxyapatite as reactive material.

4.2   *Column tests*

Figure 2 shows the curves recorded during the experiments from column experiment with the radiotracer. The results obtained from the evaluation of the data are listed in Table 1. Detailed

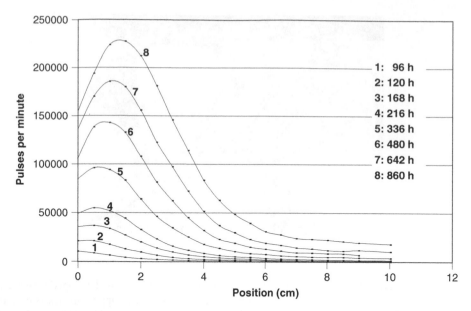

Figure 2. Activities measured in the iron column at various time intervals (experiment 1).

Table 1. Results from experiments.

	Iron (1)	Iron (2)	HAP
Velocity of the contamination front (cm/h)	19.6×10^{-3}	9.6×10^{-3}	21.2×10^{-3}
Calculated breakthrough after	766 h	5189 h	944 h
Calculated maximum concentration of uranium on reactive material (mg/kg)	3550	3019	2916
Retardation factor, R (see text)	620	1100	573

information to the experiments performed can be found elsewhere (Simon et al. 2003, Simon et al. 2004).

Having the velocity of the contaminant front the break-through of the contamination through the layer of reactive material in the column can be calculated. The experimental data enable the extrapolation of the maximum uranium concentration on the reactive material iron or HAP. These values are also listed in Table 1. From these data the distribution coefficient K_d can be calculated by division with the concentration of the uranium containing solution. From the distribution coefficient K_d the retardation factor R can be calculated (Simon and Meggyes 2000) as;

$$R = 1 + \left(\frac{\rho}{\theta}\right) K_d \qquad (1)$$

with ρ representing the bulk density and θ the porosity.

For average conditions ρ/θ is 6 kg/l (Appelo and Postma 1993). Retardation calculated by dividing velocity of uranium solution by velocity of the contaminant front resulted in similar values.

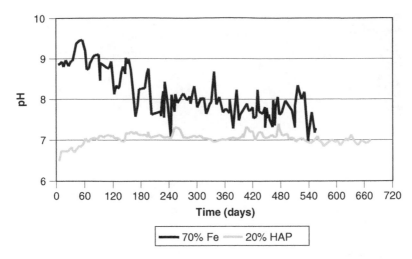

Figure 3.   pH of the effluent from long-time column experiments.

Both elemental iron and HAP are able to retain uranium from groundwater. In the case of elemental iron the reductive precipitation of uranium is accompanied with an increase in pH (see Reactions 1 to 3).

$$Fe \quad \rightarrow \quad Fe^{2+} + 2e \qquad \text{(Reaction 1)}$$

$$2H_2O + 2e \quad \rightarrow \quad H_2 + 2OH^- \qquad \text{(Reaction 2)}$$

$$O_2 + 2H_2O + 4e \quad \rightarrow \quad 4OH^- \qquad \text{(Reaction 3)}$$

Long-term column experiments with iron as a reactive material showed that the pH increased initially by 2 pH units from pH 7 in the feed water to pH 9 in the column effluent. But after about 2 to 3 months the effluent pH decreased and stabilised around pH 8 after 7 to 8 months of operation (see Figure 3). In contrast to this the pH of the effluent from columns with HAP as reactive material remained nearly unchanged.

Long-term experiments were carried out without radioindicator but with artificial groundwater containing calcium and magnesium. As the solubility of calcite and dolomite decreases with increasing pH, calcium precipitation occurs within iron columns. Alkaline earth elements are almost completely removed from the artificial groundwater with more than 800 mg of carbonates formed for every litre that passes through each column. This accumulation of solid material could lead to a loss of permeability and finally to a failure of the system. These precipitates were found after dismantling test columns and test boxes after operation with uranium containing artificial groundwater as displayed in Figure 4.

Flow conditions chosen in the column experiments represent an acceleration compared to natural conditions. The velocity in the column was around 5 pore volumes per day, i.e. 2.5 m/d, which is approximately five times faster than natural groundwater velocities. A permeable reactive barrier designed with a proportional design as in these experiments would persist for more than three years with the same pollutant concentration at natural velocities. The period of proper functioning could even be longer. Morrison found that the removal efficiency increased again after a breakthrough in his column experiment after 3000 pore volumes when the flow velocity was decreased by a factor of 10 (Morrison et al. 2001).

The results with HAP are in agreement with the results of a technical demonstration at Fry Canyon in Utah using HAP which is promising (Naftz et al. 1999). In contrast to the reaction

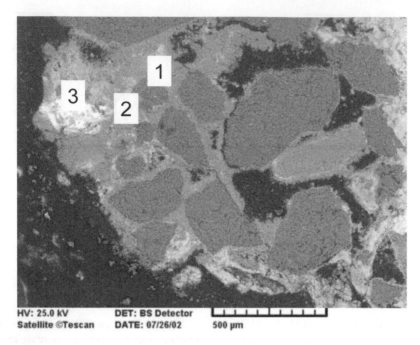

HV: 25.0 kV   DET: BS Detector
Satellite ©Tescan   DATE: 07/26/02   500 µm

Figure 4.   Image from secondary electron microscopy (SEM) from sand/iron mixture. (Item 1 is a sand particle, 2 is mainly precipitated $CaCO_3$, 3 is an iron particle. The surface is covered with $Fe(OH)_2$ and $FeCO_3$ precipitation. Result from Dr. A. Debreczeni and I. Gombkötő, University Miskolc, Hungary.)

of lead with HAP where hydroxypyromorphite $Pb_{10}(PO_4)_6(OH)_2$ as sparingly soluble mineral is formed after dissolution of HAP (Chen and Wright 1997, Ma et al. 1993) uranium removal might occur via sorption on HAP surfaces. In the non-mined Coles Hill uranium deposit phosphate mineral precipitation has been verified (Jerden and Sinha 2003). However, it was stated that the formation of autunite group minerals require a longer time interval. Such a transformation was described by Sowder et al. (1996) for schoepite $((UO_2)(OH)_2)$.

Stability of sorption complexes or mineral phases is essential to obtain the desired behaviour of a PRB system using HAP for uranium removal. Evidence exists that under normal conditions uranium remains immobile (Giammar 2001). Proper functioning of a PRB system depends on the adequate design. With the data found in this study it is possible to calculate the amount of reactive material necessary for contaminant removal. However, lessons learned from the demonstration at Fry Canyon show that site characterisation and groundwater flow patterns are the most important parameters. If groundwater does not completely pass the reactive medium, removal efficiency is poor. The cost of HAP as a reactive material is comparable to that of granular iron and is around 350 US$ per ton (EnviroMetal, Conca et al. 2000). Lower prices for HAP are feasible if waste materials, e.g. fishbone or phosphatic clay (Singh et al. 2001) containing HAP are used.

Using the radiotracer method the concentration of uranium retained in the columns can easily be measured without interfering with the flow regime or dismantling the apparatus. Investigation of the contaminant's spatial distribution in the reactive media would be difficult without a radioactive tracer. This method provides a powerful tool to elucidate uranium precipitation and adsorption within the reactive media while maintaining undisturbed flow conditions.

The results obtained using this method will help provide further information about processes within barriers and the quantity of reactive material needed to treat a given amount of contaminated water with known chemical properties.

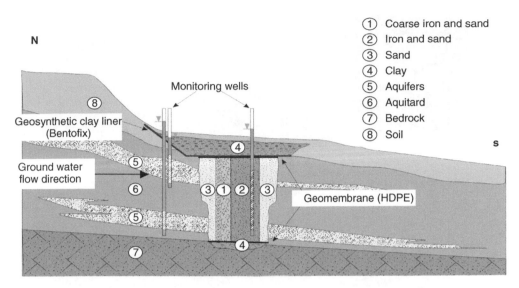

Figure 5.   Cross-section of experimental permeable reactive barrier near Pécs, Hungary (Csővári et al.).

5   CASE STUDY SITES

5.1   *Hungary*

The primary PEREBAR case study site was an area in Southern Hungary contaminated by uranium mining – a region which became part of the European Union in 2004. Uranium contaminations of up to 1000 µg/l were detected in the groundwater at the field test site selected from three potential locations.

Following preliminary in-situ column experiments using cylinders filled with reactive materials in existing wells, an experimental PRB was constructed with a length of 6.8 m, a thickness of 2.5 m and a depth of 3.8 m. The PRB consists of two different zones containing iron/sand mixtures: zone I was 50 cm thick with a low content of coarse elemental iron (12% by volume or 0.39 t/m³, grain size 1–3 mm), and zone II was 1 m thick with a higher content of fine elemental iron (41% by volume or 1.28 t/m³, grain size 0.2–3 mm). The total mass of elemental iron installed as reactive material was 38 t, of which 5 t was coarser material (Figure 5). Twenty four monitoring wells have been installed directly in or in the near vicinity of the PRB. There are wells upstream, in the first iron zone, at different depths in the second iron zone, downstream of the barrier. At one cross-section there is a line of wells across the barrier for monitoring of precipitate formation. The wells are made of pipes of 1 cm diameter for groundwater sampling. Positions of the wells are shown in Figure 6.

The first few months of operation have demonstrated very convincingly the remediation effect of the PRB system. Data obtained in a monitoring well 15 m downstream of the barrier show that the experimental PRB removes uranium very efficiently from the local groundwater (Figure 7). The concentration dropped from approx. 1,000 µg/l early 2002 to about 100 µg/l or even less after the start-up the PRB. Further long-term monitoring will evaluate changes in reactivity and hydraulic characteristics.

5.2   *Austria*

On the second case study site, on a former industrial site in Brunn a.G., Austria, a PRB system based on activated carbon was tested using periodical water sampling, in addition to routine monitoring

Figure 6.   View of the experimental permeable reactive barrier with monitoring wells near Pécs, Hungary (Csővári et al.).

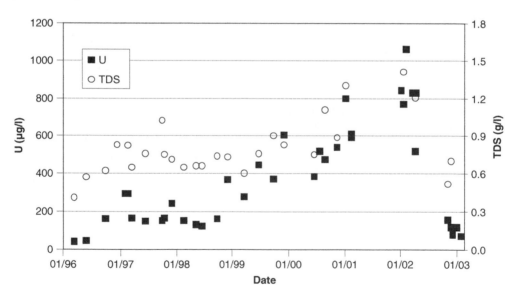

Figure 7.   Uranium concentration and total dissolved solids (TDS) at a monitoring well of the Pécs, Hungary experimental PRB. (Note a sharp drop in uranium concentration during 2002 after the barrier's start-up, Csővári et al.).

(Niederbacher 2001, 2002). The data showed that the system was functioning as planned, with all target contaminants (mainly hydrocarbons) being removed by the activated carbon. The performance of the system was also proved by tracer tests. Sampling and activated carbon analysis showed carbonate precipitation on the surface of carbon aggregates but no signs of impairment.

170

6 SUMMARY AND CONCLUSIONS

The project's results include, inter alia, improved knowledge on reactive materials for use in PRBs and their capability for uranium removal from contaminated groundwater, on geochemical processes taking place in those materials, on methods to study these processes, and on methods to define the lifetime of PRBs. Further deliverables are a newly developed material for uranyl sorption, generic expertise on the technique of electrokinetic fences, and a fully operational, pilot-scale experimental PRB using elemental iron. The results will be compiled in a book to be published by Elsevier in 2004 (Roehl et al., in preparation).

The results contribute to improving the quality of life by enhancing remediation of a large number of contaminated sites in Europe and in particular in CEECs (Central and Eastern European Countries). PEREBAR's achievements help establish the PRB technique as an accepted, scientifically sound and cost-effective tool for passive groundwater remediation.

REFERENCES

Appelo C A J and Postma D (1993) *Geochemistry, Groundwater and Pollution*, A.A. Balkema Publishers.

Birke V, Burmeier H and Rosenau D (2003, *Design, Construction and Operation of Tailored Permeable Reactive Barriers*. In: Prokop G., Bittens M., Cofalka P., Roehl K.E., Schamann M. and Younger P. (eds.), Summary Report of the 1st IMAGE-TRAIN Advanced Study Course "Innovative Groundwater Management Technologies". Tübinger Geowissenschaftliche Arbeiten (TGA), C68, 064–94.

Bryant D E, Stewart D I, Kee T P and Barton C S (2003) *Development of a Functionalized Polymer-Coated Silica for the Removal of Uranium from Groundwater*. Environmental Science and Technology, 37, pp4011–4016.

Chen X B and Wright J V (1997) *Effects of pH on Heavy Metal Sorption on Mineral Apatite*, Environmental Science and Technology, 31 (3), pp624–631.

Conca J L, Liu N, Parker G, Moore B, Adams A and Wright J (2000, *PIMS – Remediation of metal contaminated waters and soil*, 2nd International Conference on Remediation of chlorinated and recalcitrant compounds, Monterey, CA, USA, Wickramanayake, G. B., Gavaskar, A.R. and Alleman, B.C. (Editors), Batelle (Publisher).

Csővári M, Csicsák J., Földing G and Simoncsics G (2005) *Experimental Iron Barrier in Pécs, Hungary*. In: Roehl K.E., Meggyes T., Simon F.-G. and Stewart D.I.: Long-term Performance of Permeable Reactive Barriers. Elsevier, Amsterdam.

Czurda K, Huttenloch P, Gregolec G and Roehl K E (2002) *Electrokinetic techniques and new materials for reactive barriers*. In: Simon F.-G., Meggyes T. and McDonald C. (eds.), Advanced Groundwater Remediation – Active and Passive Technologies, Thomas Telford, London, pp173–192.

EnviroMetal Technologies Inc., www.eti.ca

EPA (1998) *Permeable reactive barrier technologies for contaminant remediation*. U.S. EPA Remedial Technology Fact Sheet, EPA 600/R-98/125, 102 p.

EPA (1999) *Field applications of in situ remediation technologies: Permeable reactive barriers*. U.S. EPA Remedial Technology Fact Sheet, EPA 542-R-99-002, 122 p.

Gardner R P and Ely R L (1967) *Radioisotope measurement applications in engineering*, Reinhold Publishing Corp., New York.

Gardner R P, Guo P, Ao Q and Dobbs C L (1997) *Black box gauges and anlyzers*, Applied Radiation and Isotopes, 48, 1273–1280.

Gavaskar A R, Gupta N, Sass B M, Janosy R J and O'Sullivan D (1998) *Permeable Barriers for Groundwater Remediation*. Batelle Press, Columbus, 176 p.

Giammar D (2001) *Geochemistry of uranium at mineral-water interfaces: rates of sorption-desorption and dissolution-precipitation reaction*. California Institute of Technology, Ph. D. Thesis, Pasadena, CA, USA.

Jerden J L and Sinha A K (2003) *Phosphate based immobilization of uranium in an oxidizing bedrock aquifer*, Applied Geochemistry, 18, 823–843.

Krestou A, Xenidis A and Panias D (2004) *Mechanism of aqueous uranium(VI) uptake by hydroxyapatite*. Minerals Engineering, 17, pp373–381.

Ma Q Y, Traina S J and Logan T J (1993) *In Situ Lead Immobilization by Apatite*, Environmental Science and Technology, 27 (9), pp1803–1810.

Morrison S J, Metzler D R and Carpenter C E (2001) *Uranium Precipitation in a Permeable Reactive Barrier by Progressive Irreversible Dissolution of Zerovalent Iron,* Environmental Science & Technology, 35 (2), pp385–390.

Naftz D L, Davis J A, Fuller C C, Morrison S J, Freethey G W and Feltcorn E M (1999) *Field demonstration of permeable reactive barriers to control radionculide and trace-element contamination in groundwater from abandoned mine lands,* Toxic Substances Hydrology Program – Technical Meeting, Charleston, South Carolina, USDA, US Geological Survey (Publisher), Conference Proceedings Vol. 1 (Contamination from hardrock mining), pp281–288.

Niederbacher P (2001) *Monitoring long-term performance of permeable reactive barriers used for the remediation of contaminated groundwater.* SENSPOL Workshop "Sensing Technologies for Contaminated Sites And Groundwater", May 9–11, 2001, University of Alcalá, Spain, pp162–174.

Niederbacher P (2002) *AR&B System Brunn a. G. – Site Adapted Solution of the PRB Concept.* In: Prokop, G. (ed.), Proc. 1st IMAGE-TRAIN Cluster Meeting Karlsruhe, November 7–9, 2001, Federal Environment Agency Austria, Vienna, pp32–38.

Roehl K E and Czurda K (2002) *PEREBAR – A European Project on the Long-Term Performance of Permeable Reactive Barriers.* In: Prokop, G. (ed.), Proc. 1st IMAGE-TRAIN Cluster Meeting Karlsruhe, November 7–9, 2001, Federal Environment Agency Austria, Vienna, pp23–31.

Roehl K E, Meggyes T, Simon F-G and Stewart D I (2005): *Long-term Performance of Permeable Reactive Barriers.* Elsevier, Amsterdam.

Schulze D, Heller W, Ullreich H and Segebade C (1993) *Instrumental analysis of inactive tracers by photon activation,* Journal of Radioanalytical and Nuclear Chemistry, 168, pp385–392.

Simon F-G and Meggyes T (2000) *Removal of organic and inorganic pollutants from groundwater using permeable reactive barriers – Part 1. Treatment processes for pollutants.* Land Contamination & Reclamation, 8, pp103–116.

Simon F-G, Biermann V, Segebade C and Hedrich M (2004) *Behaviour of uranium in hydroxyapatite-bearing permeable reactive barriers: investigation using 237U as a radioindicator,* The Science of the Total Environment, 326 (1–3), pp249–256.

Simon F-G, Segebade C and Hedrich M (2003) *Behaviour of uranium in iron-bearing permeable reactive barriers: Investigation using 237U as a radiotracer.* The Science of the Total Environment, 307, pp231–238.

Singh S P, Ma Q Y and Harris W G (2001) *Heavy metal interactions with phosphatic clay: sorption and desorption behaviour,* Journal of Environmental Quality, 30 (6), pp1961–1968.

Sowder A G, Clar S B and Fjeld R A (1996) *The effect of silica and phosphate on the transformation of schoepite to becquerelite and other uranyl phases,* Radiochimica Acta, pp74, 45–49.

Vidic R D (2001) *Permeable Reactive Barriers – Case Study Review.* GWRTAC Technology Evaluation Report, TE-01-01, 49 p.

Xenidis A, Moirou A and Paspaliaris I (2002) *A review on reactive materials and attenuation processes for permeable reactive barriers.* Mineral Wealth, 123, 35–48.

Geotechnical and Environmental Aspects of Waste Disposal Sites – Sarsby & Felton (eds)
© 2007 Taylor & Francis Group, London, ISBN 978-0-415-42595-7

Geotechnical and environmental assessment of contaminated sites under migration of polluting components

A.B. Shandyba

Sumy National University, Ukraine

ABSTRACT: At the present time there is growing scientific concern about the predictive procedures currently available for environmental assessment of contaminated sites. After consideration of the various approaches and geodata that may be involved, the stagnant zones model has been developed. Three stages of leaching are defined for use in the specialized model for soil-geochemistry mapping.

1 INTRODUCTION

The latest PC-generation provides the necessary computational facility for geodata management covering some aspects of environmental protection connected with ecology forecasting and soil-geochemistry mapping. At the present time, the main aim was to propose a system that was easy to use with complex geodata from controlled environmental subject areas (Rasig, 1996). Although there may be theoretical reasons for preferring one model over another, the choice is often made on much more pragmatic grounds. For example, the interrupted leaching process cannot be described without correct definition of the boundary conditions and the mass-transfer parameters for the uninterrupted process. Usually only some integral parameters are accessible by common analytical methods.

One of the most interesting and useful approaches to solving the mass-transfer equations was originally advanced by Verygin and Oradovskaya (1960). In this method uninterrupted leaching flow moves in soil pores from a soil surface to a ground water table under supposition of continuously soluting components. Both numerical and analytical solutions exist for certain cases. The most difficult part of the procedure is determining the hydrodynamical conditions of migration under periodical unregulated precipitation. Bresler et al. (1982) developed a two-dimensional model with the advantages of proposed numerical procedures in organizing and realizing the data from an actual pollution/migration process.

Typically leaching at contaminated sites consists of three stages; a time-lag (delay) associated with the movement of polluted particles of water to the control point, active leaching with decrease of contaminant concentration, and the clean stage without polluting components in stagnant zones. Obviously, the leaching time-lag at the control points will depend on the distance from the starting-line (the watershed).

2 INTERRUPTED MIGRATION

Having collected numerous experimental correlations of interrupted migration under natural hydro-geological and weather conditions it can be seen that the prevailing share of a hydraulic transfer process along a relief gradient is in accordance with the ground relief (Kremlenkova, 1993; Shandyba et al, 1997). Also it is possible to produce good numerical assessment of the technical issues involved for matters pertaining to the environment, agriculture and landscape planning

(Saarela, 1997). At the same time there has been only limited development of geodata management for linking attribute data to initial and border conditions of soil-water systems. The hydraulic simulation model with stagnant zone utilization applied to mass-transfer in porous media was described in a previous publication (Shandyba, 1995).

However, the proposed method needs to be adapted to interrupted leaching under unregulated rainfall or snowmelt with intensity h_i (mm/hr) and duration τ_i(hr). According to the aforementioned model the average concentration of pollutant in leaching water particles is

$$C = C_0 + \frac{1}{B\varepsilon}(C_0' - C_0)[1 - \exp(-Bkst)] \exp\left(-\frac{ks}{\varepsilon_1 H}\sum_i h_i \tau_i\right) \qquad (1)$$

where; C_0 – the initial concentration of pollutant in leaching water particles,
       C_0' – the initial concentration of pollutant in stagnant zone particles,
       ε and ε_1 – the dimensionless specific volumes of leaching water and stagnant zone particles respectively,
       $B = 1/\varepsilon + 1/\varepsilon_1$
       $t = \int_0^x dx/V$ – time of contact between water and stagnant zone particles,
       x   – distance from watershed to control point along a ground relief gradient line,
       V   – filtration velocity at distance x,
       k and s   – coefficient and specific surface of mass-transfer under leaching, respectively,
       H   – effective thickness of leaching layer.
It is important to remember the physical condition

$$\tau \leq \sum_i \tau_i \qquad (2)$$

To identify the model for the predictive aims and soil-geochemistry mapping a reasonable way is to find the integral leaching parameter

$$K = \frac{ks}{\varepsilon_1 H} \quad [1/\text{hr m}] \qquad (3)$$

3 METHODOLOGY

Leaching tests were conducted in the specialized washing column with distilled water ($C_0 = 0$). There are two regimes of testing:

$$t = \text{constant} \qquad (4)$$

$$\sum_i h_i \tau_i = \text{constant} \qquad (5)$$

For one soil-water sample the thickness of contaminated soil was constant ($\tau = $ constant) but it was possible to vary the intensity and duration of watering. Parameter K was then defined by a minimum of two experimental points (Figure 1).

$$K = \frac{\log \frac{C_1}{C_2}}{\theta_2 - \theta_1} \qquad (6)$$

where; C_1 and C_2 are concentrations of pollutant in leaching water after two precipitations,
$\theta = \sum_i h_i \tau_i = $ overall duration of precipitation,
The best correlation was found by ordinary statistical treatment of the experimental tests.

Analogically, the mass-transfer parameter Bks is by variation of the thickness of soil in the column under the condition $\sum_i h_i \tau_i = $ constant.

The leaching process may be controlled using EC conductometry.

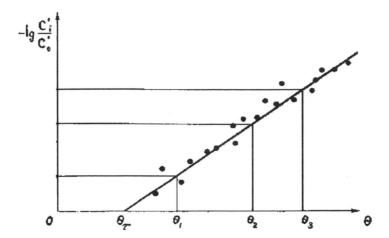

Figure 1.   Experimental results for variation of leachate concentration with time.

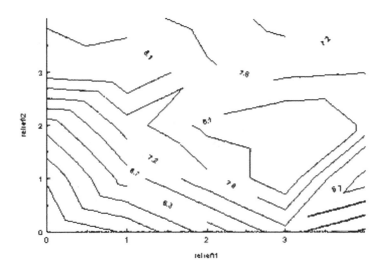

Figure 2.   Indicative topography of site modelled.

4   ENVIRONMENTAL ASSESSMENT

A very important application of the investigation was to predict the pollution area as an ecological consequence of the migration of chemical substances within soil-water systems. The results of modelling a real landscape (Figure 2) under realistic weather conditions are presented as maps of the residue relational pollution (Figures 3 and 4) for the average concentration fields. This soil-geochemistry mapping demonstrates great changes of concentration isolines and these are due to ground relief, geotechnical properties of soil and precipitation intensity.

Bresler et al. (1982) and Kremlenkova (1993) showed experimental results of the characterisation of waste disposal sites. Various natural soils, plants and landscape, and geochemical agents were considered as accumulating for the technogenic soil pollution. The content of dangerous elements depends not only the technogenic source intensity, but also on soil texture, redox conditions, ground relief slope and location of soil in the system of elementary landscapes.

175

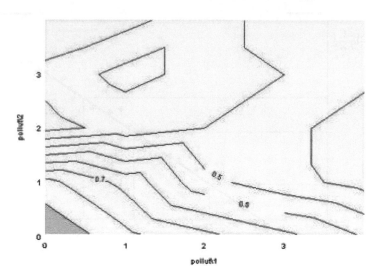

Figure 3.   Residual pollution map.

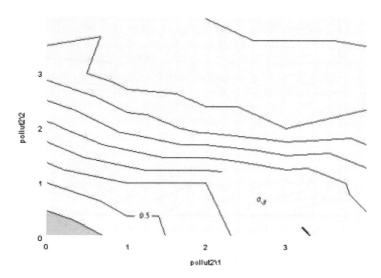

Figure 4.   Residual pollution map.

The integer loss function for limiting contamination may be defined as the relation (Shandyba, 1995):

$$f = \frac{\int\limits_{L}\int\limits_{T} rqCxdLdT}{M_oA} \tag{7}$$

where; r   – hydraulic transfer parameter;
   q   – rainfall intensity [m^3/Ha.S];
   dL – element of control border L;
   dT – element of control time T;
   M_o – initial component content [Kg/Ha];
   A   – control area [Ha].

176

5 CONCLUSIONS

The paper concerns the problem of ecological forecasting evaluated with soil-geochemistry mapping. The simulation models of mass-transfer and migration under leaching incorporate various equations to describe pollution areas in space and time. The leaching process may be controlled with conductometry of the characteristic points. The volumetric coefficient of mass-transfer is determined as the slope of the correlation line. The best correlation of the integral leaching parameter is found by ordinary statistical treatment of the experimental data.

REFERENCES

Bresler E, McNeal B E and Carter D L (1982) *Saline and Sodic Soils*, Springer-Verlag Publishing, Berlin.
Kremlenkova N P (1993) *Redistribution in soils of non-chernozemic zone*, Pochvovedenie, N 9, Moscow, pp 87–93.
Rasig H (1996) *Editing Environmental Data with a PC-based GIS,* Technology and Environment, N 2, UTA International, pp190–192.
Saarela J (1997) *Hydraulic approximation of infiltration characteristics of surface structures on closed landfills*, Finnish Environment Inst. Publ, Helsinki, 139p.
Shandyba A B (1995) *Ecology forecast for migration of the chemical substances into ground and surface water*, Fresenius Env. Bulletin, N 4 , Basel, Switzerland, pp80–85.
Shandyba A B, Vakal S V and Chivanov V D (1998) *Pollution removal by washing of contaminated soils*, Proc of the GREEN2 International Conference, Thomas Telford Publ, London, pp313–316.
Verygin N N and Oradovskaya A E (1960) *Leaching theory under filtration around hydrotechnical constructions*, VODGEO Publ., Moscow, 39p.

Geotechnical and Environmental Aspects of Waste Disposal Sites – Sarsby & Felton (eds)
© 2007 Taylor & Francis Group, London, ISBN 978-0-415-42595-7

Experimental study of reactive solute transport in fine-grained soils under consolidation

G. Tang, A.N. Alshawabkeh & T.C. Sheahan
Northeastern University, Boston, USA

ABSTRACT: This paper presents results of a study on reactive solute transport in fine-grained soils under consolidation. Lead was used as the reactive tracer. Saturated kaolin specimens were prepared with the upper-half containing Lead and the lower-half without Lead. The consolidation tests were conducted under constant stress of 25 kPa with single drainage at the top in the consolidometer (oedometer). Samples for chemical analysis were collected from the surrounding solution during the test and from the pore water of the specimen after it was sliced into four quarters and centrifuged. Comparing the test results for Lead with similar test results using a non-reactive tracer shows that a significant amount of Lead is transported out of soil due to consolidation-induced advection. Consolidation could significantly change the adsorption-desorption equilibrium and increase the advective transport of Lead, at least near the drainage surface during the early stage. Consolidation may not have a notable influence on hydrodynamic dispersion.

1 INTRODUCTION

The influence of consolidation on the transport of contaminants and their fate is an area that has received increasing interest in recent years. Consolidation coupled contaminant transport is critical for applications such as capping of contaminated sediments, dewatering of contaminated sludge and consolidation of clay liners. In addition to hydrodynamic dispersion, the consolidation process results in a significant advective component at early stages at the boundary that can cause considerable transport of contaminants through the soil and boundary.

Consolidation generally decreases the water content and porosity, increases the effective stress and changes the contaminant concentration in clays. These changes could have a significant impact on transport parameters such as hydraulic conductivity, hydrodynamic dispersion coefficient, distribution coefficient of contaminants, retardation factors etc. Decrease of water content usually leads to a decrease of hydraulic conductivity due to the decrease of porosity. Results from consolidation tests on kaolin soil samples (Sheahan and Alshawabkeh, 2003) showed that Coefficient of consolidation (C_v) increases nearly linearly with consolidation pressure increase from 25 kPa to 100 kPa. On the other hand hydraulic conductivity (k_v) decreases when the pressure increases from 25 kPa to 50 kPa and little change was noted for increase of pressure from 50 kPa to 100 kPa. Clay compaction practice means that soil fabric, rather than porosity, is likely to control the hydraulic conductivity (Holtz and Kovacs, 1981). Decrease of porosity usually increases the tortuosity so the hydrodynamic dispersion coefficient is supposed to be reduced. Zhang et al (2003) studied effect of stress on solute transport using bench-scale column tests and centrifuge column tests. The soil samples were made of 10% sodium bentonite, 50% Ottawa sand, 15% fine sand and 25% medium sand by weight. The 10.5 cm height soil sample was saturated from the bottom under a hydraulic gradient of 35 and a constant consolidation pressure of 68.95 kPa. After saturation 100 mg/L Cadmium solution at pH = 5 was used to penetrate the soil sample from the bottom. The Cadmium concentration of effluent samples and soil sections was analyzed. Test results showed that stress state may influence the hydrodynamic dispersion but it has less effect on the adsorption of Cadmium in saturated soils. It is important to consider the effect of the stress when determining transport parameters. Mazzieri et al (2002) studied the effect of consolidation on contaminant transport parameters of compacted clays (kaolin) by column tests at effective stresses of 50 kPa and 415 kPa. A solution with concentration of 6 g/L $MgCl_2.6H_2O$ was used to permeate the soil as a "pollutant" solution. From the tests it was found that the retardation factors were scarcely influenced by the effective stress and the diffusion/dispersion coefficients for Mg^{2+} and Cl^- showed three-fold and two-fold reductions, respectively, when the effective stress was increased from 50 to 415 kPa. These tests studied the impact of consolidation on transport parameters after conclusion of consolidation. Tang et al (2005) conducted an experimental study into non-reactive solute (Bromide and Chloride) transport during consolidation, which is similar to this study on reactive solute transport. Test results show that consolidation-induced advection was the dominant transport mechanism only in the early stage of consolidation near the drainage surface. The time and range was dependent on applied consolidation stress, drainage length, porosity, hydraulic conductivity and molecular diffusion coefficient. The effect of consolidation on hydrodynamic dispersion may not be significant. The behaviour of reactive solutes such as heavy metals and organic chemicals is very important since it represents real contaminants and also accounts for reactivity.

Usually contaminant transport is modelled by a coupled advection-diffusion equation model. Reible (1996) assumed that consolidation-induced advection in a cap and underlying sediment happens instantaneously after the cap is placed and the problem is uncoupled. A capping analysis program was developed and used to assess the impact of consolidation on contaminant transport for contained aquatic dredged material disposal facilities (Ruiz et al, 2002). The results showed that the flux due to consolidation was about three to five times that of diffusion over the first 30 years when consolidation was pronounced. The results clearly highlight the significance of consolidation on cap effectiveness. Smith (2000) described one-dimensional dispersion-advection transport in a deformable saturated soil with linear, equilibrium controlled sorption. Analytical solutions were derived for a quasi-steady-state contaminant transport problem. The results of the analysis indicate that for the case of contaminant transport through a consolidating soil with relatively small contaminant sorption, accounting for the contaminant on the solid phase can make a small difference to the estimated contaminant distribution, and a significant difference to the steady-state mass flux through a consolidating soil. Other coupled time-dependent consolidation-induced advection and diffusion models were developed by Peters and Smith (2002) and Van Impe et al (2002) and Alshwabkeh et al (2004) and demonstrated the role of consolidation in contaminant transport. Because the seepage velocity induced by consolidation is very high (theoretically infinite) near the drainage surface in the beginning and decreases to zero near no flow boundaries or at the end of consolidation, it is difficult to get stable and accurate numerical solutions to a time-dependent advection-diffusion equation during the early stages of consolidation (Li et al, 1992).

Ideally adding a retardation factor or applying a more complicate Freundlich isotherm or Langmuir adsorption isotherm (Fetter, 1993) to the coupled advection-diffusion model could make the reactive solute transport problem manageable in some simple cases. In reality it is much more challenging to deal with in addition to the transport because;

1. migration of reactive solute could significantly change the mechanic properties and cation exchange capacity of the soil;
2. consolidation could influence the adsorption-desorption equilibrium.

2   EXPERIMENTAL WORK

The samples were prepared by mixing a kaolin powder with Lead solution or brackish water. The reason for selecting kaolin was for its moderate properties in terms of its cation exchange capacity, which determines the adsorption properties of reactive tracers. Lead (Pb^{2+}) was selected as the reactive tracer because it is a typical heavy metal contaminant in soil and groundwater. Brackish water, which was prepared by mixing synthetic sea water with distilled water at a ratio of 1:1, and lead nitrate (Pb(NO$_3$)$_2$) were used to prepare Lead solution at concentration of 750 mg/L. Brackish water was used because it can provide other cations to compete with Lead for adsorption onto kaolin particles so that a significant amount of the Pb^{2+} can be transported out of the soil samples. Otherwise, most of the Lead will be adsorbed onto soil particles due to the cation exchange capacity of kaolin and the extremely low mobility of Pb^{2+} in soil (Fetter 1999). Nitric acid (HNO$_3$) was added to the Lead solution to lower the pH to about 3.5 so that no precipitation could happen.

Because of its high ionic strength, brackish water and Pb^{2+} can significantly change the properties of soil samples. Average values for water content (w), porosity (n), Coefficient of consolidation (C$_v$) and hydraulic conductivity (k$_v$) from tests with deionized water, 960 mg/L bromide solution, brackish water and brackish water with 750 mg/L Pb^{2+} at pH = 3.5 are listed in Table 1. From the table it can be seen that as ionic strength increases, water content required for practical soil mixing increases due to the decrease of the double layer thickness, and so do porosity and hydraulic conductivity.

To make sure the water and Lead could be uniformly distributed and adsorption-desorption equilibrium could be reached the samples were cured for at least 12 hours before testing. Chemical analysis of pore water obtained by centrifuging soil samples before testing shows that the Lead concentration ([Pb^{2+}]) in the pore water (C$_d$) is about 76 mg/L. The distribution coefficient (k$_d$) is calculated to be about 6.6 (mg/L)/(mg/kg).

Batch tests were conducted to determine the adsorption isotherm of Pb^{2+} onto kaolin in deionized water, brackish water and synthetic sea water at different pH. The results showed that the adsorption isotherm is sensitive to pH and ionic strength. The k$_d$ values are shown in Table 2 – for further details refer to Sheahan and Alshawabkeh (2003). The k$_d$ from batch tests is different from the consolidation test. The reasons could be;

1. different water soil ratio with 0.7 ml : 1 g in the consolidation test and 30 ml : 3 g in the batch test;
2. samples were cured at least for 12 hours in the consolidation tests while samples were stored for at least 24 hours with frequent shaking by hand in accordance with EPA procedures for the batch tests (Sheahan and Alshawabkeh 2003).

It has been pointed out that batch tests tend to overestimate the actual adsorption and the results could be the upper limits (Mazzieri et al, 2002).

Table 1. Soil properties with different pore water.

Pore water	w (%)	N	c$_v$ (10^{-8} m^2/s)	k$_h$ (10^{-7} m/s)
Deionized Water	50	0.58	2.4	7.0
Bromide Solution(960 mg/L)	55	0.60	2.5*	9.0*
Brackish Water	70	0.65	–	–
Brackish Water with Pb^{2+}	74	0.66	9.0**	47.0**

* c$_v$ and k$_h$ are consolidation test results with half soil sample with bromide solution on the top and half soil sample with deionized water in the bottom.
** c$_v$ and k$_h$ are consolidation test results with half soil sample with brackish water with Pb^{2+} on the top and half soil sample with brackish water in the bottom.

Table 2. k$_d$ for Pb^{2+} on Kaolin[(mg/L)/(mg/kg)](Sheahan and Alshawabkeh 2003).

Solution	pH = 7	pH = 5	pH = 3
Deionized Water	–	600	333
Brackish Water	60	60	24
Synthetic Sea Water	11	27	19

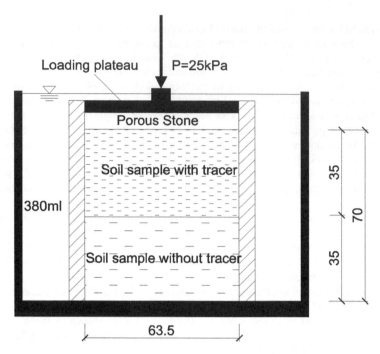

Figure 1. Schematic consolidation test setup.

The consolidation test setup is presented in Figure 1. The lower half of the cell was filled with soil without Lead first, then separated by a saturated filter paper and the upper half of the cell was filled with soil with Lead. A saturated filter paper, a saturated porous stone and a loading plate were placed above the specimen to provide single drainage on the top. The bottom of the cell was sealed. The whole cell was submerged in distilled water. Nitric acid was added to lower the pH to about 2 so that Pb^{2+} that migrated with the consolidation-induced advection and hydrodynamic dispersion from the top of the soil into the surrounding water would not precipitate. Samples were collected from the surrounding water during the test. Water was added to compensate for the water loss from the surrounding solution due to evaporation and sampling. The volume of the surrounding solution was about 380 ml.

The consolidation test was performed in accordance with ASTM D2435. One step constant load of 25 kPa was applied through loading plate and lasted for about 3 days. At the end of testing, the specimen was sliced into four quarters and a centrifuge was used to extracted pore water to analyze the Lead concentration to evaluate the final distribution across the specimen. The Coefficient of consolidation C_v was calculated using the log time method and verified by the square root of time method according to ASTM D2435. The hydraulic conductivity (k_v) was calculated from C_v by the equation

$$k_v = c_v m_v \gamma_w \tag{1}$$

where m_v is the coefficient of volume change and γ_w is the unit weight of water.

Chemical analysis of Pb^{2+} was conducted using a Varian SpectrAA-220 Atomic Absorption (AA) spectrometer. A Fisher Accumet pH meter was used to check the pH value in the solution and pore water. The displacement was measured automatically by a LVDT sensor and data acquisition system.

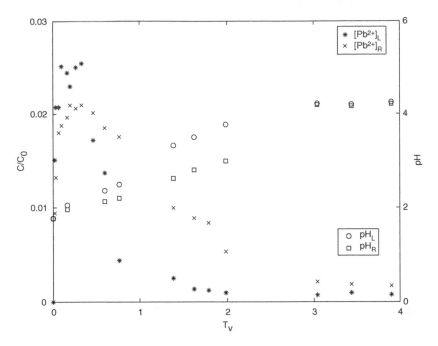

Figure 2. $[Pb^{2+}]$ and pH in the surrounding water during consolidation.

3   RESULTS AND ANALYSIS

Two sets of history curves labeled 'L' and 'R' for Lead concentration ($[Pb^{2+}]$) together with pH in the surrounding water are presented in Figure 2 in terms of the dimensionless consolidation time factor T_v and C/C_0 (where $T_v = c_v t/H_{dr}^2$, t is time, H_{dr} is the length of drainage path, C is the $[Pb^{2+}]$ in the surrounding water and C_0 is the initial $[Pb^{2+}]$ in the solution used to prepare the soil samples). The results show that C/C_0 increases and reaches a maximum at about $T_v = 0.5$ and then decreases to almost zero while pH increases continuously from 1.8 to 4.2 until $T_v = 3$. Results from similar consolidation tests with non-reactive tracer (bromide, Br^-) are shown in Figure 3 for comparison. C/C_0 from bromide tests increases continuously and becomes constant after $T_v = 1$. This suggests that the $[Pb^{2+}]$ in the solution should have similar trends. It is not likely that the Lead migrates back to soil samples because;

- at $T_v = 0.5$ there should be still be some flux from soil into the surrounding water, lead migration due to advection should increase lead in surrounding water;
- lead concentration in the soil is higher than in the surrounding water, migration of lead due to diffusion should increase lead concentration in the surrounding water;
- the loading plate and porous stone could retard that migration.

Consequently the decrease could be due to chemical reactions in the surrounding water. From Figure 2 it can be noted that the decrease $[Pb^{2+}]$ may be related to the increase of the pH. It can be seen that $[Pb^{2+}]_L$ decreasing faster than the $[Pb^{2+}]_R$ coincides with pH_L being a little higher than pH_R after the peak. Possible chemical reactions include;

- lead precipitation;
- single replacement reaction or reduction of Pb^{2+};
- adsorption to metal surface or oxide of metals.

183

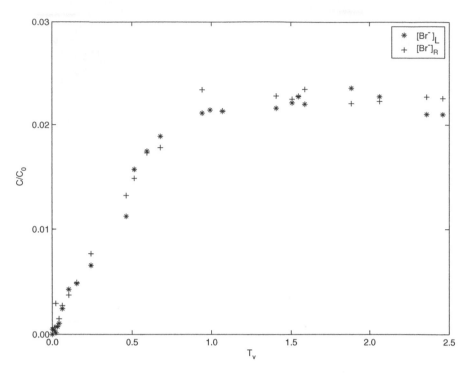

Figure 3. [Br$^-$] In the surrounding water during consolidation.

Lead precipitation may include (Snoeyink and Jenkins, 1980; Sawyer et al, 1994):

$$Pb(OH)_2(s) = Pb^{2+} + 2OH^- \quad k_{sp} = 2.5 * 10^{-16} \tag{2}$$

$$PbCl_2(s) = Pb^{2+} + 2Cl^- \quad k_{sp} = 10^{-4.8} \tag{3}$$

$$PbSO_4(s) = Pb^{2+} + SO_4^{2-} \quad k_{sp} = 10^{-7.8} \tag{4}$$

The typical chloride(Cl$^-$), sulfate(SO$_4^{2-}$), bicarbonate(HCO$_3^-$) concentration in sea water are 19000, 2700, and 142 mg/L, respectively (Snoeyink and Jenkins, 1980). The chloride, sulfate, bicarbonate etc from the brackish water used to prepare soil samples and hydroxide could precipitate Lead in the surrounding water even though pH is as low as 3–4 and calculation suggests the contrary.

Considering the standard electrode potentials E^0 (Snoeyink and Jenkins, 1980):

$$Pb^{2+} + 2e^- = Pb(s) \quad E^0 = -0.13 \text{ volt} \tag{5}$$

$$Fe^{2+} + 2e^- = Fe(s) \quad E^0 = -0.44 \text{ volt} \tag{6}$$

For Fe(s) + Pb^{2+} = Pb(s) + Fe^{2+}, E^0 = 0.44 − 0.13 = 0.31 Volt > 0, so the single replacement reaction or reduction reaction can proceed because there is iron present in the metal parts in the test setup.

The pH increase could be due to the transport of the higher pH pore water from the soil sample into the surrounding water, the erosion of the metal parts in the test setup and possibly the open carbonate system(Snoeyink and Jenkins, 1980). About 20 ml pore water at pH of about 4 was drained from the soil into the surrounding water. That could help to increase the pH. Erosion of the metal parts in the test setup (like the reaction Fe + 2H$^+$ = Fe^{2+} + H$_2$(g)) is noted and could consume some protons and partly account for the increase. The open carbonate system can affect the pH but not much at such a low pH. Further research is needed to determine exactly the reason behind the decrease of [Pb^{2+}] and increase of pH.

Table 3. Profile of lead and pH distribution from top to bottom ($T_v = 4.7$).

Location	$[Pb^{2+}]_{I,L}$	$[Pb^{2+}]_{I,R}$	$[Pb^{2+}]_{F,L}$	$[Pb^{2+}]_{F,R}$	$pH_{I,L}$	$pH_{I,R}$	$pH_{F,L}$	$pH_{F,R}$
Top	76.0	75.6	44.1	47.9	3.25	3.22	3.74	3.87
Upper	76.0	75.6	31.6	45.2	3.25	3.22	4.14	3.94
Lower	0	0	3.0	2.9	6.20	6.30	4.16	4.08
Bottom	0	0	1.2	0.2	6.20	6.30	4.20	4.50

$[Pb^{2+}]_{I,L}$, $[Pb^{2+}]_{F,L}$ is the initial and final concentration of lead for test 'L'; $[Pb^{2+}]_{I,R}$, $[Pb^{2+}]_{F,R}$ is the initial and final concentration of lead for test 'R'; $pH_{I,L}$, $pH_{F,L}$ is the initial and final pH in (pore) water for test 'L'; $pH_{I,R}$, $pH_{F,R}$ is the initial and final pH of (pore) water for test 'R'. The unit for concentration is mg/L.

Table 4. Profiles in the soil sample(mg/L)(from top to bottom).

Location	Initial Br^-	Br^- ($T_v = 2.7$)	Br^- ($T_v = 3.8$)
Top	961.6	713.8	669.1
Upper	961.6	572.3	532.2
Lower	0.0	295.2	356.2
Bottom	0.0	222.6	263.2

The concentration curves could be the lower bounds for the Pb^{2+} transported out of the soil. The results from the reactive test can be compared with those of the nonreactive test when T_v is not large. In Figure 2 C/C_0 is about the same as, or a little higher than, that in Figure 3 for T_v less than 0.5. Unlike nonreactive solute, for which the concentration of solute in the pore water (C_d) is almost the same as C_0 because of no adsorption, C_d (76 mg/L) is only about 10% of C_0 (750 mg/L) for Pb^{2+} because of adsorption. The Pb^{2+} transported out of the drainage surface due to advection is much more higher (about an order of magnitude) than what is expected by assuming C_d to be the $[Pb^{2+}]$ in the drained pore water. Several factors could contribute to that unexpected high values:

- The water content(74%) for Pb^{2+} test is higher than that for the bromide (55%), the settlement for Pb^{2+} test is about 9 mm and Br^- is about 6 mm, this means about 35% more volume of pore water was drained from Pb^{2+} test than from Br^- test at the same T_v or consolidation ratio.
- The lower pH (1.7) in the surrounding water in the beginning could be of help but this is not likely due to the retardation of the porous stone, loading plate, and the advection from the soil to the water.
- $[Pb^{2+}]$ in the drained pore water is much higher than C_d.

Only the last reason could account for such a big difference. So it is highly possible that the consolidation induced-advection or high seepage velocity near the drainage surface in the early stage or the effective stress could significantly change the adsorption-desorption equilibrium and increase the mobility of Pb^{2+}.

The profiles of $[Pb^{2+}]$ and pH in the pore water across the specimen in the beginning and in the end are presented in Table 3. Profiles from bromide tests are listed in Table 4 for comparison. The values represent the average for four quarters from the top to the bottom of a specimen. It can be seen from Table 3 that the pH profile in the end is almost uniform while there is big Lead concentration gradient. The pH increase in the bottom of the specimen could be due to the high mobility of H^+.

Unlike the nonreactive test results, in which an average of about 14% of bromide was transported into surrounding water due to advection and 25% of the bromide was transported to the lower half due to diffusion at a T_v of about 3 (Tang et al, 2005), from Table 3 it can be seen that a significant amount of Lead was transported out of soil through the drainage surface and a detectable amount of Lead wasonly found in the lower half of the specimen at a T_v of 4. Assuming $k_d = 6.6$ and that adsorption-desorption equilibrium is reached for the whole specimen, the percentages of Lead remaining in the upper half and transported to the lower half could be roughly estimated to be about 55% and 3%, respectively. So adsorption does significantly retard the migration of lead to the bottom due to diffusion.

4 CONCLUSIONS

Tests were conducted to evaluate the effect of consolidation on transport of Lead in soils during the consolidation process. In comparison to similar test results with nonreactive solutes, the consolidation test results show that a significant amount of Lead is transported out of soil due to consolidation-induced advection. The high seepage velocity near the drainage surface in the early stages of consolidation could significantly change the adsorption-desorption equilibrium and increase the mobility of Lead. However, the retardation of the diffusion transport might not be influenced. Difficulties arise when the pH is lowered and ionic strength is increased to increase the mobility of Lead in soils because of the low buffer capacity of kaolin and the high sensitivity of the adsorption isotherm of Lead to pH and ionic strength.

REFERENCES

Alshwabkeh A N, Rahbar N, Sheahan T C and Tang G (2004) *Volume Change Effects on Solute Transport in Soils under Consolidation,* Proceedings of Geo2004, Amman, Jordan, ASCE.
Fetter C W (1999) *Contaminant Hydrogeology(second edition),* Prentice-Hall, Upper Saddle River, NJ.
Holtz R D and Kovacs W D (1981) *An Introduction to Geotechnical Engineering,* Prentice-Hall, Englewood Cliffs, NJ.
Lo I M C, Zhang J, Hu L and Shu S (2003) *Effect of Soil Stress on Cadmium Transport in Saturated Soils,* Practice Periodical of Hazardous, Toxic, and Radioactive Waster Management, 7(3), pp170–176.
Li S, Ruan F and McLaughlin D (1992) *A Space-Time Accurate Method for Solving Solute Transport Problems,* Water Resources Research, 28(9), pp 2297–2306.
Reible D D (1996) *Appendix B: Model for Chemical Containment by a Cap,* Guidance for In-Situ Capping of Contaminated Sediments, www.epa.gov/glnpo/sediment/iscmain
Mazzieri F, Van Impe P O and Van Impe W F (2002) *Effect of Consolidation on Contaminant Transport Parameters of Compacted Clays,* Environmental Geotechnics (4th ICEG) '02, Swets & Zeitlinger, USA, pp 183–188.
Peters G P and Smith D W (2002) *Solute transport through a deforming porous medium,* International Journal for Numerical and Analytical Methods in Geomechanics, 26, pp 683–717.
Ruiz C E, Schroeder P R, Palermo M R and Gerald T K (2002) *Disposal facilities shortage generates innovative dredged material management and contaminant flux evaluation solutions,* Dredging Research, Vol.5, No. 2.
Shackelford C D (1990) *Diffusion of Contaminants Through Waste Containment Barriers,* Geotechnical Engineering, Transportation Research Record 1219, pp169–182.
Sheahan TC and Alshawabkeh A N (2003) *Practical Aspects of a Reactive Geocomposite to Remediate Contaminated, Subaqueous Sediments,* Soil and Rock America 2003, Proceedings of the 12th Pan-American Conference on Soil Mechanics and Foundation Engineering, Cambridge, MA, Vol. 2, pp1417–1422.
Smith D W (2000) *One-dimensional Contaminant Transport through a Deforming Porous Medium: theory and a solution for a quasi-steady-state problem,* International Journal for Numerical and Analytical Methods in Geomechanics, 24, pp 693–722.
Sawyer C N, McCarty P L and Parkin G F (1994) *Chemistry for Environmental Engineering (4th ed),* McGraw-Hill, New York.
Snoeyink V L and Jenkins D (1980) *Water Chemistry,* Wiley, New York.
Tang G, Alshawabkeh A N and Sheahan T C (2005) *Experimental Study of Nonreactive Solute Transport in Find-Grained Soils Under Consolidation,* Proceedings of Geo-Frontiers 2005, Austin, Texas.
Van Impe P O, Mazzieri F, Van Impe W F and Constales M (2002) *A Simulation Model for Consolidation and Contaminant Coupled Flows in Clay Layers,* Environmental Geotechnics (4th ICEG) '02, Swets & Zeitlinger, USA, pp189–194.

Geotechnical and Environmental Aspects of Waste Disposal Sites – Sarsby & Felton (eds)
© 2007 Taylor & Francis Group, London, ISBN 978-0-415-42595-7

Groundwater protection by preventive extraction at an industrial waste deposit – history and evaluation

C. Cammaer
SLiM vzw, Diepenbeek, Belgium

L. De Ridder
Umicore, Olen, Belgium

T. Van Autenboer
SLiM vzw, Diepenbeek, Belgium

ABSTRACT: Some 20 years ago hydraulic isolation with a central pumping well was proposed to protect the groundwater underneath an industrial waste deposit close to an existing deposit of mixed domestic and industrial waste at Olen (Belgium). A major advantage was the simplicity of the system. The system is still in use and necessary, as the steady increase of salts in the central well and the improvement of downstream groundwater quality indicate. However, a large number of additional wells and several hydrogeologic studies prove the aquifer to be far more complex than previously thought. Extensive pumping tests confirmed earlier studies and stressed the importance of the production wells of the refinery. Mathematical modelling proved the central pumping sufficient to reverse the downstream groundwater flow underneath most of the deposit. It also provided a better understanding of the local aquifer which is crucial in the remedial actions now being planned at and around the refinery.

1 AN INDUSTRIAL WASTE DEPOSIT AND A VULNERABLE AQUIFER

1.1 *The original state of the deposit*

Aerial photographs from the early 1960's show the early deposit as a pile of rubble in wetland close to the canal (Figure 1). The wetland to the North of the canal is caused by infiltration of the canal onto the clayey topsoil. The refinery uses some 2.5 M m^3/yr groundwater as process and cooling water.

Until the 1980's the deposit consisted of a mixture of domestic waste, filter residues, slag and probably some contaminated materials from a non-ferro refinery. The refinery (Umicore Olen) has a long and complex history. Established in 1908, the company moved to Olen in 1912 to manufacture chromium salts. Cobalt production started in 1925, copper in 1928, germanium in 1953. The plant was best known (in Belgium) for radium (started in 1920). With the rich uraninite deposits from Katanga and sound management, 97% of the world production was soon manufactured in Olen (i.e. 40 g/y). With a rich industrial past (no roasting), problems with wastes and groundwater are to be expected; colourful chromium residues used as a landfill, artificial hills with iron oxides (along with traces of other metals) due to the cobalt industry and various precipitates and slag as a result of the copper refinery. Waste material from the production of radium was at first of limited value but became strategic in WW-II. There is still a small amount of radioactive waste at the Olen site causing a disproportionately big administrative problem. Groundwater contamination did not spread very far downstream of the refinery thanks to the local geology and to the high pumping

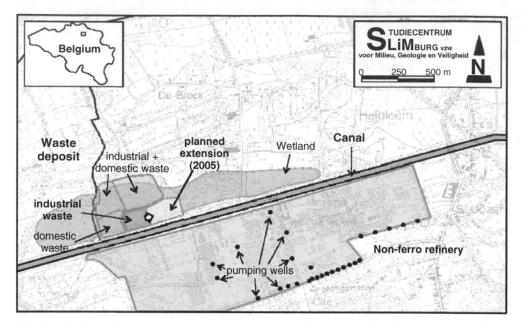

Figure 1.   Study area with the canal between the refinery and the waste deposit (situation in 1984 with mixed domestic and industrial wastes).

rate in several groundwater wells. Sanitation of the different sites is now planned and groundwater remediation under study.

The groundwater in the area is shallow and can easily be contaminated by a landfill without protection, as indicated by the chlorine content in some downstream wells. In the 1980's some concern was voiced about the influence of waste on the groundwater quality. A regulatory administrative body dealing with waste was created by the Flemish government in 1981. Administrative and technical directives started to flow freely with regulations logarithmically expanding. Domestic and industrial waste were no longer to be dumped together and a hydrogeologic study became mandatory. The high chlorine content did cause some concern and some prevention of further infiltration of leachate deemed necessary.

1.2   Local hydrogeology

The water bearing formation consists of 80 m glauconite sands of Miocene age (Diest sands) covering an up to 60 m thick impermeable Boom Clay (Figure 2). The Quaternary sands on top are up to 3 m thick in this area. With a shallow groundwater table the aquifer was classified as "very vulnerable". It is intensively exploited in a large part of Flanders with the local refinery as a major user.

Most of the sandy aquifer can be considered as homogeneous with rather coarse glauconite sands and a hydraulic conductivity (k) between 6 to 14 m/d. Observation wells drilled around the waste area at first confirmed this. A two week pumping test showed a slight difference in head losses for observation wells penetrating the top zone and the deeper ones. But as a whole the local hydrogeology was believed to be fully unconfined. The groundwater table seemed to be regular.

1.3   Hydraulic isolation

Hydraulic isolation was advocated. Leachate from the waste deposit reaching the water table would be removed and the dispersion of groundwater already containing high concentrations of

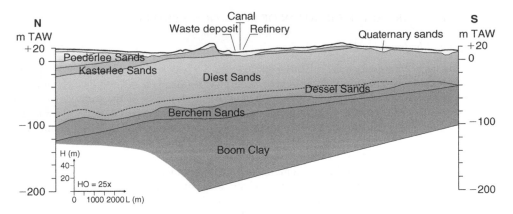

Figure 2. Geological cross section (after the official geological map and cross sections of the area).

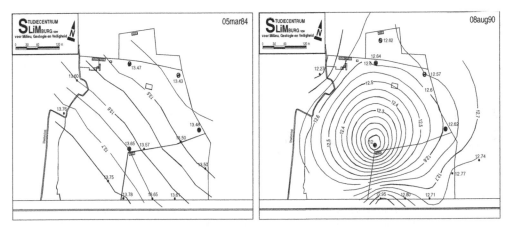

Figure 3. Piezometric maps showing local groundwater table before pumping (March 1984) and with central pumping well in operation (August 1990, 30 m³/h).

salt reduced. The installation of the system was simple and its functioning could easily be controlled by equally simple head measurements. There was a lack of experience with artificial liners with a limited life expectancy estimated to be maximum 25 years.

The very low permeability of the filter residues and chemical precipitates would also limit percolation. Leaching to the groundwater was expected to decrease as the thickness of the waste increased (up to 10 m). The produced groundwater could to be used in the plant or if necessary treated and discharged – there was excess capacity at the water treatment plant). The local lay-out was favourable and the economic aspects good as compared with passive protective systems.

The installation and operation of the system was convincingly simple and geologically sound. A hydrogeological study including a two-week pumping test determined the optimum pumping rate. The results further showed that the necessary groundwater lowering could be obtained with one central pumping well and a minimum pumping rate of 15 m³/h. Maps of the hydraulic head indicated that all water permeating through the wastes would be captured not only from the industrial waste but also from the domestic deposit (Figure 3). The influence of the water production at the refinery was clearly identified but also underestimated.

2 TWENTY YEARS AFTER CREATION OF ISOLATION SYSTEM

2.1 *Impact of hydraulic isolation*

The system has been in continuous operation for 20 years. The salt content (chlorine and sulphate) did show a regular increase (the high salt content prohibited its use in the refinery) which indicated that leachate and some metals were removed (Figure 4). The quality of the wells downstream improved and migration of the old salt plume was thought to be stopped or seriously retarded (Figure 5 – showing the canal, the plume with salt groundwater downstream and the pumping well planned to control leachate from the industrial waste site).

Hydraulic isolation is clearly effective but must be continued. The quality of the water of the central well decreased as salts and some metals increased thereby eliminating its use in the hydrometallurgy. The expected decrease in leachate was not effective and local anomalies in the ground water flow and quality, e.g. hanging water table, were discovered. A more thorough investigation of the hydraulic isolation was required.

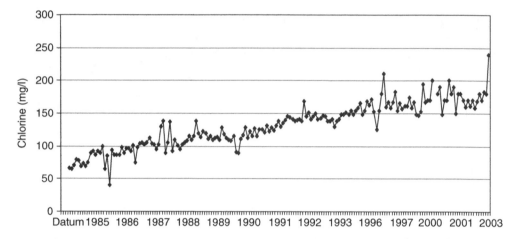

Figure 4.   Increasing chlorine content in produced water since hydraulic isolation started.

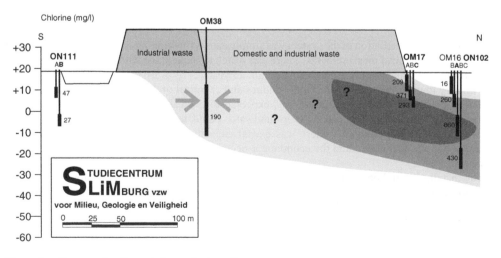

Figure 5.   Cross section through the waste deposit.

190

2.2 *Additional information about local hydrogeology*

Since 1990, with several hydrogeological studies, the number of observation wells in the neighbourhood increased considerably. Some 200 wells with corresponding lithology, measurements of hydraulic head and chemical analysis provided a better understanding of the local shallow aquifer. Originally the water-bearing formation was believed to be homogeneous but the new data showed that it was much more complex. A top zone, up to 15 m thick, was found to be heterogeneous, finer grained as a whole and containing clayey lenses.

The clayey zone has been observed in other localities further east, but its importance varies considerably. In one area a thick zone of very fine sandy and clayey layers forms an impermeable separation between a shallow aquifer and the deeper water bearing formation, which is protected from polluted surface water or leachate. In other localities the clayey zone is insufficiently developed or too irregular to protect the groundwater. At the Olen site the clayey character of this top zone changes gradually thereby creating a transition from a fully unconfined aquifer in the west through semi-unconfined to fully confined in the east (the latter with a separate shallow unconfined aquifer on top). The clayey top zone and the shallow aquifer indicate that the emphasis of remedial active will be to deal with the shallow groundwater. Because this clayey top zone and the shallow water table complicate the 'official' aquifer vulnerability, it was understood that study of the local hydrogeology would be crucial in the process of deciding on the best available technique to prevent further contamination.

A gradual east-west transition from a confined to an unconfined aquifer was found to be valid as a general picture. It was used in steady state conditions, but locally large scale pumping of the groundwater at the refinery was found to have a much more important impact than previously thought. Figure 6 shows a detailed E-W cross section with the well developed clayey area in the east and the resulting differences in water levels above and under the clayey zone. The waste deposit is downstream (to the north) of the western end of the profile.

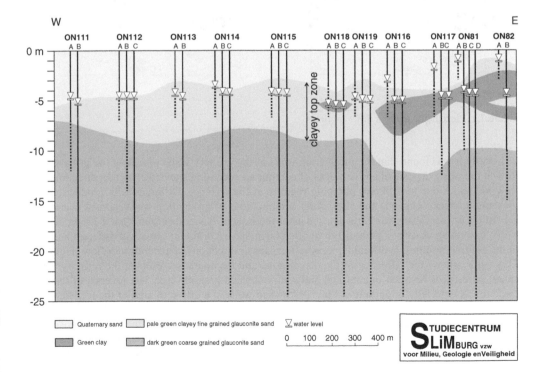

Figure 6.   Cross section along the canal (about 2 km).

191

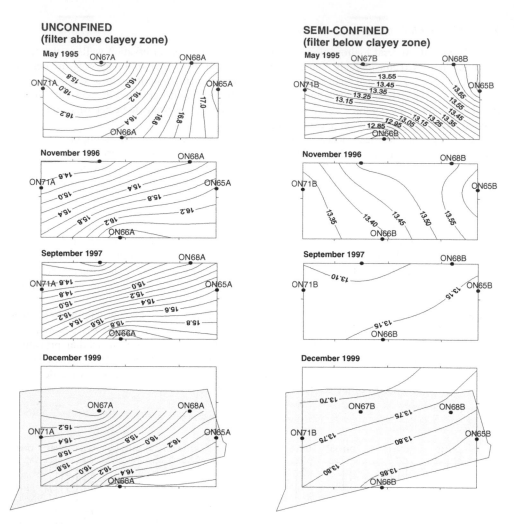

Figure 7.   Piezometric maps at different times.

Locally it was found that the production wells at the refinery also controlled the groundwater flow at the waste deposit site. Five weeks of continuous recording showed changing directions and flow velocities due to different pumping regimes at the refinery. Differences in production rate (day-night, week-weekend) and alternating combinations of production wells (spread over a vast area of the refinery) were easily recognised in the recordings. It was also found that these production wells, extracting water at a great depth of 50 to 80 m, have a different impact in different horizons of the aquifer, resulting in different piezometric maps or directions of groundwater flow (Figure 7). This figure shows the influence of the clayey zone and the operating wells at the refinery which cause differences between unconfined and semi-confined groundwater. The reversal of the flow direction is especially important in May 1995 when wells in a different area of the refinery were taken into production.

The local groundwater, previously believed to be fully unconfined at the waste deposit site, now showed signs of being semi-unconfined in stress conditions. It was felt that the production wells could be intensifying the impact of the central well hydraulic isolation (Figure 8). It also became clear that controlling and measuring pumping well efficiency was not as straightforward as previously estimated. Infiltration from the canal could be important as shown by the much

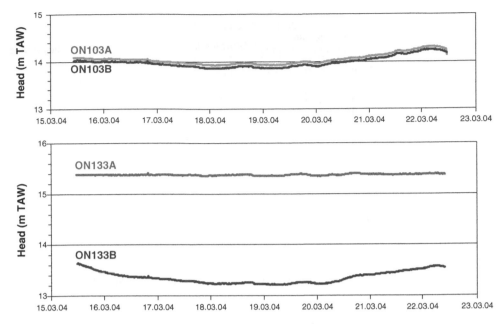

Figure 8.   Different impact west (ON103) and east from the dumping site (ON133).

higher gradient near the waterway. This was confirmed by rapid changes in the water temperature of the canal due to discontinuous discharges of cooling water, reflected in the shallow wells close to the canal. Figure 8 shows the impact during a normal working week at the refinery, showing a maximum effect around Wednesday (full production) and a minimum on Sundays. East of the waste deposit (confined area) there is a big difference in piezometric heads above (ON133A, red) and below (ON133B, blue) the clayey zone. More to the west, in the unconfined part of the aquifer, this effect has gone (ON103A/B).

3   MODELLING GROUNDWATER FLOW

3.1   *The regional model*

Modflow96 was used to simulate the groundwater flow for a vast region surrounding the non-ferro refinery.

When constructing the conceptional model, special consideration was given to the east-west transition (confined-unconfined). This was accomplished defining a separate model layer (2) with gradually changing hydraulic characteristics (such as sand, clayey sand, clay). This layer with a constant thickness is used in the model to control a spatial difference between two extremes: one unconfined aquifer or two aquifers (unconfined and confined) – Figure 9. The pumping wells for the refinery are screened much deeper than the central pumping well. The thickness of layer 2, used to simulate the clayey zone, is kept the same and with the same horizontal hydraulic conductivity for the whole model area, while its vertical hydraulic conductivity varies.

As there are many observation wells in the area, some nested wells with screens at different depths, the steady state calibration was based upon some 200 head measurements. The difference in hydraulic head in multiple wells in the confined area was considered important. A maximum difference of 0.1 m between measured and observed data was tolerated. Calibration was mainly focused on the spatially varying hydraulic characteristics of model layer 2 (the clayey zone).

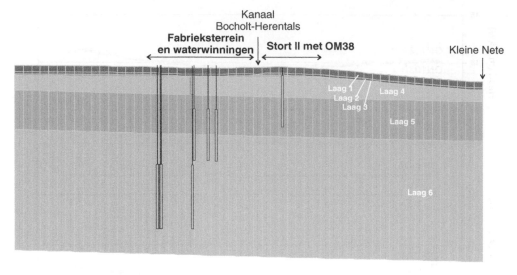

Figure 9.   North-south section showing the 6 model layers used to model local hydrogeology.

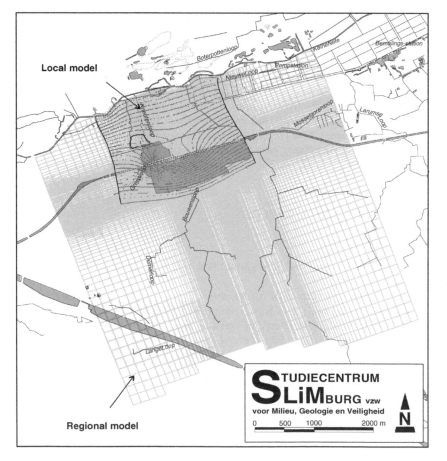

Figure 10.   The grid of the regional model with the grid of the local model superimposed.

A second calibration was carried out for transient flow using the data of a two week pumping test, thereby fine tuning the connection between the 6 different model layers. A second set of piezometric heads measured in a different season was used to validate the model.

3.2 *Conversion of the original model to a local one*

The regional model, once operational, was converted into a local one to refine the grid around the deposit (Figure 10). The boundaries of the model were a small river in the north and a watershed in the south. The grid in Figure 10 is oriented according to the groundwater flow direction. This local model was calibrated again, for two pumping tests (1984 and 2003). Both pumping and recovery tests used the central pumping well (hydraulic isolation) and all available wells were measured for several weeks.

The results confirmed the initial calculation of the hydraulic characteristics. For the calibration pumping data for the central well and production wells were used as input stress periods in the transient local model (3 hour interval for several weeks).

4 APPLICABILITY OF THE MODEL

The calibrated model illustrates the effect of the different pumping combinations for all wells operating simultaneously in 1984 (Figure 11), under changed conditions in 2005 (Figure 12) and for a hypothetical situation without production wells at the refinery (Figure 13). A film loop was helpful to evaluate these data with simulations every 3 hours over several weeks.

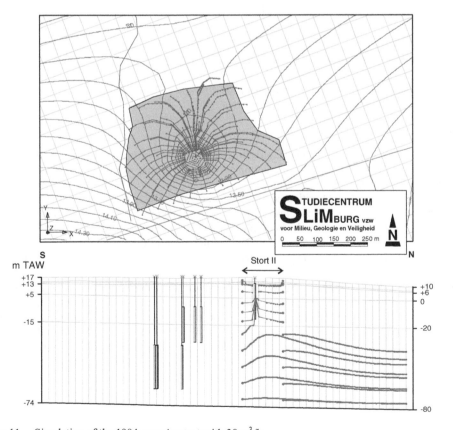

Figure 11.   Simulation of the 1984 pumping test with 30 m³/h.

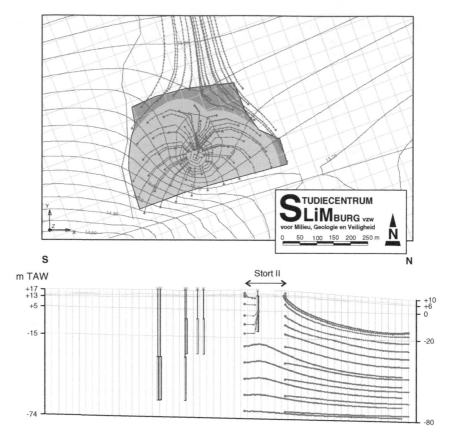

Figure 12.   Simulation of the 2005 pumping with 20 m³/h.

As the 2003 pumping test involved collecting data for equally long periods with and without the central pumping well in action, the interaction between the central well and production wells at the refinery became clear.

The pumping test of 1984 with a central pumping rate of 30 m³/h was reconstructed with the model and did prove the early conclusions to be correct (Figure 11). Particle tracking (Modpath) and resulting path lines confirm the hydraulic isolation with one central pumping well: its impact covers the entire dump. The analytical calculations however used for optimising the pumping rate (to 15 m³/h) were found to be ambiguous since the impact of production wells was not fully realised at that time.

The impact of the central pumping well was measured far beyond the confines of the dump. The local model however showed that with a pumping rate of 20 m³/h the flow is not always reversed underneath the entire dump (Figure 12). Particle tracking (Modpath) and resulting path lines show the effect of the production wells at the refinery, part of the shallow groundwater under the dump leaves the area. Deeper groundwater migrates freely. Leachate underneath the northern part of the dump site (historical mix of domestic and industrial waste) is allowed to migrate elsewhere – attracted to the production wells during the week, released to the natural groundwater flow during the weekend.

The model also illustrated that the production wells are an important factor controlling the general groundwater flow in a large area at and around the refinery. Pumping large quantities of water limits, retards or even reverses the natural flow of contaminated groundwater (Figure 13).

196

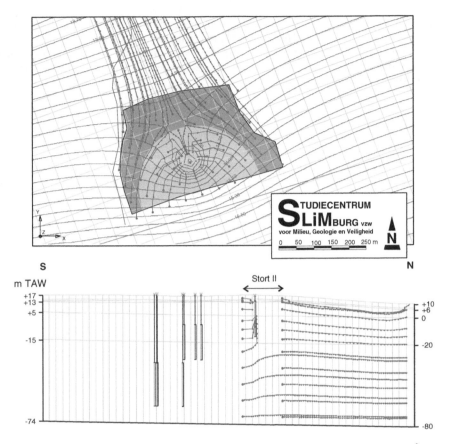

Figure 13. Simulation of a hypothetical situation with the effect of hydraulic isolation (20 m³/h) without production wells.

Particle tracking (Modpath) and resulting path lines show that in this case the regional groundwater flow dominates and the central pumping well is insufficient to cover the dump site.

5 CONCLUSIONS

Geology can be simple on a millimetre or kilometre scale or when there are too few data to contradict each other. The Olen study clearly showed that interpretation becomes increasingly complex as more wells are drilled, more water levels are measured and more samples taken and analysed

Groundwater models are biased and based upon a simplified concept of the geology and hydrogeology and must therefore be considered as a simplified description of aquifer behaviour. The main advantage however is that they demand a lot of data to be collected and the local hydrogeology to be fully understood. More data makes it possible to calibrate and evaluate the model, thereby making the model, if successful, more reliable to use for predictions. Detailed hydrogeologic investigations are necessary and should be regional, not restricted to the dump site. In the Olen case the model clearly and unequivocally illustrated that the production wells played an important role governing and restricting pollution transport. Besides the need to continue the preventive pumping, the continued use of groundwater at the refinery was strongly advocated.

Preventive pumping and hydraulic isolation can provide effective groundwater protection, as long as some local conditions are met; a homogeneous aquifer, a suitable local layout with excess water treatment capacity, a policy in which long term provisions can be included. Use of the water is probably only possible on paper. Hydraulic isolation remains a well established remediation technique. Although the present EU regulations make clay layers, artificial liners and drains mandatory beneath waste dumps, the technique might still be used for temporary protective measures or for areas where such infrastructure is not possible.

ACKNOWLEDGEMENTS

The studies and tests referred to have been financed by Umicore Olen. Research was carried out by the authors (C. Cammaer, T Van Autenboer) then at the Limburgs Universitair Centrum and more recently at SLiM vzw. L. De Ridder is the Environmental Affairs Manager of the Umicore Olen Site in Belgium.

Geotechnical and Environmental Aspects of Waste Disposal Sites – Sarsby & Felton (eds)
© 2007 Taylor & Francis Group, London, ISBN 978-0-415-42595-7

Thermal properties of bentonite under different conditions – simulation of real states

R. Vasicek

Czech Technical University, Prague, Czech Republic

ABSTRACT: Research on thermal properties of bentonites is one of a number themes connected with the research of the construction of the sealing around canisters containing high-level radioactive waste in a deep repository. Heat from the deposited waste can cause cracks within the surrounding system which is intended to contain radionuclides. This paper describes the results of laboratory tests undertaken to investigate water movements, thermal conductivity and expansion/shrinkage of bentonite mixes within a specially designed apparatus.

1 INTRODUCTION

The high-level radioactive waste repository in Czech Republic is planned to be situated in the granitic massif at a depth of 500 m or more. Bentonite will be used as the sealing material in the repository.

The bentonite sealing is one part of the multibarrier system, which consists of metal canister, bentonite sealing and host rock. This multibarrier system should protect the environment against penetration of radionuclides from the radioactive waste.

The stored canisters and their bentonite sealing (buffer) will be exposed to water inflow from the surrounding (host rock) all the time. During the initial stage – after placing of canisters – the buffer and host rock will be exposed to the elevated temperature generated by spent nuclear fuel. This initial temperature depends on the cooling time of a canister in the interim storage. It should not exceed 90°C because the long-term thermal loading (more than approx. 90°C) can change the material properties. It also causes vapour generation, which significantly complicates all predictions of the behaviour of the whole system. The accumulated heat also can cause create cracks and opening of the junctures between highly compacted bentonite blocks, which are planned as structural elements of the buffer. This is connected with the increased possibility of the radionuclide penetration, therefore good heat dissipation is required.

2 BASIC PARAMETERS

Thermal conductivity λ [W/mK], specific heat capacity c [J/kgK] and thermal diffusivity a [m^2/s] are the main determinant quantities for thermal-technical calculations, which can characterize material properties. Heat capacity can be also expressed as the capacity of unit volume – thus one obtains volume heat capacity c_ρ [J/m^{3}K]. Relations between parameters are shown in Equations (1) and (2), where ρ is density of material.

$$a = \frac{\lambda}{c.\rho} \qquad (1)$$

$$c = \frac{c_\rho}{\rho} \qquad (2)$$

All the foregoing material parameters are dependent on the material state. Generally, changes of temperature do not have not significant influence on values of the thermophysical parameters in the expected range of temperature. Change of the water content (w) has the most significant effect on porous material. Of course, the thermal properties are also influenced by the dry density of the material.

Thermal conductivity is the parameter describing the heat flow through the solid material. It is affected most significantly by changes of water content. The knowledge of this reaction is the required input information for mathematical modelling. Therefore the research was focused specifically on the changes of thermal conductivity.

3 REQUIRED VALUES OF THERMAL CONDUCTIVITY OF BUFFER MATERIAL

In consequence of the requirement for good heat dissipation by the buffer to the surrounding rock, a high value of thermal conductivity of the buffer is needed. It would be useful to express the required value of thermal conductivity. It would be possible to estimate it from Equation (3) which defines the total heat flux Q through the boundary for cylindrical geometry (expected shape of the deposit place) – L is the length of the cylinder, λ is thermal conductivity; T_i, and T_0 are the temperatures on inner and outer boundaries; r_i and r_0 are the radii of inner and outer boundaries.

$$Q = A.q = 2.\pi L\lambda \frac{T_i - T_o}{\ln(r_o/r_i)} \qquad (3)$$

Unfortunately, due to the relatively early phase of the project of the deep repository, it is not possible to use either the correct data or the approximate dimensions of the deposit place and the heating capacity of the canister with the spent nuclear fuel. The known values are only the required temperature limit (90°C in buffer near the canister) and the value of thermal conductivity of the host rock. Thermal conductivity of the buffer should be at the same or higher level not to be the "brake" for the transferred heat. Possible junctures between blocks and the cracks in the blocks are not taken into account because these tests deal only with material properties not with the whole sealing system.

Average thermal conductivity (at 60°C) of granitic massif is 2.61 W/mK (Ikonen, 2003) and determines the comparative value for the buffer.

4 SELECTION AND DESCRIPTION OF METHODOLOGY, CONDITIONS AND DEVICES

The ability to detect the thermal conductivity and the heat capacity, ease of use and rapidity of measurement were the basic requirements for the devices which were to be used. Hence the application of a dynamic measuring method. Other factors which determined the choice of apparatus were; the ability to prohibit swelling and to record the swelling pressure, the ability to add water to the sample at high temperatures (up to 200°C). Sample preparation technology was taken into account in the design.

The device satisfying the foregoing specific requirements had to be developed by Applied Precision, Bratislava (SK). It was impossible to use any standard equipment. A dynamic impulse method with a point source of heat is used. The heat pulse flows into the "half-space" shaped sample and there is continuous scanning of the increase and decrease of the temperature wave.

The advantages of the method used were mentioned above when describing the apparatus requirements. One difficulty is the necessity to provide a sufficiently smooth sample surface for ideal contact with the measuring probe. This requirement is mostly fulfilled by compacting samples

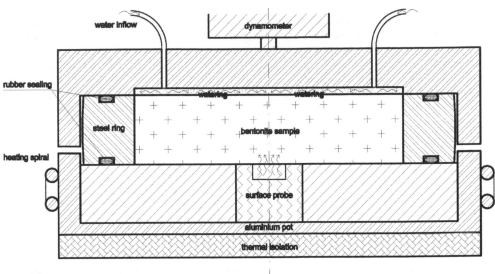

Figure 1. Test chamber.

in a steel form. For the material tested a sufficient size of the "half-space" is a cylinder with 50 mm diameter and thickness 20 mm. Another, but less problematic difficulty, was the necessity of achieving a uniform sample temperature before measuring. This was managed by software and put as the starting condition of the measuring process.

The saturation of the sample is guaranteed by the watering system. It consists of a water tank, water heater, permeable carborundum plate in contact with the upper surface of sample in the measuring chamber, valves for isolating the part of the watering system in which the water is heated. Of course the system also consists of the pipes connecting all parts. Heating of the sample is through the heating coil around the "base" (the aluminium pot with the measuring probe inside). This probe is in contact with the lower surface of the sample (Figure 1).

The saturation process is managed by free water level – approximately 1 m above the sample for the tests at laboratory temperature. During tests at higher temperatures the samples are saturated by the water heater which sucks water in from the tank. The whole watering circuit is isolated by valves.

The "Basic" measurements (Vasicek 2004) which were a consequence of selection of the most suitable material mixture for further tests were undertaken using the device 'ISOMET'. This applies the same measuring method as APT-P01 test chamber but only allows measurement of the thermal conductivity and volume heat capacity at laboratory temperature (approximately 25°C), under atmospheric pressure and on unsaturated samples (without pressure sensing). Surface probes were used for the highly compacted blocks and a needle probe was used for the bulk samples. The spread of measured values were:– for thermal conductivity λ; $\pm(10\%$ of reading $+ 0.005$ W/mK), for specific volume heat capacity c_ρ; $\pm(15\%$ of reading $+3 \times 10^3$ J/m^{3}K).

These "Basic" tests give data on the dependence of thermal conductivity on dry density of sample and on the water content. The admixtures added to change material properties were graphite (for elevation of λ) and silica sand (for reduction of swelling pressure). The range of mixtures was from 70–90% bentonite (Ca bentonite from Rokle locality, ground, not activated – RMN), 0–10% silica sand (locality Provodin) and 0–20% graphite (locality Netolice, portion of carbon 94–96%, ash content maximum 6%, grain size less than 0.025 mm).

On the basis of past findings the chosen admixture was 85% RMN Bentonite, 10% sand and 5% graphite for further tests under expected real conditions.

5   MEASUREMENT UNDER EXPECTED CONDITIONS

Temperature in the buffer depends on the heat output of the canister holding the spent nuclear fuel. This value is very much influenced by the cooling time during interim storage, by the amount of the fuel in the canister and also by the geometry of the repository (interaction between the canisters).

The design of the canister and the geometry should satisfy the aforementioned requirement, i.e. not to exceed 90°C in the buffer.

The test in APT-P01 permits modelling of the real conditions expected in the buffer, especially near the canister surface. The complex tests are very time consuming because of very low permeability of bentonite – approximately 1 month is needed to achieve a saturated state within a sample.

5.1   *Tested material*

As a consequence of the "Basic phase" and other activities at the Centre for Experimental Geotechnics certain materials were chosen for further tests in the APT-P01 chamber:

a) Mixture Rokle bentonite +10% sand +5% graphite – the chosen material for Mock-up-CZ test and for detailed examination
b) "FEBEX" bentonite – natural Ca Spanish bentonite – for comparison with other materials
c) MX80 bentonite – Na bentonite, the most-widely known one – measured for comparison with other materials

Most samples were prepared by compacting mixtures to dry density of approximately $1770 \, kg/m^3$. Only two samples were cut from imported, previously-compacted, blocks containing FEBEX.

5.2   *Types of test*

All tests permitted recording of temperature, thermal conductivity, volume heat capacity, swelling pressure and vapour pressure. The four types of test that could be conducted were:

1) At elevated temperature, without adding water to the sample – these tests are not so time consuming
2) At laboratory temperature, with addition of water to the sample
3) At elevated temperature, but under 90°C (limit for vapour evolution), with addition of water to the sample
4) At elevated temperature – above 90°C, with addition of water to the sample

Because of the conditions expected in the waste deposit, the most common test type was case (3), i.e. a temperature of approximately 80°C and addition of water to the sample. Since better sample saturation and thermal conductivity is expected at a temperature around 25°C case (3) tests give results which are on the 'safe side'.

6   RESULTS – DESCRIPTION OF INITIAL AND FINAL STATES

6.1   *Initial states*

These were;

- Water content: 0.5–14.7%
- Thermal conductivity: 0.7–1.0 W/mK
- Dry density: $1630–1820 \, kg/m^3$

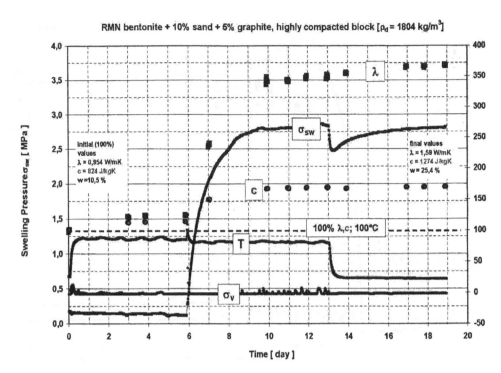

Figure 2. Evolution of swelling pressure (σ_{sw}), vapour pressure (σ_v), temperature (T), thermal conductivity (λ) and heat capacity (c).

6.2 *Final states*

- *Vertical distribution of water in sample in fully saturated state*
  The core sample (diameter 30 mm, height 30 mm) was taken from a fully saturated sample by drilling after testing. Afterwards it was cut into three parts – in accordance with its position in the testing chamber, i.e. upper, middle or bottom part. All cut samples were dried at 110°C for 24 hours and their water content was determined. For example, the water content distributions for two samples taken from a mixture of 75% Rokle bentonite +10% sand +5% graphite were;
  – 26.5, 24.7, 24.1, 25.1 (upper portion, midle portion, bottom portion, average)
  – 25.2, 24.7, 24.1, 24.7 (upper portion, midle portion, bottom portion, average)
  There was no significant difference between the upper and bottom layers (it was approximately 1%). Therefore it can be assumed that the whole sample was fully saturated after test.
- *Water Content (mass)*
  The water content of fully saturated samples was approximately;
  a) For mixture Rokle bentonite +10% sand +5% graphite −25.2% (average from 5 samples)
  b) MX80 and FEBEX bentonites −30.2% (from 1 and 2 samples respectively)
  Two later mentioned materials are only bentonite, e.g. without addition of sand and graphite which decrease the final water content. All tests were finished by dropping the temperature to the laboratory level.
- *Thermal Conductivity*
  Thermal conductivity values of fully saturated samples were obtained as follows:
  a) Mixture of Rokle bentonite +10% sand +5% graphite –
  $\lambda = 3.51$ W/mK (at temperature of approximately 85°C, range of $\lambda = 0.01$ W/mK) from three samples not subject to thermal loading. There are no significant differences between observed values of thermal conductivity for temperatures of 85°C and 25°C – the difference is approximately 0.2 W/mK as illustrated in Figure 2.

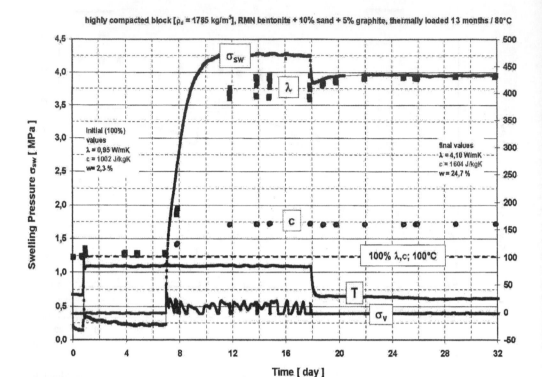

highly compacted block [ρ_d = 1785 kg/m³], RMN bentonite + 10% sand + 5% graphite, thermally loaded 13 months / 80°C

Figure 3. Evolution of swelling pressure, vapour pressure, temperature, thermal conductivity and heat capacity.

$\lambda = 4.10$ W/mK was observed for two samples of the mixture containing Rokle bentonite which were thermally loaded for 13 months at 80°C (Figure 2). The increase of thermal conductivity was about 0.6 W/mK (i.e.17%). This increase seems to be caused by long-term thermal loading. Verification of this phenomenon is a task for future tests.

b) FEBEX bentonite –
  $\lambda = 1.9$ W/mK (from 2 samples)
c) MX 80 bentonite –
  $\lambda = 1.6$ W/mK (from 1 sample)

- Swelling Pressure
  a) Mixture of Rokle bentonite +10% sand +5% graphite:
    – unheated samples $\sigma_{sw} = 4.0, 4.1$ MPa (2 samples)
    – samples heated at 85°C $\sigma_{sw} = 3.9, 4.3$ MPa (2 samples), 2.9 MPa

    There are no observed significant differences between values of swelling pressure for temperatures of 85°C or 25°C.

  b) FEBEX bentonite:
    – unheated sample $\sigma_{sw} = 3.0, 3.9$ MPa (2 samples)
    – sample heated at 85°C $\sigma_{sw} = 3.2$ MPa (1 sample)

    The value of the heated sample is within the range for unheated samples.

  c) MX 80 bentonite:
    $\sigma_{sw} = 3.5$ MPa (1 sample)

- Influence of Expansion/Shrinkage on Pressure of Sample

  Generally it was observed that changing temperature had a strong influence on changes of sample pressure – this is caused by thermal expansion/shrinkage of material.

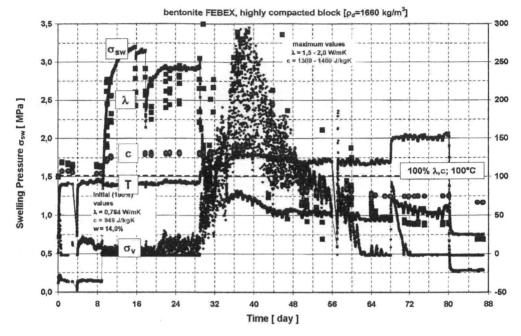

Figure 4. Evolution of swelling pressure, vapour pressure, temperature, thermal conductivity and heat capacity.

The decrease of swelling pressure caused by shrinkage when the temperature dropped from 85°C to 25°C was more significant than the effect of additional saturation of a sample. The average difference between the initial and the final values of the sample pressure was 0.3 MPa (Figure 3).

When the temperature dropped from 150°C to 25°C the sample pressure decreased from 1.0 MPa to 0.0 MPa (FEBEX bentonite, at an initial water content of 14% – Figure 4). The zero value results from loss of contact between the sample and the dynamometer due to drying shrinkage.

7 CONCLUSIONS

In their initial state, before testing, none of the samples achieved the comparative value of thermal conductivity of the host rock (2.6 W/mK) – dry or naturally moist material had lower thermal conductivity (0.7–1.0 W/mK). However, a thermal conductivity values equivalent to the rock was quickly achieved and significantly exceeded during saturation of all samples made from the mixture of 75% Rokle bentonite +10% sand +5% graphite (a value up to 3.5 W/mK was obtained). Further research is needed to determine the exact water content at which the mixture exceeds the comparative conductivity of the rock. Also verification of the influence of long-term thermal loading on thermal conductivity should be done.

Unlike the mixtures, the samples made from natural bentonites did not achieve the rock thermal conductivity in any state (the maximum obtained was 1.9 W/mK).

The values of thermal conductivity for mixtures were significantly higher than expected. The prediction was based on extrapolation of the curve relating thermal conductivity and water content (Vasicek 2004). The average value resulting from tests under 'real conditions' was 3.5 W/mK against the expected 2.1 W/mK – i.e. more then 65% higher. A similar phenomenon was observed during the Ophelie Mock-Up test (Dereeper et al 2002). In that case it was considered that additional heat transfer processes (convection and evaporation/ condensation) may be occurring – but that material was exposed to a temperature higher than 80°C.

ACKNOWLEDGEMENTS

The Thermophysical research has been supported by Grant No. 103/02/0143 (Czech Science Foundation CTU 0404411, Czech Technical University).

REFERENCES

Dereeper B., Verstricht J. and Gatabin C. (2002). *The Mock-up OPHELIE: A large scale backfill test for HLW disposal.* (CEA). The 6th International workshop on Design and Construction of Final Repositories, 11–13 march 2002, Brussel, Belgium, 5 p.
Ikonen K. (2003). *Thermal Analyses of Spent Nuclear Fuel Repository*, VTT Processes,
Vasicek R. (2004). *Impact of Admixtures on Thermal Properties of Bentonite,* Distec 2004 – International Conference on Radioactive Waste Disposal; Kontec Gesellschaft für technische Kommunikation mbH, Tarpenring 6, 22419 Hamburg, Germany, 558 p.

Geotechnical and Environmental Aspects of Waste Disposal Sites – Sarsby & Felton (eds)
© 2007 Taylor & Francis Group, London, ISBN 978-0-415-42595-7

Low Leachability of As, Cd, Cr and Pb from soils at a contaminated site in Walsall, West Midlands

R. Wesson, C. Roberts & C. Williams
University of Wolverhampton, Wolverhampton, England

ABSTRACT: In order to improve understanding of the environmental availability of trace metals, leaching tests were carried out on soil samples from a contaminated site in Walsall, West Midlands. Leachability of the metals was determined using ICP-MS after agitating the samples with de-ionised water for 24 hrs. Additionally, physical and chemical parameters of the soils were determined to enable the assessment of the association of the target metals with soil components. Whilst high levels of trace metals were found to be present on the site, only a small proportion of the total metal flux was found to be leachable. This was a consequence of the age of the waste on the site with the most labile fraction of the trace metals having been already leached out. The available data indicates that the trace metals are associated with Fe oxides and organic matter. In view of the nature of these soil components, it is thought to be unlikely that the leachability of the trace metals will increase without significant changes in site conditions taking place.

1 SOIL CONTAMINATION

1.1 Introduction

Legislation with regard to contaminated land has traditionally been based on the total amount of the contaminant present in the soil. However, the potential for the contamination of local aquifers or surface water by the leaching of contaminants from the soil is of prime concern. For this reason leaching tests have been developed which aim to demonstrate the potential solubility of the contaminants in question (Forstner, 1993; Lewin et al, 1994; Sahuquillo et al, 2002). This is a move from assessing the bioavailability of the metals in question to evaluating their environmental availability. This is arguably more significant from the point of view of assessing the risks that contaminants on site pose to the wider environment than analysing the quantity available to a single species. This is especially true when attempting to quantify the risk to ground and surface water resources. A number of factors have an influence on metal solubility in soils, and these factors are likely to have a synergistic effect.

1.2 Sorption of metals

The ability of soils to adsorb metals onto the solid phase plays a major role in their availability and is influenced by the mineralogical composition of the soil. The colloidal fraction i.e. those particles less than $2\,\mu m$ in size are particularly important due to there high surface charge and surface area (Zhuang and Yu, 2002). This fraction includes the clays and is frequently referred to as the clay fraction (Rowell, 1994). Clay minerals provide some degree of attenuation of heavy metals present as a result of precipitation and exchange processes (Alloway and Ayres, 1997). Trace metals may also be sorbed by or co-precipitated with hydrous oxides of iron (Sharma et al, 2000; Turner, 2000). Hydrous oxides of iron such as hematite and goethite, exhibit very low solubilities and consequently may be important in controlling the solubility of trace metals. Whilst iron oxides commonly provide a coating on clay surfaces thus enhancing the binding of metals to this fraction. Cave and Hamon (1997) found that hydrous oxides may also be associated with coarser minerals.

1.3 *Influence of organic matter*

Organic matter has an important role in governing the solubility of heavy metals in soils due to its high cation exchange capacity. Organic matter can be characterised into three fractions:

1. Humin which is insoluble throughout the pH range
2. Humic acid which is insoluble in solutions below pH 2
3. Fulvic acid that is soluble throughout the pH range (Spark et al, 1997).

The role of humic substances in enhancing or retarding contaminant mobility is not solely dependent on how it is defined. (Wu et al, 1999) emphasise the importance of organic matter in controlling the solubility of heavy metals. The metal ions form soluble complexes with organic matter which are the dominant form of dissolved metals in many natural waters. Sauve et al (2000) state that the majority of dissolved metal is found in the form of metal-organic complexes. The solubility of the humic substances affected by other soil factors such as pH and calcium concentration.

Warwick et al (1998) suggest that the influence organic matter on contaminant mobility is largely pH dependent with enhancement at lower pH and inhibition at higher pH. The author's research also suggests that humic acid retards metal mobility despite being present in the mobile phase as a result of humic acid coating grains of sand in the packed column that was being utilised. Fein et al (1999) suggest that the pH dependent behaviour is due to the humic substances under acidic to neutral conditions sorbing to positively charged mineral surfaces whilst under neutral to alkaline conditions they exist as negatively charged ligands in solution.

2 METHODOLOGY

The site being investigated is situated to the south of Walsall near to the A34 and has an area of 0.15 hectares, as illustrated in figure 1. The site has small streams adjacent to I and these form part

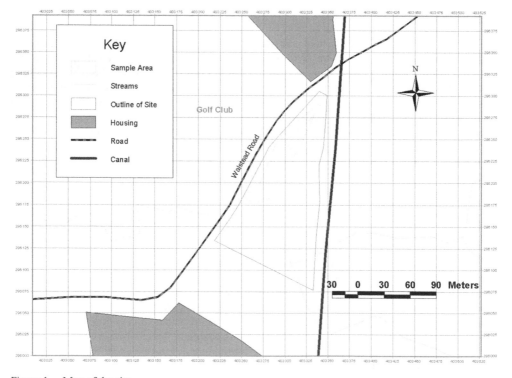

Figure 1. Map of the site.

of the Trent River catchment. IRL carried out a site investigation in 1992 and it was found that the site was filled to a depth of 3.0 metres with "black ashy sand with some clinker, brick, refuse and metal". Landfilling was completed prior to 1962.

2.1 Sampling

A grid was produced for the site in advance of taking samples using Arcview GIS giving sample distances of 10 m. The sample points co-ordinates were programmed into a Trimble XRS GPS to allow easy location on each site. Samples were taken from the top 10 cm of the soil profile and sealed in plastic bags prior to transport to the laboratory.

The samples were dried at 40°C prior to sieving to 2 mm. 0.5 g of sample was digested with 10 ml of aqua regia and 2.5 ml of H_2O_2 in a Milestone Ethos 900 Microwave Digester. Following digestion, the samples were filtered and made up to 50 ml. Additional digestions were carried out using 0.25 g of sample in 2 ml HF, 2 ml of HNO_3 and 1 ml of H_2O_2 using the same microwave digestion apparatus. Following digestion 1.2 g of boric acid was added to the vessel and the samples made up to 100 ml. Analysis was carried out using ICP-OES. It was found that greater amounts of As, Cd and Pb, were extracted using aqua regia whereas in the case of Cr and Fe hydrofluoric acid extracted the greater amount.

2.2 Leaching tests

The leaching test utilised the procedure described in Lewin et al (1994). This utilised 10 g of sample in 100 ml of de-ionised water, which was agitated for 24 hours on an orbital shaker at 100 rpm. The samples were then filtered with a 0.45 μm filter using vacuum filtration. Following this period the samples were acidified with 5 ml of HNO_3 and refrigerated with a non acidified aliquot reserved for UV/VIS analysis. Analysis of arsenic, cadmium, chromium, iron and lead was carried out by ICP-OES and ICP-MS.

2.3 Particle size analysis

Particle size analysis was carried out on the fraction that passed a 2 mm sieve. As organic matter when present in significant quantities is liable to skew the results it was necessary to remove this component prior to further analysis. Approximately 50 ml of soil was added to a 250 ml glass beaker and wetted with distilled water. Aliquots of hydrogen peroxide were added to the soil sample until visible reactions ceased. The beakers containing the soil samples were then placed in a water bath at 70°C and further aliquots of hydrogen peroxide were added to the soil sample. The beaker was covered with a watch glass and the water bath was maintained at a temperature of 70°C. Aliquots of hydrogen peroxide were added until there was no further visible reaction. The samples were kept in the water bath until excess liquid was evaporated.

Particle size analysis was carried out using a Malvern Mastersizer Long Bed laser granulometer with a MS-17 automated sample presentation unit. To prevent errors due to flocculation of the particles, each sample had a few drops of sodium hexametaphosphate added as a dispersal agent. Three sub-samples were taken for analysis from each sample. Each sub-sample was analysed in triplicate and three sub-samples were taken from each sample for each fraction analysed.

2.4 Organic content

Total Organic Carbon (TOC) was analysed by thermo-gravimetric analysis. The loss between 200–600°C was recorded as the TOC content of the soil as described by (Leinweber and Schulten, 1992). Dissolved Organic Matter (DOM) in aliquots of the leachate that had not been acidified was measured by UV/VIS absorbance at 254 nm. This has been found by a number of researchers to be well correlated with dissolved organic matter as measured by TOC analysers (Fairhurst et al, 1995; Deflandre and Gagne 2001). As the latter authors point out, calibration is difficult with this method as no single organic compound can be used so the absorbances reported represent relative not absolute values.

3 RESULTS

3.1 *Metallic contaminants*

The levels of all four metals exceed the CLEA Soil Guideline values (SGV) for all classified land uses except for commercial/industrial with lead above the SGV for this use in some samples (Table 1).

High levels of As, Cd, Cr, and Pb occur on the site (Figure 2). In all cases the actual amount of the trace metals which are leachable is very low (Figure 3). As the waste present on the sites has

Table 1. CLEA Soil Guideline Values.

| | Soil Guideline Value (mg/kg dry weight soil) | | | | | |
| | | Cd | | | | |
Land use	As	pH6	pH7	pH8	Cr	Pb
Residential with plant uptake	20	1	2	8	130	450
Allotments	20				130	450
Residential without plant uptake	20		30		200	450
Commercial/Industrial	500		1400		5000	750

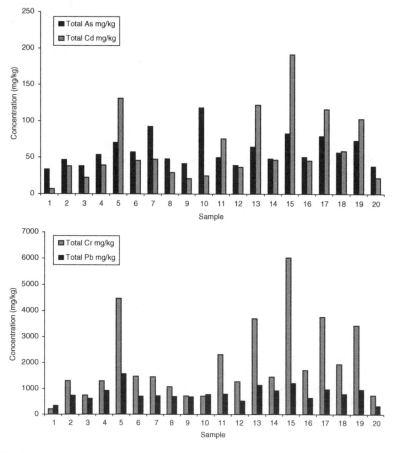

Figure 2. Total trace metals.

210

been in situ for at least two decades, much of the labile fraction is likely to have been leached from the sites (Abollino et al, 2002). This indicates that the metals present on both sites must be present in soil components that bind them strongly enough to prevent them being leached by water.

All four trace metals demonstrate a significant correlation with Fe. In the case of Cd and Cr this association is particularly strong, whilst As and Pb demonstrate a weaker but still statistically significant relationship (Table 2). Significant correlation between Fe and the metals being studied in this programme have been shown to indicate an association of the metals with iron hydroxides surfaces in the soil (Voigt and Brantley 1996; Sharma, Rhudy et al. 2000). It is equally possible that the association of the trace metals with iron may indicate that all five elements were present in some of the original waste that was disposed of on site. Due to the lack of records pertaining to the origin of the waste it is not possible to state what the origin was, but it is likely that some industrial waste found its way onto the site. The statistical relationship does not allow differentiation to be made between binding to iron hydroxides and waste origin. However, initial results from sequential extractions carried out on the soils (Figure 4). This indicates that Fe is present mainly as oxides indicating that the trace metals are associated with iron hydroxides.

3.2   *Influence of particle size*

Cd, Cr and Pb demonstrate a statistically significant correlation with soil particles in the <0.5 μm range (Table 2). This indicates that they are bound to the fine clays which due to their relatively large

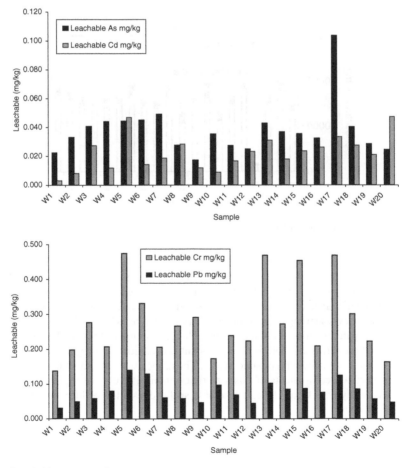

Figure 3.   Leachable trace metals.

Table 2. Correlations between total metals and soil variables (correlations discussed in the text are highlighted)

	Fe		<0.5 um		TOC	
	r	sig	r	sig	r	sig
As	0.511	0.021	−0.262	0.264	0.853	0.000
Cd	0.850	0.000	0.556	0.013	0.106	0.665
Cr	0.845	0.000	0.569	0.011	0.084	0.732
Pb	0.560	0.010	0.467	0.038	0.256	0.275
Fe	1.000		0.338	0.145	0.433	0.057

	8–16 um		16–31 um		31–60 um	
	r	sig	r	sig	r	sig
As	0.504	0.024	0.815	0.000	0.780	0.000
Cd	0.050	0.838	0.299	0.214	0.384	0.104
Cr	0.044	0.859	0.283	0.240	0.363	0.126
Pb	0.358	0.121	0.419	0.066	0.488	0.029
Fe	0.085	0.723	0.336	0.147	0.415	0.069
TOC	0.656	0.002	0.772	0.000	0.746	0.000

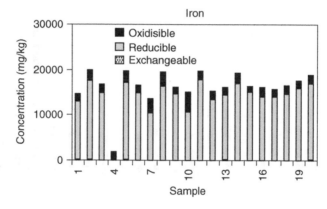

Figure 4. Sequential extraction of Fe.

surface area have a high cation exchange capacity and makes them important sorbents for heavy metals (Droppo and Jaskot 1995). Lead additionally shows a positive correlation with particles in the 31–60 μm size range, which is the coarse silt granulometric class. Arsenic demonstrates a good correlation with particles in the 4–60 μm size range and a very strong correlation with TOC. TOC additionally shows a strong correlation with this size range, which indicates that the arsenic is bound to organic matter coating the soil particles. The correlation of Pb with the coarse silt cannot be explained in these terms, as there is no significant correlation with organic matter or the other silt groups that are associated with organic matter. The association of lead with this size range may result from particulate components of this size that contain Pb, or coatings on this particle size class that specifically sorb lead. The data available so far does allow a differentiation to be made between either scenario.

3.3 *Contaminant leachability*

Table 3 shows that leachable Cd is correlated with both pH and DOM. The relationship with pH is negative indicating that lower pH is related to lower cadmium solubility. This indicates that

Table 3. Correlations between leachable trace metals, dissolved organic matter and electrochemistry.

	DOM		pH		eh	
	r	sig	r	sig	r	sig
As (mg/kg)	0.184	0.438	−0.162	0.495	−0.010	0.967
Cd (mg/kg)	0.497	0.026	−0.550	0.012	−0.177	0.456
Cr (mg/kg)	0.195	0.411	−0.223	0.344	0.066	0.781
Pb (mg/kg)	0.151	0.524	−0.046	0.848	0.095	0.689

should the site pH fall, the mobility of cadmium would be enhanced. The correlation with DOM indicates the complexation of cadmium by the soluble fraction of the humic substances. The lack of correlation between the other trace metals and these two parameters is interesting. It would be expected that lower pH would also enhance the mobility of As, Cr and Pb if Cd solubility shows a pH dependency. It would be also be expected that these metals also would undergo complexation with DOM. There is a lack of correlation between redox potential and trace metal leachability. This indicates that the variations in redox potential on this site are not enough to significantly influence trace metal mobility.

4 CONCLUDING REMARKS

The proportion of metals leachable at from the soil samples is a small percentage of the total metal loading. Whilst iron hydroxides play an important role in binding the trace metals that occur on site other soil components are significant. The fine clay fraction is significant in the case of Cd, Cr and Pb, but arsenic shows a stronger relationship with the silt fraction and organic matter. With the exception of Cd the leachability of the trace metals does not seem to be significantly influenced by variations in pH. Cd is additionally the only metal associated with DOM indicating the formation of complexes with organic ligands in solution. The next phase of this project will investigate the association with other soil components and assess how these may influence the leachability of the contaminants present on site.

REFERENCES

Abollino O, Aceto M, Malandrino M, Mentasti E, Sarzanini C and Barberis R (2002) *Distribution and mobility of Metals in contaminated sites. Chemometric investigation of pollutant profiles,* Environmental Pollution 119, pp177–193.

Alloway B J and Ayres D C (1997) *Chemical principles of Environmental Pollution,* London, Blackie.

Cave M and Harmon K (1997) *Determination of Trace Metal Distributions in the Iron Oxide Phases of Red Bed Sandstones by Chemometric Analysis of Whole Rock and Selective Leachate Data,* Analyst 122, pp501–512.

Deflandre B and Gagne J P (2001) *Estimation of dissolved organic carbon (DOC) concentrations in nanoliter samples using UV spectroscopy,* Water Research 35(13), pp3057–3062.

Droppo I G and Jaskot C (1995) *Impact of River Transport Characteristics on Contaminant Sampling Error and Design,* Environmental Science & Technology 29, pp161–170.

Fairhurst A J, Warwick P and Richardson S (1995) *The Effect of pH on Europium-Mineral Interactions in the presence of Humic Acid,* Radiochimica Acta 69(2), pp103–111.

Fein J B, Boily J, Kubilay G and Kaulbach E (1999) *Experimental Study of Humic Acid Adsorption onto Bacteria and Al-oxide Mineral Surfaces,* Chemical Geology 162, pp33–45.

Forstner U (1993) *Metal speciation – general concepts and applications,* International Journal of Environmental Analytical Chemistry 51(1–4), pp5–23.

Leinweber P and Schulten H R (1992) *Differential thermal analysis, thermogravimetry and in-source pyrolysis-mass spectrometry studies on the formation of soil organic matter,* Thermochimica Acta 200, pp151–167.

Lewin K, Bradshaw K, Blakey N C, Turrell J, Hennings S M and Flavings R J (1994) *Leaching tests for assessment of contaminated land: Interim NRA guidance,* R&D Note 301, Bristol, Environment Agency.

Sahuquillo A, Rigol A and Rauret G (2002) *Comparison of leaching tests for the study of trace metal remobilisation in soils and sediments,* Journal of Environmental Monitoring 4(6), pp1003–1009.

Sauve S, Hendershot W and Allen H E (2000) *Solid-Solution partitioning of Metals in Contaminated Soils: Dependence on pH, Total Metal Burden, and Organic Matter,* Environmental Science and Technology 34(7), pp1123–1131.

Sharma V K, Rhudy K B, Cargill J C, Tacker M E and Vazquez F G (2000) *Metals and grain size distributions in soil in the middle Rio Grande basin, Texas USA,* Environmental Geology 39(6), pp698–704.

Turner A (2000) *Trace Metal Contamination in Sediments from U.K Estuaries: An Empirical Evaluation of the Role of Hydrous Iron and Manganese Oxides,* Estuarine, Coastal and Shelf Science 50, pp355–371.

Voigt D E and Brantley S L (1996) *Chemical fixation of arsenic in contaminated soil,* Applied Geochemistry 11, pp633–643.

Warwick P, Hall A, Pashley V, Van Der Lee J and Maes A (1998) *Zinc and Cadmium Mobility in Sand: Effects of pH, Speciation, Cation Exchange Capacity (CEC), Humic Acid and Metal Ions,* Chemosphere 36(10), pp2283–2290.

Wu J, West L J and Stewart D K (1999) *Influence of humic acid on the electromobility of copper in contaminated soil. Geoenvironmental Engineering,* Contaminated ground, Fate of pollutants and remediation, Thomas Telford.

Zhuang J and Yu G-R (2002) *Effects of surface coatings on electrochemical properties and contaminant sorption of clay minerals,* Chemosphere 49, pp619–628.

Tailings

Geotechnical and Environmental Aspects of Waste Disposal Sites – Sarsby & Felton (eds)
© 2007 Taylor & Francis Group, London, ISBN 978-0-415-42595-7

Self-weight consolidation behaviour of Golden Horn (Haliç) dredged material

S.A. Berılgen
Trakya university, Yildiz-Istanbul, Turkey

P. Ipekoglu & M.M. Berılgen
Yildiz technical university, Yildiz-Istanbul, Turkey

ABSTRACT: Determination of in situ deformation behavior of contaminant soils such as dredged materials, mine tailings and sludge has become an important problem in geotechnical engineering in past decades. In the research reported this paper, a model tank test was conducted in the laboratory in order to gain an understanding of the in-situ self-weight consolidation behavior of Golden Horn dredged material. In this test a cylindrical tank measuring 1 m in height and 0.80 m in diameter was used. The slurry material used in the tests was slurried to 300% water content which matches the initial water content of the pumped dredged material in the storage site. Pore pressure and settlement measurements were then taken during the tests by a computerized data acquisition system. Numerical analyses were carried out for prediction of time dependent behaviour of the soil column in the tank under its self-weight.

1 INTRODUCTION

Storage and reclamation of storage lands for highly compressible dredged materials, mine tailings, and sludge has presented the geotechnical engineering field with an important economical and environmental challenge in recent years. Storage of dredged materials obtained from the bed sediments of contaminated undersea environments and industrial and urban waste materials is an issue that needs urgent attention by the scientific community. The storage of waste sludge and reclamation of these storage areas requires economic and environmentally-friendly solutions to the current problem. There are many international examples of such well managed reclamation and storage of sludge and waste materials in countries such as U.S.A and Japan. Istanbul, the largest city in Turkey and one of the major metropolitan areas in the world, cleaned one of its environmentally polluted areas – the Golden Horn (Haliç) – by dredging 5 million m^3 of the bed sediments and pumping the resulting sludge to the storage areas near the now-defunct rock quarries behind a dam built in Alibey district. The utilisation of the land lying over the storage areas of Golden Horn dredged material is socially and economically very desirable. It is crucial to analyze the characteristics of the sludge material, including its behaviour under self-weight and/or under surcharge loading. Results from this study would shed light on the extent of expense and planning required for land reclamation and rehabilitation projects. This article contains results from studies that are focused on determining the self-weight consolidation behaviour of the dredged material coming from Golden Horn which has been placed in the Alibey storage area.

2 THE GOLDEN HORN

It is generally accepted that the formation of Bosphorus and Golden Horn started approximately 8000 years ago with the flooding of ancient river valleys with the waters of the Mediterranean.

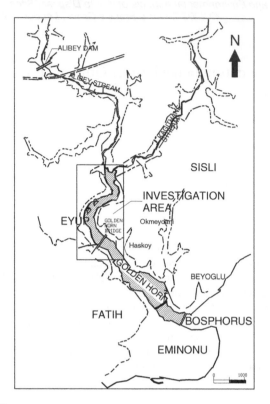

Figure 1.   The Golden Horn.

Archeological findings indicate that these environmental changes have taken place concurrently with the start of human dwellings in the area. The thickness of deposits in the Golden Horn indicate a very high rate of deposition (approximately 7 m/1000 years), which is believed to be due to the elevation caused by young tectonic movements, (Özaydin Yildirim, 1997).

The Golden Horn is located in a topography which falls towards the South. Alibey and Kagithane streams flowing southwards reach Golden Horn at its north end where it is about 450 m wide, then it extends first in southwest direction to Eyüp from that point on it continues in the southeast direction towards Bosphorus (Figure 1).

Bed elevations in the Golden Horn are higher than those in the Bosphorus and therefore the Golden Horn can be considered as a suspended valley. The rise due to tectonic movements caused the formation of three ridges in the sea bottom. It is postulated that the third ridge sets a boundary for rapid deposition and filling, and the sections of Golden Horn beyond this point are under the influence of active marine and current environment.

The rapid growth of population in the Golden Horn area due to industrialization starting from early 1950's and uncontrolled and untreated discharge of municipal and industrial waste until very recent years have led to an unacceptable situation. The upstream part of Golden Horn became almost completely filled. Water depth in this region was generally reduced to less than 1 m, and large areas were turned into swamps. Because of this situation, Alibey and Kagithane streams could no longer flow freely into the Golden Horn, the navigation became impossible even for small vessels, and due to lack of sufficient oxygen a very polluted environment with dense odour was created (Figure 2).

The organic content of the sea bottom deposits are determined to vary between 3% to 6%, and samples are generally classified as sandy clayey silt (MH) according to Unified Soil Classification System, (Özaydin et al, 1997).

Figure 2.   Golden Horn before rehabilitation works.

3   REHABILITATION WORKS

The Golden Horn Rehabilitation Project envisaged the dredging of sea bed sediments in the upstream half to provide a minimum water depth of 5 m. Considering that large areas near the upstream end were completely filled, and to make the project more feasible and manageable, it was proposed that in these parts not the total area of the Golden Horn is dredged but only wide channels are formed by dredging for Alibey and Kagithane streams. Only beyond a certain point would the whole surface area be dredged to give a 5 m water depth. Not only did this allow reduction of the volume of material to be dredged, but also provided some storage area for dredged materials on the sides of the dredged channels.

After evaluation of all possible alternatives it was decided that the bulk of dredged material was to be transferred via a pipeline to an abandoned crushed rock quarry pit and stored there by means of two rock fill dams. Some of the dredged materials were planned to be stored behind berms formed along the newly opened channels and the low lands along the shoreline.

Experimental investigations have shown that sea bed deposits dredged from the Golden Horn could attain the characteristics of natural soft-medium stiff cohesive soils upon consolidation. It is postulated that due to surface evaporation and consolidation, the storage areas for dredged material could be converted to fill areas suitable for recreational facilities and very much needed green areas in the centre of the city, (Özaydin et al, 1997).

4   LABORATORY INVESTIGATION OF SELF-WEIGHT CONSOLIDATION BEHAVIOUR OF GOLDEN HORN (HALIÇ) DREDGED MATERIAL

This study was undertaken with two main goals in mind:

- Determination of consolidation behaviour over time (as well as the time it takes for the storage of dredged material to reach the eventual equilibrium state).

219

- To study the storage settlement and self-weight consolidation behaviour of recent Haliç dredged material under laboratory conditions.

4.1 *Experimental model tank and Haliç dredged material consolidation behaviour*

A cylindrical experimental model tank (Figure 3) measuring 100 cm in height and 80 cm in diameter was constructed in order to study the time dependent self weight consolidation behaviour of high water content Haliç dredged material – Ipekoğlu, 2004.

Haliç dredged sludge as acquired from the storage location had 95% water content. In order to simulate the original conditions at the time of disposal, water was added to bring the water content up to 300%. This mix was then stored in sealed barrels for 2–3 weeks and was periodically stirred in order to avoid segregation of the materials inside the mix and remove air bubbles from the mix, thereby assuring the mix to be homogeneous throughout the barrel. This way the sample was thoroughly saturated with water. In order to accurately model the storage environment, 0.8% salt-water was added. Eventually a material mix with a density of 1.17 g/cm^3 was obtained. Using a pump this material was transferred into the model tank. This mix was then allowed to consolidate under its self-weight inside the experimental setup. The index properties obtained from the experimental setup for the sludge material received from the storage reservoir can be seen in Table 1. According to these data, the material was identified via the USC (Unified soil classification system) system as a high plasticity organic silt soil (MH).

Figure 3. Model tank.

Table 1. Haliç dredged material index properties.

Plastic Limit	50%
Liquid Limit	75%
Plasticity İndex	25%
Specific Gravity	2.72 Mg/m^3
Organic Material Content	12.6%

During the experiment, the height of the material was continuously recorded by visually observing the consolidation of the sludge material inside the model tank. Throughout the experiment, water was allowed to drain from the bottom of the tank through a gravel layer on top of which a geotextile fabric was placed. This setup allowed for drainage in the system, which helped the consolidation of the sludge material, much like in its original storage setting. Experimental data for consolidation with time is presented in Figure 4 (Ipekoğlu, 2004).

4.2 *Pore pressure measurements*

Pore pressure measurements were taken at various depths as high water content dredged soil settled under its self-weight in the model tank. For this purpose, sludge material measuring 96 cm in height was placed inside the model tank and several ceramic piezometers were placed inside the sludge at depths of 2, 20, 40 and 60 cm from surface. Pore pressure inside the piezometers were measured through transducers and recorded via a computer controlled data acquisition system. One end of the ceramic piezometers was connected to the tube carrying the water while the other end was connected to the transducer. Four different length tubes were then properly secured inside the tank with metal wires, one end connected to the piezometer and the other to the transducers taking the measurements outside the model tank. It is essential for the tubes and the piezometers to be filled with water prior to the beginning of the experiment. Transducers are then connected and sealed so that the entire system is airtight (Figure 5). This way the whole experimental setup was saturated with water (Ipekoğlu, 2004).

Figure 6 shows experimental results for pore pressure against time behaviour at various depths during a self-weight consolidation test in the model tank (Ipekoğlu, 2004).

5 NUMERICAL ANALYSIS

In the model tank, the height of the slurry material was continuously recorded in order to obtain experimental data for consolidation behaviour of the mix. To numerically analyse the same consolidation behaviour under self-weight, CS2 software, which employs a non-linear finite strain solution algorithm, was used (Fox and Berles, 1997). The model that is considered in this program uses the following vertical effective stress-voids ratio (σ_v'-e) and

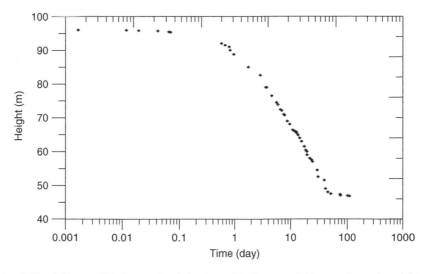

Figure 4.   Self-weight consolidation vs. time behaviour of dredge material in experimental model tank.

Figure 5.   Position of the transducers with respect to the model tank.

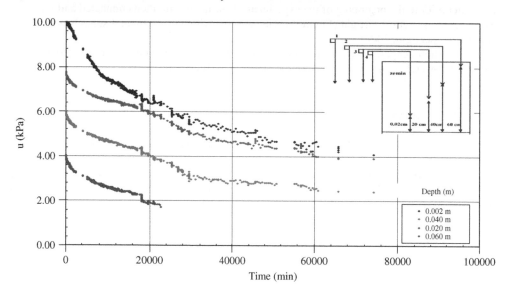

Figure 6.   Pore pressure variation with time.

permeability-voids ratio (k-e) relationships in order to solve the time dependent large strain consolidation problems:

$$e = A(\sigma' + Z)^B \qquad (1)$$

$$k = Ce^D \qquad (2)$$

The material parameters A,B,C,D and Z were obtained from the Haliç dredged slurry at a voids ratio equal to that the material in the model tank ($e_0 \cong 8$) in a 'Seepage Induced Consolidation Experiment' setup (Somogyi, 1979). Results of this experiment can be found in Table 2.

The experimental results for the consolidation behaviour under self-weight for pore pressure against time were then compared and contrasted with the numerical results obtained from the non-linear finite strain CS2 program – these are presented in Figures 7 and 8. While the mix inside the model tank was consolidating under its self-weight, water was removed from the tank in two

Table 2. Material parameters from 'Seepage Induced Consolidation Experiment' ($e_0 = 8.0$, σ in kPa, k in m/s).

$e = A(\sigma' + Z)^B$	A	2.55
	B	−0.28
	Z	0.01
$k = Ce^D$	C	3.0E-07
	D	0.80

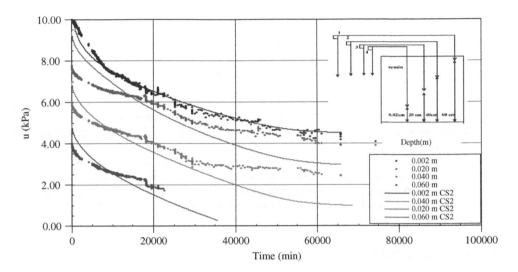

Figure 7. Comparison of pore pressure-time behaviour from analysis and experiment.

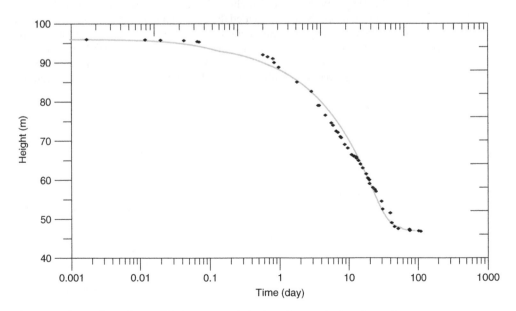

Figure 8. Comparison of consolidation vs. time behaviour from analysis and experiment.

ways; through drainage at the bottom, through removal of water from the top following sedimentation of the mix. It can be observed from Figures 7 and 8 that there is good agreement between the experimental data and numerical results. The experiment for the consolidation of the mix lasted for 109 days at Yildiz Technical University's Geotechnical Laboratories. However, pore pressure results were obtained only for 16 days from first transducer and the remaining transducers provided data for only 51 days due to air bubbles that formed inside the tubes that were connected to the piezometers.

6 CONCLUSIONS

In this study, behaviour of the Haliç dredged material that is being stored at the disposal site was experimentally investigated. Samples dentical to Haliç dredged material in its pumped state were placed inside the model tank that was built for this specific study. This mix had a water content of 300% and an initial voids ratio of $e \cong 8.00$ and was then allowed to consolidate under its own weight in the laboratory for 109 days. Using the independently-determined experimental material parameters from the 'Seepage Induced Consolidation Experiment' (Berilgen, 2004), consolidation against time and pore pressure against time behaviour of the Halic dredged sludge was predicted and modelled through numerical analysis tool CS2. This predicted behaviour was later compared with the experimental observations from the laboratory. Predictions and experimental results were in good agreement.

Valuable experience for prediction of high water content dredged soil consolidation was gained from this experimental research. It is clear that by using proper soil samples, appropriate experimental setup and application of modern numerical analysis tools, the soil behaviour at storage sites can be realistically predicted and rehabilitation and reclamation of these lands can thus be achieved in planned times.

ACKNOWLEDGMENTS

This investigation was funded and supported by Yildiz Technical University Research Fund, Project No. 21.05.01.0101 and the Municipality of Greater Istanbul. The Authors wish to thank both YTU Research Fund and the Municipality of Greater Istanbul. The Authors greatly appreciate the assistance of Prof Fox (Ohio State University) who provided the computer code CS2 and Prof. Znidarcic (University of Colorado) who encouraged the production of a seepage- induced test system and provided supporting documents. In addition, the Contractor for Golden Horn Rehabilitation Project, Southern Maryland Dredging Company and Gülermak A.Ş., kindly provided samples of dredged materials and field monitoring data. The Authors gratefully acknowledge their cooperation and assistance.

REFERENCES

Berilgen S A (2004) *Consolidation Behavior of Soft Clays,* Ph.D. Thesis (in Turkish), Institute of Science, Yildiz Technical University.
Fox P and Berles J D (1997) *CS2: A Piecewise-Linear Model for Large Strain Consolidation,* International Journal for Numerical and Analytical Methods in Geomechanics Vol: 21, 453–475.
Ipekoğlu P (2004) *Sludge Disposal Area Rehabilitation,* Ph.D. Thesis (in Turkish), Institute of Science, Yildiz Technical University.
Özaydin K, Yildirim S and Yildirim M (1997) *Haliç Rehabilitation Project – Feasibility Report: Fourth Interim Geotechnical Report*, Technical Report (in Turkish), Yildiz Technical University.
Özaydin I K and Yildirim M (1997) *The Golden Horn: Its Formation, Deterioration and Hopes for Rehabilitation*, Special Volume Honoring Prof. Vedat Yerlici, Bogazici University, Istanbul, pp. 195–210.
Somogyi F (1979) *Analysis and Prediction of Phosphatic Clay Consolidation: Implementation Package*, Technical Report, Florida Phosphatic Clay Research Project, Lakeland, Florida.
Znidarcic D and Liu J C (1989) *Consolidation characteristics determination for dredged materials,* Proc. 22nd Annual Dredging Seminar, Centre for dredged studies, Texas A&M Univ., College Station, Tex., pp45–65.

Geotechnical and Environmental Aspects of Waste Disposal Sites – Sarsby & Felton (eds)
© *2007 Taylor & Francis Group, London, ISBN 978-0-415-42595-7*

Geotechnical investigation of tailings dams with non-destructive testing methods

A. Brink, M. Behrens, S. Kruschwitz & E. Niederleithinger
BAM, Berlin, Germany

ABSTRACT: The environment and human beings are highly endangered by unsafe mine tailings facilities. The most important factor for safety is the stability of the tailings dams. Within the scope of a European research and technical development project ("Sustainable Improvement in Safety of Tailings Facilities", Tailsafe) geophysical non-destructive testing methods are used to investigate tailings dams to determine build up, condition, water content and water flow. The main techniques applied are geoelectrics (Spectral Induced Polarization – SIP) and georadar (Ground Penetrating Radar – GPR). In the paper measurements and results from a test site located in Germany are presented.

1   INTRODUCTION

Unsafe mine tailings facilities pose considerable risk to the environment and human lives. Tailings are the fine residue of the milling process from the mining industry. Large amounts of them are produced and must be disposed. Due to the processing they normally appear in slurry form. Therefore large tailings ponds are required to store them. Usually these ponds are surrounded by dams built by the upstream method.

In particular, a tailings dam's stability is essential for safety of the tailings facilities. In the majority of cases the dams are composed of the coarser part of the tailing material itself. Only a thin shell at the outermost part of the dike is constructed on the basis of geotechnical considerations and provides therefore some established structural integrity.

A key factor for a dam's safety is the water which seeps through the dam towards its free face. Deficient water management is one of the main causes of accidents and hazards emanating from tailings facilities. Nearly every dam failure can be attributed to springs forming at the downstream dam face, dam overflow or dam erosion.

2   THE TAILSAFE PROJECT

In 2002 a European research and technical development project ("Sustainable Improvement in Safety of Tailings Facilities", TAILSAFE) was founded which focuses investigation on the structural parameters of tailings dams stability, their measurement and their evaluation as regards risk factors, with particular attention to;

- the stability of bodies of fine material and their liquefaction and mobilisation behaviour
- the special risks inherent when such materials include toxic or hazardous wastes
- authorisation and management methods and procedures for tailings ponds and dams

Within the project, those parameters relating to the structural stability of tailings ponds and dams will be investigated, and methods for their measurement and monitoring developed further, to increase knowledge and technology levels in the fields of;

- materials parameters of tailings, with special reference to their heterogeneous and sometimes hazardous nature

- flow-deformation behaviour of tailings as fine hydraulically-placed non-plastic materials
- probabilistic approaches for stability analysis
- structural and lithological parameters of tailings impoundments
- physical conditions and fluid content, pressures and movement inside the dams
- non-destructive/geophysical methods for dam structure assessment
- whole-process water management by further developing paste technology

As a first approach the geophysical non-destructive testing methods geoelectrics and georadar are used to investigate tailings dams to determine build up, condition, water content and water flow.

3 EXPERIMENTATION

3.1 Geoelectrics (Spectral Induced Polarization – SIP)

Electrical methods for corrosion and moisture detection are well established in Civil Engineering (Schickert et al, 1999). But due to the fact that the electrical resistivity is depending on many factors (matrix material, porosity, salt content, moisture, temperature and others) data interpretation is often difficult. The same problem arises in the use of electrical methods in geophysics. During the last decades much effort was spent on the use of more sophisticated electrical methods. Most promising is the SIP method, which is used for example to discriminate minerals in mining geophysics.

Most electrical methods use separate electrodes for current injection and voltage measurements to avoid the influence of contact resistances (Figure 1). The apparent resistivity ρ_a is calculated by

$$\rho_a = K \cdot \frac{U}{I} \tag{1}$$

K (configuration factor) is dependent on the electrode arrangement at the surface. The apparent resistivities ρ_a are integral values over a certain volume in the subsurface dependent on the electrode arrangement and the true resistivity distribution. Large electrode distances lead to large penetration depths and large volumes. The resistivity ρ at a certain point in the subsurface has to be reconstructed

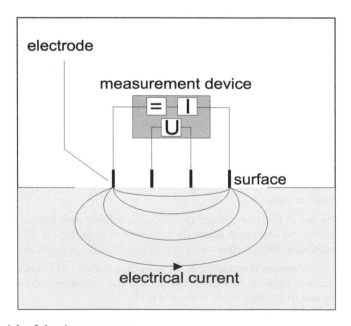

Figure 1.   Principle of electric measurements.

from large data sets of apparent resistivities ρ_a from different positions and penetrations depths via model calculations or inversion schemes.

The electrical resistivity r is a complex-valued quantity and consists of a real and an imaginary component. The real component is mainly influenced by the water content of a medium. Therefore it is often used as an estimate of the medium's porosity or the salt content of the pore fluid. The imaginary resistivity component (often referred to as phase) contains information on the chargeability behaviour of a medium. Typically, media with narrow pore spaces (like silts and clays) show high chargeability, media with larger pore spaces (like sands and gravels) exhibit lower chargeabilities.

3.2 *Georadar (Ground Penetrating Radar – GPR)*

The radar technique is based on the generation of short electromagnetic impulses and the recording of their reflections and backscattering at layer boundaries and objects. Two antennae (one transmitting, one receiving, together in one housing in most cases) are moved on the ground surface (Figure 2).

Radar traces (time series of signals at the receiving antenna) are recorded at a short distance interval controlled by an internal clock or a survey wheel. The radar traces along a profile are plotted as radargrams (vertical slices, B-Scans). If many profiles are recorded in the area under investigation time slices (horizontal slices, C-scans) can be generated. In most cases colour codes or grey scales are used to present the data instead of plotting each single trace.

A lot of application parameters have to be selected carefully for successful use. The main criterion is the selection of the optimum antenna frequency. High frequencies, e.g. 1.5 GHz, lead to high resolution (cm-range) but low penetration depth (20–50 cm typical). Low frequencies, e.g. 200 MHz, have low resolution (10 cm-range) and high penetration depth (more than 5 m in dry sandy soil). Resolutions and penetration depth also depend strongly on the medium (concrete, sand, clay) and moisture. Penetration is often very poor in wet media and clay. Other parameters to be selected are measurement time, distance between traces, samples per trace, amplifier settings, etc. An experienced operator is needed. After data acquisition several stages of data processing are necessary. Among them are gain control, bandpass filtering, static corrections and 2D or 3D display parameters.

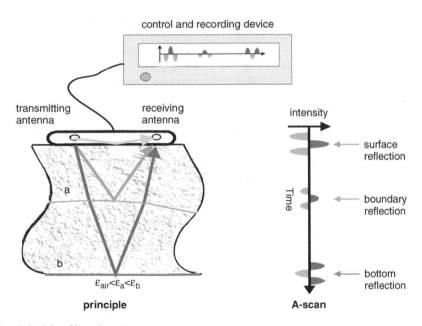

Figure 2.   Principle of impulse radar.

Figure 3. Measurement profiles and sampling points at Münzbachtal.

4 MEASUREMENTS

SIP and GPR measurements were performed at an abandoned tailings facility with two dams at Münzbachtal in Saxony (Germany), which is one of the test sides for the TAILSAFE project. Figure 3 shows a view from above of the tailings pond and the profiles that were measured along the two dams (P1 and P2 for the lower dam, P3 for the upper dam) and across the pond area (P4). The Profiles P1 and P2 are 90 m long and the other two have a length of 50 m.

4.1 *Geoelectrics*

On all profiles geoelectrical measurements were done with 25 electrodes at intervals of 3 m on the dams and 2 m on the plateau. Current was induced with a maximum voltage of 200 V at frequencies from 156 mHz to 500 Hz. Figure 4 shows the SIP measurement on profile P2 at the lower dam.

Results for profile P1 at the lower dam and profile P4 on the plateau are shown in Figures 5 and 6 respectively.

In the diagram for the resistivity it can been seen that the first two metres of the dam are very dry and that there are some inhomogeneities in the centre part of the dam.

The resistivity for profile P4 shows alternating layers of wet silt and dry sand up to a depth of 6 m. Wet tailings can be assumed beneath. With the phase only two different zones can be determined, but the wet, fine grained tailings can be distinguished from the covering coarser sand too.

4.2 *Georadar*

Radar measurements were carried out with two different antennae (200 and 500 MHz) and twice for each profile. An example is contained in Figure 7 which shows the readings at profile P2. The first

Figure 4.   SIP measurements at Münzbachtal.

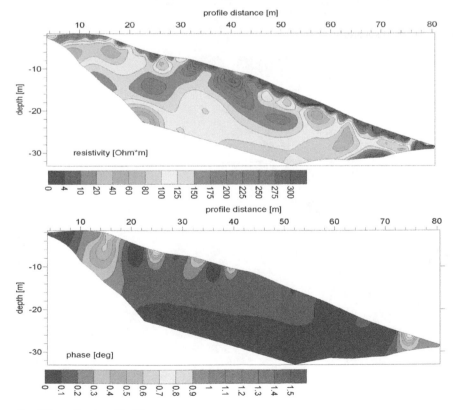

Figure 5.   SIP results at profile P1.

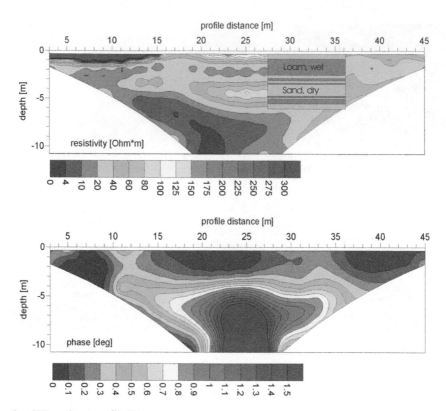

Figure 6.   SIP results at profile P4.

Figure 7.   GPR measurements at Münzbachtal.

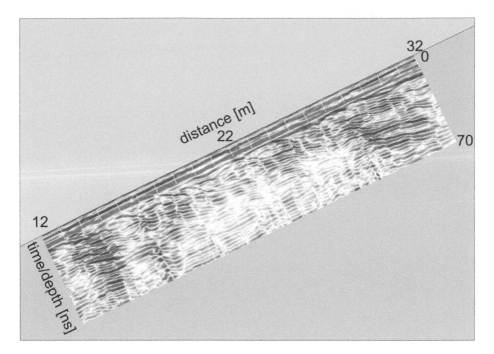

Figure 8.   GPR results at profile 2.

measurements were done by pulling the antenna along by hand. After these first tryouts a winch was used to improve the results with a more continuous movement over the profiles. Figure 8 shows the results of the GPR field tests at profile P2.

5   CONCLUSIONS

As tailings dam stability is the critical aspect for the safety of mine tailings facilities it is necessary to know as much as possible about a dam's structure. As shown in this paper geophysical non-destructive testing methods like SIP and GPR can be suitable for the structure analysis. While measurements and data analysis are still ongoing the potentials and limitations of the techniques cannot be defined yet. So far, layers of different materials, for example silt, sand or clay, in combination with the water content can be determined (Figure 6 for example). Further investigations will show whether it is possible or not to detect the phreatic surface within the dam, which is one of the key factors for the stability analysis.

ACKNOWLEDGEMENTS

The Authors would like to thank BAM for the opportunity to undertake research into this topic and to publish the findings.

REFERENCE

Schickert G, Henschen J, Krause M, Maierhofer C, Weise F, Wiggenhauser H and Borchardt K (1999) *ZfPBau-Kompendium*, http://www.bam.de/service/publikationen/zfp_kompendium/welcome.html

Geotechnical and Environmental Aspects of Waste Disposal Sites – Sarsby & Felton (eds)
© 2007 Taylor & Francis Group, London, ISBN 978-0-415-42595-7

Tailings impoundments – A growing environmental concern

J. Engels & D. Dixon-Hardy
Department of Mining, Quarry and Mineral Engineering, University of Leeds, UK

ABSTRACT: Technological advances in recent years have meant that it is now possible to mine lower grade mineral deposits more economically. Developments in large mobile equipment and mineral processing techniques have played an important role in achieving this. However, these low grades result in large waste fractions that need to be stored by a method that meets environmental legislation that is becoming more stringent as decades go by. The most common method of storage is a surface impoundment which is raised as the volume of tailings requiring storage increases. Failures worldwide have sparked concerns over the design, construction and management of these impoundments. Many failures occur due to poor construction techniques and the management of the water within the impoundment. Liquefaction events, overtopping, and piping are all aided by the degree of saturation of the tailings and embankment materials. Reducing the water content should, in theory, reduce the risk of failure and this is one reason for the recent development of paste technology. This paper looks at the types of surface disposal methods currently available and past examples of particular failures.

1 INTRODUCTION

Tailings have to be stored in a suitable location as to prevent minimal impact on the surrounding environment. Today there are several techniques available to provide this safe storage. In the past tailings were dumped without any lining systems or disposed of into the local water course. This still occurs today, and even the world's largest mining companies discharge direct in to rivers and seas. The Ok Tedi mine in the Philippines and the Lihir gold mine in Papua New Guinea dispose of their tailings and waste rock directly into the local water course.

Failures during the 1960's triggered concern for the environment and earthquake induced failures brought about the liquefaction phenomenon, which is still to this day unpredictable and misunderstood. Today a wide variety of site factors are assessed when a proposed impoundment is to be built. Some of these include:

- Distance and elevation relative to the mill
- Hydrology
- Topography
- Geology

2 RECENT DEVELOPMENTS

The last thirty years has seen a dramatic technological advancement in tailings disposal. Thickened, paste, backfilling, and dry stacking technologies are being tried and tested at various mines around the world. Backfill has been used successfully for a number of years allowing ore-bearing pillars to be removed, thus increasing productivity of a mine. However, the use of dewatering technology and binders generate higher costs compared to conventional surface impoundment storage.

Figure 1.   Cross-section of a typical thickened disposal operation (modified from Falconbridge Ltd).

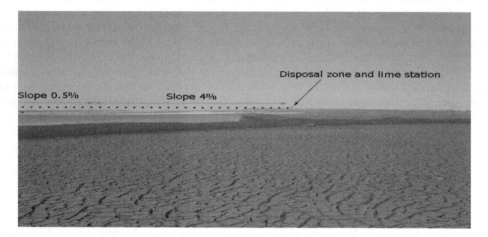

Figure 2.   View across the impoundment showing the distinct conical pile.

2.1   *Paste disposal*

Surface thickened disposal was first developed back in 1973 at the Falconbridge-owned Kidd Creek Metallurgy Plant in the town of Timmins, Ontario, Canada. The site has a Central Thickened Discharge (CTD) system that creates a conical pile of tailings (see Figure 1). The vast volume of tailings is concentrated in the conical pile resulting in low surrounding embankment walls. Multiple discharge points (risers) can be developed, e.g. Bulyanhulu Gold Mine in Northern Tanzania, if the tailings saturation is very low resulting in a high-angled conical pile.

More than 100 million tonnes of base metal sulphide tailings are stored at the Kidd Creek thickened site, and by the time the site is expected to close in 2023 more than 130 million tonnes will be stored. The radius of the conical pile (Figure 2) is 1.2 km and the height of the cone is 25 m. The height of the cone increases by 0.2 m/yr and by closure the height is expected to be 29 m.

The poor topography, hydrology and soil conditions in the Kidd Creek area prompted the need for an alternative to a large conventional impoundment:

- The site is situated on high ground surrounded on three sides by Porcupine River tributaries.
- High dykes would be unsuitable for the area and the risk of instability would be increased due to the extremely poor soil conditions.
- Predominantly cold climate conditions would only allow a short window for dam construction.

- The tailings are finely ground (45 μm) and imported materials would have to be used for dyke construction, thus increasing costs.

Surface paste and thickened disposal has many advantages and disadvantages over conventional surface impoundment storage:

Advantages:

- Costs associated with pumping and recycling of mine water are reduced.
- With no ponded water held in the impoundment, if a failure occurs there is no liquid to aid the flow of tailings.
- The embankments of the impoundment need minimal raising due to the disposed tailings ceasing their flow near the centre of the impoundment, thus forming the conical pile.
- Problems with Acid Mine Drainage (AMD) are reduced if correct management methods are in place. The correct disposal technique is to spread the paste evenly in the impoundment covering a previous layer before AMD can generate.
- The shallow angle of the conical pile (typically 1–4%) combined with the interbonding of the paste layers through filling of previous layer dissemination cracks with new paste make the impoundment very stable.
- A paste impoundment is far more stable in areas of seismic activity and far more resistive to liquefaction events compared to a conventional impoundment.
- Decant towers used to prevent overtopping of conventional impoundments can fail under the ever increasing weight of tailings. This increases the risk of failure. Paste disposal do not require decant (only for surface water runoff collection zones) as there is no ponded mine water.
- Progressive restoration of the impoundment can be carried out during operation.
- There is no erosion of the embankments through groundwater movement and natural seepage of the impoundment. Hence migration of tailings effluent into the environment is reduced dramatically.
- A paste impoundment can be made very stable by using thin layers of paste. As the old layers naturally desiccate and crack the overlaying fresh paste fills the cracks forming a bond between the two layers.
- Reduced costs of insurance, rehabilitation and closure.
- The paste impoundment can be reconfigured to accept other types of paste waste. This is one method of gaining revenue from the disposal operation that costs mining companies millions of US dollars.

Disadvantages:

- High capital and operating costs of high rate dewatering equipment and the use of synthetic or organic flocculants.
- Once the tailings have been deposited they are exposed to the elements as they are not covered with water. Hence, there is no barrier to prevent rapid oxidation of the tailings and the generation of AMD. If subsequent disposal layering is not managed correctly then AMD can be a serious problem.
- Paste disposal is a relatively new method of tailings storage and is unproven compared to the conventional impoundment, even if it is problematic (Fourie, 2003).
- Higher cost of pumping the paste tailings.

2.2 *Large conventional impoundments*

Conventional impoundment storage is one of the cheapest methods of surface storage (excluding river disposal) and is still the most common and preferred method. It is cheaper for mines producing high volumes of waste as no binders or dewatering have to be used. The tailings are pumped straight from the processing plant to cyclones that raise the embankment(s) with the sand fraction in the tailings. It is not common for a tailings embankment to be built to its full height prior to disposal, unlike water retention dams. The idea is to progressively raise the embankment so as to avoid having

Figure 3. The L-L embankment at Highland Valley Mine tailings impoundment (courtesy of Teck Cominco Ltd).

a very low freeboard which can raise the phreatic surface of the embankment, thereby increasing the risk of failure.

The slimes are disposed of behind the embankment with the fines settling out the furthest away from the spigot. Three basic designs are adopted when constructing and raising dams, i.e. for upstream, downstream, and centreline.

The Highland Valley Copper Mine in British Columbia, Canada has one of the world's largest tailings impoundment. Two embankments (the L-L and the H-H) contain the tailings in a cross-valley design called the Highland Pond (Figure 3). The L-L dam is built using the centreline technique (Figure 4) and is a zoned earth fill structure. The downstream face of the embankment has compacted cycloned sand berms to aid seismic resistance. The H-H dam at the other end of the impoundment is only 57 m high and consists of an earth and rock filled centreline structure (Scott, 2001).

The processing plant produced an average of 133,040 tonnes of tailings a day in 2003 which is pumped to the Highland Pond 7 km away. The L-L embankment is raised using the sand fraction of the tailings and is currently 128 m high. By the time operations cease in 2009 the impoundment is expected to be 170 m high and the impoundment will store an estimated 1.3 billion tonnes of tailings. As the sands are deposited, mechanical equipment is used to compact the embankment before additional layers of sand are added. Hydraulic cells are also used to raise the downstream side of the embankment and aid with consolidation of the sands (see Figure 3).

The stability of a tailings embankment is determined by the method of raising (Figure 4). The most common type used is the upstream design which raises the embankment towards the centre of the impoundment. The build material is deposited on the slimes. The centreline method raises the embankment on the slimes and also on the downstream face. For the downstream design the downstream face is raised and the advancing toe moves away from the impoundment.

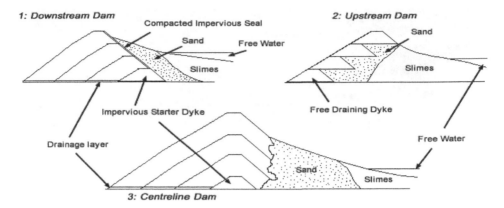

Figure 4. Methods of raising tailings embankments.

It is not surprising that the majority of tailings impoundments that fail are of the upstream design. However, the majority of tailings impoundments around the world are upstream, mainly due to the low volumes of build material and the low cost. Decant failure, storms causing overtopping, embankment erosion, piping and liquefaction are common modes of failure of upstream embankments. All these failures are due to water mismanagement. The well documented Merriespruit and Los Frailes tailings failures were caused by ponded water mismanagement.

Downstream designed embankments are the most stable but are more costly owing to the amount of material required to raise the embankment. Costs further increase if materials are not readily available locally. In these situations and where an upstream design is not suitable (for example in seismic areas) then other methods of tailings storage maybe investigated.

3 DRY STACKING

Dry stacking of tailings has been developed due to advancements in large capacity pressure and vacuum filters. The tailings are dried out to less than 20% water content (depends on the specific gravity) allowing them to be transported by conveyor/truck to the disposal site. The tailings are spread out and compacted to increase the density of the stack. The deposit is much more stable than a paste disposal site.

3.1 Advantages

a. The high density of the tailings reduces the overall volume of the storage facility required (less land required).
b. Surface stability is much safer than conventional wet deposition and allows for quick plantation.
c. The dry stack can be raised to heights which would not be economical with conventional impoundments.
d. Groundwater contamination is reduced.
e. Useful in arid climates where water conservation is an issue.
f. For cold climates dry stacking prevents pipe freezes and frosting problems with conventional impoundments.
g. Binders (cement) can be used to increase the stability of the deposit.
h. Low seepage levels from the stack.

3.2 Disadvantages

a. Higher costs than conventional impoundment storage.
b. Largely unproven technology and disposal method.
c. Problems with dust and tailings becoming airborne.

237

Figure 5.   Dry stacking operation (courtesy of Amec, Earth and Environmental).

3.3   *La Coipa site*

This is a silver/gold operation located in the Atacama region of Chile. Each day 18,000 t of tailings are dewatered by belt filters, conveyed to the stack area and deposited by a mobile radial stacker (Figure 5). Vacuum filters are used to dewater the tailings in the hope of retaining dissolved gold from the process solution. The high rate of dewatering helps to lower the saturation of the tailings stack which helps in water conservation and stability in the high seismic area (Martin and Davies, 2002).

4   HYDRAULIC MINING OF TAILINGS

Processing technology has advanced over the last twenty years and waste from old mining operations can now be economically exploited. In parts of South Africa and Australia the reprocessing of tailings has helped to remove old impoundments adjacent to neighbouring towns and reprocess the in-situ gold so allowing storage of tailings in a more suitable and safer location. The majority of reprocessed tailings are old gold mining waste which can now be exploited thanks to the development of carbon-in-pulp (CIP) and carbon-in-circuit (CIC) processing (Cuevas and Jansson, 2002).

4.1   *The technique*

English China Clay first developed hydraulic mining monitors back in the early 1960's. They are still used at the English China Clay sites to erode the soft rock and Kaolin. Figure 6 shows a typical hydraulic monitor.

These monitors are now used on old tailings impoundments to help reprocess and move the waste to a more suitable location. A typical flow sheet for the reprocessing of tailings is shown in Figure 7. The monitor uses high pressure water to erode the tailings in sections (Figures 8 and 9), washing the material downstream to be collected in a sump. When the tailings were dumped in the dam they generally segregated according to the coarseness of the material (coarsest fraction nearest the spigot) and if a particular area of a dam is too coarse for pumping then blending is required. The sump helps to mix the coarse and fine fractions and screen any large objects from clogging the pipe or slurry trench. The slurry is then pumped to thickeners and the underflow is reprocessed in the plant. The tailings are then stored in a tailings impoundment which is generally situated in

Figure 6.   A hydraulic monitor in Cornwall.

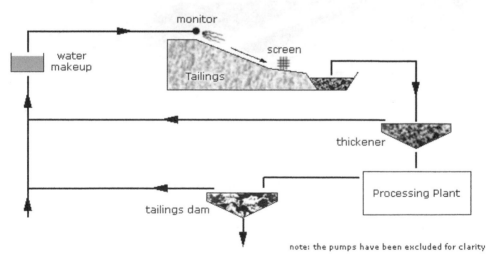

Figure 7.   Hydraulic mining of tailings.

a better location and in a more advanced engineered impoundment compared to the extraction site (Cuevas and Jansson, 2002).

To allow the tailings to be extracted, part of the embankment is deliberately breached which provides a flow path for the tailings to wash into the sump (Figure 10). As the impoundment is worked the tailings can be washed through this particular breach or through subsequent breaches that can be formed to help aid the flow of tailings to the processing plant, thus reducing pumping costs.

Figure 8.   Hydraulic mining of a tailings dam (courtesy of Fraser Alexander Ltd).

Figure 9.   Cutting into old tailings (courtesy of Fraser Alexander Ltd).

5   CASE STUDIES OF DELIBERATE BREACHING

5.1   *The Disputada mine, Chile*

The mine has three tailings dams that currently need to be moved for environmental reasons. During the move of the tailings they are reprocessed to retrieve more ore. One of the impoundments is located on the side of a valley, and the other two are situated in a steep narrow valley (Figures 11 and 12).

Hydraulic monitors were considered the only economical option as blending of the coarse fraction with the middlings and fines allows pumping of the tailings. One complication is that the tailings are mixed with rock deposited during the construction of the impoundment and the subsequent impoundment division to form roadways. Avalanche scree from the steep valley sides has also deposited a small amount of rock. The rock will have to be removed by mechanical methods that

Figure 10.   Aerial view of a hydraulically mined tailings dam (courtesy of Golder Associates).

Figure 11.   Mining into the unsaturated zone (courtesy of Anglo American, Chile).

Figure 12.   Mining the unsaturated zone upstream of Figure 11 (courtesy of Anglo American, Chile).

have to work in dry conditions. Hydraulic operations will need to be carried out in a separate area to allow the mechanical equipment to remove the large objects from the dry tailings. Dredging of the material in the dams would cause damage to the mechanical equipment and pumps, also dry load and haul is considerably more expensive than hydraulic mining. During the extraction of the tailings the screened material is pumped to a header tank almost 5 km away at a height elevation of 200 m. The tailings are then gravitational fed from the header tank to a new tailings impoundment 50 km away. It is necessary to move 19,000 tonnes per day, over an 8-month annual cycle, to make the operation a success. Mining during the winter months poses hazards from avalanches, blizzards and freezing of pipelines and mechanical equipment (Cuevas and Jansson, 2002).

   In the future mining of the saturated zone will pose a great challenge. Early indications show that floating equipment should be used to prevent a liquefaction event from occurring and risking the lives of the operators.

5.2   The Kaltails project, Western Australia

In the Boulder and Lakewood areas of the city of Kalgoorlie large gold tailings impoundments were posing an environmental concern. The Kaltails project was established to reprocess and move these tailings dumps (Figures 13 and 14). The operations ceased in September 1999 and had been ongoing for just over a decade. The tailings dumps were hydraulically worked with monitors, reprocessed and stored in an engineered impoundment located 10 km south east of Kalgoorlie. 60 million tonnes of tailings were extracted and 695,000 ounces of gold was recovered by CIC and CIP leach and absorption circuits (Anon, 1998). Hydraulically mining extended over 333 hectares of land and another 262 hectares will be used for waste rock storage from the Kalgoorlie Consolidated Gold Mines (KCGM) Superpit. The remaining 71 hectares have been contoured, seeded and re-vegetated.

242

Figure 13.   Remaining part of a waste dump being hydraulically mined (courtesy of Newmont Mining).

Figure 14.   External view of waste dump being internally eroded by a monitor (courtesy of Newmont Mining).

6   CASE STUDIES OF ACCIDENTAL BREACHING

6.1   *Mufulira mine, Zambia*

On the 25th September 1970 the Mufulira Mine in Zambia had an underground breach of a tailings dam. The night shift crew were on duty and the tailings dam above them collapsed causing nearly

243

Figure 15.   Sinkhole and processing plant.   Figure 16.   Aerial of the sinkhole.

Figure 17.   Aerial of breached embankment at Aznacóllar.

1 million tonnes of tailings to fill the mine workings killing 89 miners (Nellar and Sandy, 1973). A sinkhole opened (Figures 15 and 16) on the surface allowing surface water and ponded effluent to continue to pour into the workings.

6.2   *Los Frailes mine, Spain*

On the 25th April 1998 a tailings dam failed at the Los Frailes mine in Aznalcóllar, Spain. The failure is thought to have occurred as a result of acid seepage that passed through the embankment walls from the tailings eroding the marl foundations of the dam. The weakness in the foundations combined with the minimal length of beach (ponded water was encroaching onto the embankment) caused high stress in the foundations. This resulted in failure of the embankment material (Figure 17)

Figure 18. Aerial of embankment breach at Merriespruit.

and 4.6 million cubic metres of toxic effluent and tailings made its way into the Río Agrio and Río Guadiamar rivers.

6.3 *Merriespruit dam, South Africa*

On the 22nd February 1994 the Merriespruit tailings dam failed due to overtopping from heavy rains causing a flowslide (static liquefaction) of part of the embankment (Figure 18). The failure occurred due to poor management of the freeboard and ponded water within the impoundment. Half-a-million cubic metres of tailings flowed out of the impoundment and eventually stopped 2 km away in the town of Merriespruit. Seventeen people were killed and scores of houses were demolished (Fourie, 2003; Chu and Leong, 2003).

The impoundment is 31 m high and had problems prior to the major failure. Small slips had caused the impoundment to close temporarily, and only mine water with small amounts of tailings were deposited. The deposition of these tailings caused the ponded water to move to the opposite side of the impoundment which rendered the decant system useless. A satellite recorded the transition stages of the decant pond relocation as more tailings were deposited with the mine water. Heavy rains that fell on the day of the failure (30–55 mm in 30 mins) caused the overtopping (UNEP, 2001).

7 CONCLUSIONS

As technology progresses lower grade ores are mined are more waste is produced. For example, gold mining has typical extraction rates of 1% gold to 99% waste. The gold is generally extracted from the rock by cyanide which when spent is stored with the tailings in an impoundment. These huge volumes of tailings and toxic chemicals are the cause for concern with tailings storage today.

Paste, thickened and dry stacking of tailings are proving to be a safe solution for tailings storage. Although these are relatively new methods compared to conventional storage they are proving to provide greater stability and reducing risks of failure. This is mainly because of the low saturation of the tailings, the low dykes, and because there is no ponding of mine water. However, most of the current paste operations that store sulphide tailings generate high levels of acid due to mismanagement of the layering. One of the main reasons for this is that it can be very difficult to direct the flow of deposition and the subsequent layering, particularly if disposal operations are on a continuous uninterrupted basis, and most processing plants are.

It is now possible to mine gold from large volumes of old tailings impoundments which were created as little as 30 years ago. In the future, mining of other ore bearing tailings deposits maybe more economical than exploiting new similar grade deposits. The waste being produced and stored today from the mining industry is considerably more than that from 30 years ago. This results in a higher concentration of tailings deposits that will be an environmental burden in countries around the world. With the world's environmental concerns growing over time and environmental legislation becoming more stringent, mining of these large impoundments may be a more environmentally friendly alternative for the mining industry.

REFERENCES

Anon (1998) *Environmental report,* Newport Mining, Normandy, Australia

Chu J and Leong W K (2003) *Discussion – Defining an appropriate steady stae line for Merriespruit gold tailings,* Natural resources, Canada

Cuevas R and Jansson D (2002) *Hydraulic mining: An alternative solution*

Fourie A B (2003) *In search of the sustainable tailings dam: Do high density thickened tailings provide the solution,* School of Civil and Environmental Engineering, University of the Witswatersrand, South Africa

Martin T E and Davies M P (2002) *Stewardship of tailings facilities*

Nellar R R and Sandy J D (1973) *How Mufulira has been rehabilitated,* World Mining

Scott M D (2001) *Towards decommissioning – Highland Valley's copper tailings ponds,* Canadian Dam Association, Canada

UNEP (2001) *Tailings dams – Risk of dangerous occurrences, lessons learnt from practical experiences,* United Nations Environmental Programme, Division of technology, Paris

Geotechnical and Environmental Aspects of Waste Disposal Sites – Sarsby & Felton (eds)
© 2007 Taylor & Francis Group, London, ISBN 978-0-415-42595-7

Some legal aspects of tailings dams safety – existing authorisation, management, monitoring and inspection practices

R. Frilander, K. Kreft-Burman & J. Saarela
Finnish Environment Institute, Helsinki, Finland

ABSTRACT: Legislation and regulations on tailings dams differ in various countries; the differences are also considerable among the European countries. The heterogeneous regulations can be one of the answers to many problems concerning tailings dams. In most countries tailings dams remain outside the scope of dam safety regulations meant for water retention dams. It is one of the purposes of the TAILSAFE project to stress that tailings and other waste dams should be dealt in the future according to the same general dam safety legislation. This article contains results from the TAILSAFE project (http://www.tailsafe.com).

1 INTRODUCTION

The reasons for dam failures never remain unexplained. All the failures have been caused either by poor dam design, poor dam operation or by the combination of these two. From this point of view, well-organized and functioning dam safety legislation and regulations are of a crucial importance. ICOLD (1989) gives some recommendations on how tailings dam statutory legislation could be arranged. Amongst these are laws, commissions, registers, permit procedures for design, construction, operations and maintenance, supervisors, authorities, inspections and rehabilitation.

Golder Associates (2001) suggest that the regulations concerning tailings storage facilities should be flexible enough to accommodate variations in physical, technical and social considerations of different sites. An ideal regulatory framework would accommodate future changes as technical knowledge and community expectations increase. Regulations should also be written to proactively address potential challenges and hazards, rather than as a reaction to unacceptable events and performance. The form of the regulations can be either focused or separated. In some countries the regulatory scheme rely on specific dam safety legislation. On the other hand, there are countries where dam safety is treated as one aspect of more general legislation, for example combined with water, dams, energy or natural resources (Bradlow et al, 2002).

2 MAIN POINTS OF DAM SAFETY LEGISLATION

Water-retaining dam safety procedures are usually considered to be of a very high standard. However, tailings dams are often regarded as, so to speak, "lower level dams" and therefore not dealt with to the same high safety standards as water-retention dams, even though they pose very high risks to population and environment. Traditionally, dam safety legislation concerns only water-retention dams. The European Dam Safety Club (BETCGB, 2001) has collected existing material concerning dam legislation from the following countries: Austria, Finland, France, Germany, Italy, the Netherlands, Norway, Portugal, Romania, Spain, Slovenia, Sweden, Switzerland and the United Kingdom. It has been found out that the standards and procedures in dam safety legislation vary a lot. The criteria to be applied by dam legislation include height, volume and risk posed by the dam. Dam height varies from 3 to 15 m and volumes from $50\,000\,m^3$ to $1\,500\,000\,m^3$. The classified hazards to human and environmental safety can be distinguished as high-risk, medium-risk and low-risk dams. In general, water dam legislation gives clear orders and demands about the responsibilities and activities concerning dam safety issues. In water-retention dam legislation there are more actors with exactly defined tasks described by law. This has resulted in an unequal safety situation of the water-retention dams in comparison to tailings and other dams.

3 LEGISLATION CONCERNING TAILINGS DAMS

3.1 Mining legislation

In most cases tailings dams are operated by mining companies and supervised by State Mining Authorities. This might easily result in a situation where the State Mining Authority has lots of tasks to cope with such as mine production, etc and therefore tailings dam safety might be considered as a minor issue.

In the European Union there is no specific legislation concerning waste from mining operations. The member states have their own mining and environmental legislation. In Finland the Mining Act and Mining Decree includes a paragraph which requires that suitable and applicable parts of the dam safety guidelines should be taken into account. Both the Act and Decree are to be reformed within a few years time.

All the EU Candidate Countries went through political and economic changes in the late eighties and early nineties. In all cases new legislation was introduced in the field of mining law. The first innovative legislative ideas were usually followed by corrective actions and subsequent amendments of mining laws (Hamor, 2002). Hamor points out that the term "mining law" is often replaced by some other title, e.g. Subsurface Resources Act (in Bulgaria), Earth Crust Act (in Estonia), Law on the Subsoil (in Latvia), and Law of the Underground (in Lithuania), etc. Poland has a combined legislation on geology and mining and Slovakia has two separate acts governing the issue – one act on the protection and utilisation of mining resources and one act on the mining operation activities. However, in Hungary, Romania and Slovenia there still exist traditional Mining Acts. According to Hamor, the lack of the traditional "mining law" term in the new legislation is a sign of the acceptance and adoption of the sustainability concept that the relevant acts deal with the sustainable management of the complex geo-environment.

Tailing facilities safety is not the priority for mining safety legislation if it is considered at all. In this case, there exist considerable differences among the Candidate Countries. In Poland tailings are regulated mainly by Building Law. In Romania specific regulations on tailing ponds are covered by the law and special orders are issued by the Ministry of Water and Environment Protection and the Ministry of Industry and Resources. In Hungary a specific regulation on tailing ponds is being drafted (Hamor, 2002).

3.2 Legislation on waste and tailings

The Waste Framework Directive 75/442 (with amendment of Directive 91/156) states that the Member States must take the measures necessary to ensure that "the wastes are covered or disposed of in such a manner that they have no impact on human health or cause any environmental damage". Mining waste is included under the scope of this directive, if it is not covered by another legislation. In 2003 the Commission launched the proposal of the new directive "on the management of waste from the extractive industries". The Directive concerning Landfill of Waste (99/31/EC) set up the surveillance programme for water, leachates and gases and the results of the monitoring must be shared with authorities. This directive also applies to mining waste. Certain mining wastes are covered by the list of hazardous waste (European Waste Catalogue, decision 2001/118/EC).

3.3 Legislation concerning water

In the European Union the management of water is based on integrated management depending mainly on quality standards and limit values for emissions. Directives also concerned with tailings sites are, for example:

- Discharges into Water, Directive 76/464/CEE with other Directives on discharges of dangerous substances
- Groundwater Protection Directive 80/68
- Water Framework Directive

3.4 Legislation on environmental issues

The Environment Impact Assessment (EIA) Directive (85/337/EEC) (with amendments of Directives 97/11/EEC and 92/104/EEC) is an integral part of the laws on mining operations on most of the Union countries. The first amendment is applied to the safety and health of the workers in surface and underground mining industries. Golder Associates (2001) suggests that EIA should be undertaken and submitted to government authorities as a part of the TSF approval process. In spite of the existence of well-developed environmental legislation in the EU, the major accidents of Aznalcóllar and Baia Mare have indicated that the EU legislation does not cover all the aspects of mining activities.

3.5 Mining site closure

The objective of the decommissioning phase is to achieve the closure plan to meet specified end land uses agreed to between the regulators and the company. Closure objectives, including a specified end land use, should be stated by the government and addressed by the EIA process (Golder Associates, 2001). At the moment there are no European regulations for closed mines and closure procedures. On the other hand, mine sites can be identified under Landfill Directive (1999/31/EEC) which specifies the closure conditions.

In the Central and Eastern European candidate countries there exist detailed regulations on mining safety, but closure is not regulated in the same detail as the opening of mines. The mining license is usually based on the approval of a closure plan. The reclamation and post-closure monitoring are covered by the mine closure licensing procedure. Where there are reclamation requirements most countries have specific provisions. However, this is not the case for post-closure monitoring. This is usually laid down by the authorities in their resolutions. However, in lots of cases no safety procedures nor reclamation apply to the abandoned mines. Orphan mines do not have an operator or a legal successor and therefore they pose a high risk to the environment. In most of the candidate countries this problem is tackled in the mining act at least. The solutions vary and depend, among others, on the ownership of the given mineral commodity and the type of extraction technology. For instance, in Poland if the mine operator disappears without a legal successor, the landowner is responsible for the reclamation of open pits. In Slovenia the reclamation of orphan mines is taken care of by local communities which look for financial support from governmental funds. In Hungary the mining rights of the bankrupted mining company without a legal successor are to be announced for tender by the Hungarian Mining Office. If the transfer of the rights and obligations remains unsuccessful for a year, the licence is removed from the register and necessary measures to cover the costs of closure, reclamation, etc., are initiated. If the reserved financial guarantees of the company do not cover the total cost, the obligation of reclamation and environmental clean-up eventually is shifted to the State which is the original owner of the minerals. The status of orphan mines is not regulated in Estonia and Latvia at all (Hamor, 2002).

4 TAILINGS DAM GUIDELINES AND CODES OF PRACTICE

In 1989 ICOLD published Bulletin 74 "Tailings Dam Safety – Guidelines" which contains information about unique aspects of tailings dams and general provisions with guidance for design, construction, operation and rehabilitation.

4.1 *Finland*

The Finnish Dam Safety Code of Practice has been applied to tailings dams. The Dam Safety Code of Practice includes, among other issues, some technical recommendations concerning the stability, the drainage system, the height of the dam (including frost protection), the slope protection and the permitted location and size of trees.

4.2 *Canada*

In Canada "A Guide to the Management of Tailings Facilities" (Mining Association of Canada, 1998), presents a full life cycle management framework for tailings storage facility. Guidelines give recommendations for planning, design, construction, operation and finally for decommissioning and closure. The framework is expanded into a series of checklists addressing various stages of the tailings dam life cycle. The Guide identifies six key elements for effective implementation of the dam operation and management. These elements are:

- Management Actions,
- Responsibility,
- Performance Measures,
- Schedule,
- Technical Considerations and
- Other References.

Currently in Finland and Canada the general dam safety guidelines include aspects concerning tailings dams. The Canadian Dam Association (CDA) updated their dam safety guidelines in 1999. The Guidelines include recommendations concerning responsibilities for dam safety, scope and frequency of dam safety reviews, operation, maintenance and surveillance and also emergency preparedness. These guidelines state that conventional earthfill dams for water storages and tailings dams for mining residues are in many cases similar, for instance if design criteria for stability of the dams are concerned.

4.3 *United States of America*

In 1994 the US Environment Protection Agency (USEPA) published a Technical Report entitled "Design and Evaluation of Tailings Dams". The document is intended for government agencies assessing effects on the environment. Sections of the document are based on the book "Planning, Design and Analysis of Tailings Dams" by Steven Vick (1990). The report provides an overview of the methods of tailings disposal and the types of storage facilities. General information is presented on the design of tailings dams, including a discussion on design variables, such as site-specific factors, site location, hydrology, geology, ground water, foundations and seismicity. Water control and management are also presented, including discussions on hydrology, management of storm flows, infiltration and seepage control and tailings water treatment.

4.4 *Mexico*

The Mexican Official Standard (1997) stipulates the compulsory requirements for site selection, construction, operation and monitoring of a tailings storage facility. These requirements include:

- An environmental impact study
- Compliance with laws governing the preservation of historical or cultural heritage
- Assurance that there will be no percolation of toxic leachates to the nearest aquifer or surface water body within 300 years
- Approved plans for surface and groundwater monitoring
- Detailed characterisation of the underlying geological structure and the mechanical properties of rock formations and soil deposits

- Land surveys of the site to delineate elevations and features such as roadways and pipelines
- Compliance with civil work design standards for dams issued by the Federal Electricity Commission
- Monitoring instrumentation for a tailings facility over 50 m in height.

4.5 South Africa

In South Africa a policy of "self management" is applied which requires mines to prepare an Environmental Management Program Report (EMPR) at the planning stage. Thereafter, the requirements of the "Code of Practice for Mine Residue (SABS 0286-1998)" apply during the life cycle stages of design, construction, operation and closure. The Code of Practice addresses the life cycle of tailings facilities in terms of safety, construction, operation and environmental impact. It contains objectives, principles and minimum requirements for good practice and has the aim of ensuring that no unavoidable risks, problems and/or legacies are left to future generations. A process of continual management and continuous improvement throughout the life cycle is envisaged. The Code of Practice requires that each tailings facility be assigned a Safety and Environmental Classification. In terms of safety each tailings facility is classified as having a high, medium or low safety hazard. The tailings facility is environmentally classified according to the spatial extent, duration and intensity of its potential impacts and is considered as either "significant" or "not significant". These classifications determine the minimum requirements for investigation, design, construction, operation and decommissioning.

4.6 People's Republic of China

In the People's Republic of China tailings storage facilities must be designed and constructed in accordance with National Codes (Design Standards). Provincial authorities are responsible for issuing a license to construct and operate the tailings facility. "Code ZBJ 1-90 (1991) Design Standard – Tailings Facility for a Mine" addresses the design of a tailings facility for a mine and classifies a tailings facility into one of five classes according to storage capacity and dam height. The Code specifies the minimum factor of safety for various operating conditions. Tables are presented stipulating the minimum freeboard and storage (beach) length for different classes and types of construction (upstream and centreline), minimum crest widths and downstream slope angles.

4.7 Australia

For all mining projects in the Western Australia a Notice of Intent (NOI) has to be submitted in accordance with the "Guidelines to Help you get Environmental Approval for Mining Projects in Western Australia" (DME 3, 1998). The NOI is a document addressing the environmental issues associated with the mining project. The NOI should contain a design report that documents the design of the tailings facility. The design should be carried out in accordance with the "Guidelines on the Safe Design and Operating Standards for Tailings Storage" (DME, 1999). The design is required to take cognisance of development, operational and rehabilitation/closure conditions. The DME Guidelines set out minimum requirements in this regard.

5 AUTHORIZATION AND RESPONSIBILITIES

5.1 Authorities for dam and tailings dam safety

The purpose of the authorization is to minimize the risks concerning tailings dams. This is mainly achieved by involving different sections and levels of different ministries. However, Golder Associates (2001) suggest that all tailings storage facilities should be assessed, licensed and monitored by a single authority.

A study performed by Bradlow et al (2002) revealed that among the 22 studied countries 11 jurisdictions have designated a regulatory authority that is exclusively dedicated to dam safety. In some of these countries the specifically designated authority may share jurisdiction over some aspects of dam safety with other regulatory bodies. In 15 countries the regulatory authority deals with dam safety as part of the other responsibilities – with the exception of Australia which has a system where the regulatory framework identifies a specific individual to be responsible for dam safety issues.

In Finland the Ministry of Agriculture and Forestry has the overall supervision and guidance issuing guidelines for dam safety. The supervision of dams subject to the Mining Law is undertaken by the Safety Technology Authority of the Ministry of Trade and Industry. Regional Environment Centres are responsible for official decisions, supervision of observance of rules and regulations issued by virtue of the Dam Safety Act, excluding rescue services. Ministry of the Interior and its authorities (provincial governments and municipal rescue authorities) are responsible for rescue services and emergency action planning. The role of the Finnish Environment Institute is to give expert opinions to regional environment centres on the safety monitoring programmes for P dams (dangerous) and hazard risk assessments, to improve dam safety and participate in preparation of the dam safety code of practice.

According to Bradlow et al (2002), the authorities have many responsibilities and powers in dam safety issues. They are in charge of developing norms and standards. They have the power to issue licenses and permits. Considering the surveillance of the dam, the authorities are monitoring and conducting the inspections and/or giving the approval of the inspectors. Regulators should also establish and maintain a database on all tailings dams, operating and others (closed). Information databases are collected by making an inventory of dams and by maintaining the register. In Canada there have been quite a few recent tailings dam incidents, which were partly caused by the lack of relevant and accessible historical database and/or inadequate appreciation of that database (Martin et al, 2002).

In Finland the Regional Environmental Centres must supervise the fulfilling of all requirements on the dam safety legislation including procedures of applied permissions, surveillance for the demands of the permissions and roles in emergency situations.

5.2   *Responsibilities of persons in charge of dam safety*

The responsibilities for dam safety are shared among the different parties including dam owners, dam operators, designers (and constructors) and stakeholders. Stakeholders should be involved already during the environmental impact process of the tailings storage facility.

The primary responsibility for dam safety rests with the owner of the plant – including appropriate monitoring, maintenance and provision for emergency measures. The owner must ensure that the dam is designed by a competent and experienced person. In addition, the owner must recognize the importance of good management principles and practices. Responsible management of tailings might be expensive, but expenses are incomparably higher if the tailings facility fails and causes enormous damage to humans and the environment. The owner must implement means of detection and, if possible, repair of the defects that can occur in the dam. The owner of a dam is obliged to make himself acquainted with the regulations concerning his dam, and, on his own initiative, ensure that they are observed. The owner must also have a detailed file with all the documents concerning the dam.

As an example of the different responsibilities of a dam owner the items of dam safety for dams in general in Switzerland are the following (BETGCB, 2001):

- The control of the working order of the outlet gates and the spillway gates.
- Visual surveillance and the reading of the monitoring system. Data must be immediately analysed (control of the behaviour of the dam).
- Annual inspection by a experienced professional (regular control of the state).
- Publication of annual reports about the results of surveillance and monitoring – these reports are intended for the Authority of Surveillance.
- Expert evaluations of the dams (at least every five years) by confirmed experts in the field of dams (engineers, geologists). These evaluations include an opinion on the condition of the dam,

an analysis of its behaviour, an examination of the monitoring system with a proposed programme of monitoring. A special evaluation can be required (for instance safety in the case of flooding).
- Setting up the register of the dam.

Dam operators have to ensure that the tailings facility has an Operation Manual for guidance in tailings management. Operators must maintain the contacts with designers and most importantly they must ensure that the design work is closely combined with operators active participation. The operators have to guarantee that the dam is operated by qualified staff. For instance, in France the operator of a dam has a register for all the events, incidents, maintenance activities, etc. The operator must carry out periodical visual surveys and implement suitable monitoring. The regulation indicates (but does not impose) frequencies of inspections and monitoring. Every fault must be reported to the administration. The operating instructions for exceptional events are established by the operator and approved by the administration. The operator publishes an annual report of the surveillance, the monitoring and the operation of the dams. Every two years, the report includes a detailed analysis of the results given by the monitoring (BETGCB, 2001). In Spain the operating director must be a qualified engineer. The operating instructions have to include the following issues:

- procedures in the case of exceptional events
- programme of monitoring and periodical inspections
- information procedure for water releases
- alarm system.

These instructions are included in the file of the dam. The operator keeps up to date a register of the dam. (BETGCB, 2001).

6 MANAGEMENT OF THE TAILINGS FACILITY

The management of the tailings storage facility starts after dam construction with the commissioning and implementation of the dam (in relation to the plans and construction work descriptions etc.). After this stage the filling of the impoundment basin can start. During the tailings dam operation and functioning, including water balances, increasing the height, emergencies and decommissioning, the importance of the surveillance, monitoring and inspections can never be underestimated. The management and supervisory personnel should have sufficient experience in carrying out demanding earthworks, and the persons responsible for these works should have experience of previous works on embankment dams. The supervisory personnel and management should not be dependent on each other, and the supervisor should have the right to halt construction in case the conditions, materials used or work methods differ from those specified in the design documents.

6.1 *Operational safety manuals*

An operational safety manual (OSM) should be prepared for the documenting operation, maintenance and surveillance. The OSM should be implemented, followed and updated at appropriate intervals. The manual should also contain suitable and sufficient information in order to operate the dam safely, to maintain it in safe conditions and to monitor its performance. There are states (for example Finland and British Columbia in Canada) where operation manuals and annual inspections/ reviews by specialists are a regulatory requirement. The Canadian operation and maintenance manual contains a general information on the dam, operational procedures and emergency preparedness plan.

According Martin et al (2002), the Operation Manual (OSM) for a tailings facility might be the most important measure in implementing good stewardship practices for tailings. Recommended elements of the OSM are the following:

- Project administration, responsibilities of operation, safety and review roles of the corporation
- Design overview and key design criteria

- Tailings deposition and water management plans
- Planning requirements
- Training and competency requirements
- Operating systems and procedures
- Dam surveillance (checklists, signs of unfavourable performance, responses to unusual observations)
- Reporting and documentation requirements
- Emergency action and response plans
- Construction and quality assurance/quality control requirements
- Standard formats for status reports in certain times, performance reviews
- Reference reports and documents

6.2 *Commissioning of the tailings dam*

According to the Finnish Dam Safety Act, the construction of a dam has to be carried out in such a way that in structure and strength it meets the requirements that no safety risk would arise from either the dam itself or its use. From this point of view the commissioning inspection of a dam shall be made in such a manner that all issues relevant to dam safety are adequately considered. The quality of the dam structures shall be assessed. The commissioning inspection is the responsibility of the chief dam designer or another competent person. The commissioning inspection is based on data in the dam plans, the quality control programme and, if necessary, the hazard risk assessment.

The commissioning inspection begins with a written notification to the authority. Then the inspection continues with the necessary field inspections, which can be reviews of structures and foundations conducted during different stages of the work. The commissioning inspection is completed when all the structures are operationally ready, have been brought into full-scale use and have been approved to function as planned. At the closing of the commissioning inspection the records of the field inspections and the completion documents are collected, and a summary (final statement) and a proposal for dam qualification are compiled from them and included in the dam safety file.

6.3 *Emergency situations*

Tailings dam failures are typically caused by poor water management, overtopping, foundation failure, drainage failure, piping, erosion or earthquake. Potential effects caused by the tailings dam failures are loss of life, contamination of water supplies, destruction of aquatic habitat and loss of crops and contamination of farmland. In addition, tailings dam failure may threaten the protected habitat and biodiversity and lead to loss of livelihood. Guidance have also been given for the mining industry in raising awareness and preparedness for emergencies at local level (UNEP, 2001).

In Finland the dam safety guidelines recommend reporting and evaluating the possibility of failures and other problems endangering the structures of the reservoir; their causes and consequences. Also the plans for future prevention of such failures or other problems in the use of the dam must be documented.

According to the Dam Safety Act, the emergency action plan has to be made for a dam which in the event of an accident may manifestly endanger human life or health or seriously endanger the environment or property. The plan is based on a dam-break flood analysis (hazard risk assessment). The municipal rescue authority is responsible for the emergency action planning. The dam owner is obliged to assist the rescue authorities in drawing up the plan, to draft the relevant assessments and necessary action plans for his part. The dam owner also bears the responsibility to acquire and maintain the facilities and materials referred to in the action plan and to take other measures to safeguard people and property against the risk posed by the dam and to participate in the implementation of the action plan. The Regional Environmental Centre, when this is considered necessary, can instruct that a dam breach hazard analysis is prepared for a waste dam. A dam breach hazard analysis for a tailings or other dam can be prepared by supplementing the evaluation of dam failures (MMM, 1997).

Precautions against a dam accident are to be taken according to instructions for the tailings or other waste dams, as well as the possible harmful and toxic nature of the waste material which must be reported at all waste dams. This can be done on the form for evaluating the environmental consequences of a dam failure. For the most dangerous dams (P dams), the reports and plans for accident prevention are prepared according to instructions in the Dam Safety Code of Practice. For the assessment of a danger to human health or to the environment, the following details of the impounded substance are to be provided:

- The nature of the impounded substances, the substances harmful or dangerous for health and environment, and their content in the impounded material.
- the total volume of the impounded substance and an estimate of the amounts of harmful or dangerous components likely to be washed out into the environment in the case of an accident.

Additional material to be presented are an estimate of the magnitude of the risk of danger to human health and to the environment, an estimate of the consequences, and an action plan for the prevention of these dangers in the case of a dam accident. If there are people living so close to a dam in a potential downstream hazard area that an alarm given by the rescue services could not possibly reach them in time, the area at particularly high risk must be provided with a system capable of sounding the alarm in time. The normative time limit for an area at particular high risk is two hours from a dam failure. The need for such a system has to be ascertained by the rescue authority.

6.4 *Decommissioning and remediation of tailings dams*

The method of decommissioning the tailings facility depends on the characteristics of the impounded material. The different alternatives are tearing down the dam, surface cover structures and preventing the environmental impacts caused by the tailings facility. In every case the area need to have a surveillance programme for the environmental emission and water levels.

The mining industry in Peru has used The Mine Closure Guide prepared by Golder Associates Ltd to provide an outline of closure objectives, approaches and technical issues for the planning of closure of mines or mine facilities including tailings storage facilities. The guide contains the following issues relating to tailings storage facility:

- Perpetual disruptive forces and control technologies
- Chemical stability of soluble minerals, acid drainage and chemical reagents, and control technologies
- Design methodologies, including treatment and encapsulation
- Guidelines for the design of covers
- Closure alternatives including water management and landfill stability
- Closure plans for waste materials that address potential environmental issues, typical closure technologies and typical design elements
- Post-closure performance monitoring.

6.5 *Documentation*

In order to improve the safety of existing tailings dams keeping the documentation updated is of a great importance during dam operation. Documentation should include operating manuals, dam safety files, maintenance diaries, malfunctions reports and records/minutes of the inspections.

According to the dam legislation survey (BETCGB, 2001), dams in Germany are operated with the help of the approved documents including:

a. operating plan for the use of the water,
b. operational and maintenance instructions for the plugs,
c. operating and measuring installations and
d. instructions for dangerous situations with the required communication.

In Finland the dam owner or holder is obliged to store all the documents relevant to dam safety in a special safety file. This also includes the control and measurement documents, which must be registered in the same dam safety file. A record is made of the annual and regular inspection and included in the dam safety file. The records can be made more graphic with drawings, photographs, videos, etc. A copy of the record is also sent to the authority (regional environment centre) and, for a dangerous dam, also to the provincial government, the regional fire commander and the municipal fire authority even if these did not participate in the inspection. Amendments and supplements to the safety file are sent to the regional environment centre (MMM, 1997). In Finland according to the Dam Safety Code of Practice, structural and operational disturbances affecting dam safety must be reported. The report must present the cause of the disturbance, the investigations conducted and the measures undertaken. The disturbance report has to be delivered to the dam safety authority.

7 SURVEILLANCE, MONITORING AND INSPECTIONS

Safety surveillance, monitoring and inspections belong together when evaluating the safety of a tailings or other dam. The Canadian guidelines (MAC, 1998) suggest that periodic inspections and reviews, audits, independent checks and comprehensive independent reviews are the main part of surveillance programme. Dam monitoring involves among other things:

- monitoring the height of water or other substance impounded in the basin,
- inspection of the visible parts of the dam structures and the dam downstream area during each inspection visit,
- the observations and measurements listed in the monitoring programme and other issues relevant to the dam and
- other special items related to dam safety (especially of tailings and waste dams).

The risks to tailings impoundments are normally caused by seismic, hydrological or operational disturbances. Tailings dams are normally inspected yearly by independent experts for various safety concerns. One might suspect that this kind of evaluation represents only an instant of time and does not tell how the facility is really being operated. From these points of view testing and evaluation of possible failure scenarios should be added to the annual inspections (Glos, 1999). As a result of the surveillance to the operational manual, the description of the possible problems at the tailings dam should be documented, for example excess filling, overflow, their causes and measures for preventing similar problems in the future.

7.1 Surveillance programmes

The safety monitoring programme has to be drawn up well in advance of completion of a dam so that it can be approved for compliance before starting the filling. According to Hurndall (1998), the dam safety programme should include as a minimum

- some form of legislation, regulation and/or guidelines
- an inspection programme for dams with significant hazard potential
- an operation, maintenance and surveillance programme in place
- regular independent review of the safety of dams
- an audit programme in place.

The Finnish procedure for reducing the risks of a dam damage includes a safety monitoring programme, which has to be drafted for each dam referred to in the Dam Safety Act. The dam safety monitoring programme has to be drafted by the dam owner. The programme has to be drafted in such a manner that all the issues relevant to dam safety are subjected to surveillance and inspection. The programme may include rules concerning the monitoring proper, annual inspections and the inspections made at regular intervals (not exceeding five years). A safety monitoring programme or its amendments are approved by an authority (MMM, 1997). In Finland the safety surveillance programme of a tailings or other dam is prepared in accordance with the Dam Safety Code of Practice. The safety surveillance programme includes regular inspections (every 5 years) performed by a competent expert, annual inspections (in the intermediate years) done by maintenance personnel and surveillance between inspections according to the programme defined in the basic inspections. Special attention is to be paid to the observation of:

- intake pipes of the waste material, pumping and transfer lines and discharge pipe systems,
- inspection wells and collecting wells and
- quality of work in constructing and increasing the height of waste dams and drainage areas.

In Finland seasonal surveillance of the dam is to be carried out by the schedule required in the safety surveillance programme. In addition, surveillance is required when the structures come, or may have come, under special strain after the break-up of ice, during a flood, or because of heavy rains or a storm. During the surveillance the following observations should be made:

- Investigation of the visible parts of the dam structure
- Observation of the internal inspection galleries and wells
- Visual observation of the collection wells and discharge points of the dam filter system (function of drains and colour of seepage)
- Reading of the stand pipes, measuring weirs and other monitoring instruments
- Inspection of the drains in the downstream area and abutments
- Inspection of the inflow pipes, pumping lines and outlet channels
- Checking the inspection of the monitoring and collecting wells
- Quality control of the building and work to increase the height of the dams and the waste areas
- Evaluation of environmental impacts.

7.2   Frequency of monitoring

The frequency of the inspections depends on the dam classification. Deviations from the normative periods are allowed if a system replacing the inspections is in use, e.g. remote monitoring cameras, telemetric apparatuses and computers with the alarm systems based on them. The use and function of the replacement systems should be described in the draft monitoring programme.

The annual inspection should establish the state of the structures and changes in them visually and by means of equipment test runs. The structures should be inspected in spring or early summer after the flood and thaw. In addition, and if necessary, during floods and after exceptionally heavy rainfalls and storms, inspection visits should be made to dams which are or may have been subject to extra strain. The conditions of the structures and facilities, and for waste and tailings dams, the type of impounded material, and changes in them that affect dam safety should be established. This should be done by regular inspections (to be held at intervals not exceeding five years) with measurements, analyses of observational data, test runs of the equipment and other investigations where considered necessary.

In Sweden water retention dams are inspected by County Councils with the support of dam safety experts at the owner's expenses. The general policy of dam inspection suggested by the "Commission of Safety of Dams" includes:

- Weekly visual inspection made by the operation team.
- Annual inspection made by a qualified engineer – this inspection includes a test of the gated spillways. The report of the inspection is sent to the County Council.
- Every four years, a detailed examination should be made by an engineer independent of the local operation team. It is considered to organize an evaluation of the safety level of the dams every fifteen years (BETCGB, 2001).

7.3   Annual inspection

The annual inspection of a dam should be performed when the soil is not frozen. In the course of the annual inspection the measurements and observations made during the year are reviewed.

Taking into account the changes that have occurred, the operational state of the measuring devices is checked and the parts of the dam and the associated facilities requiring repair are investigated in the field. In the annual inspection special attention should be paid to checking the condition, performance and alarm systems of the dam. In addition, the checking measures taken by the operating personnel during sudden heavy rainfall are to be reviewed. The filling and discharge channels of the basins and associated structures should be inspected when the spring flood has subsided. Correspondingly, the structures and facilities of waste dams, such as the inflow and outflow systems, are inspected once a year. A record is made of the annual inspection and test runs and included in the safety file of the dam owner.

7.4 *Regular inspection*

In Finland a regular inspection is made at intervals not exceeding five years. The date of the first regular inspection is counted from the date of the commissioning inspection. A representative of the dam owner or holder, authority and a competent person participate in the regular inspection. In the regular inspection the agenda is usually as follows:

- the compiled observational data and other results are gone through
- repairs made and the reasons for them are checked
- structures are inspected as considered necessary and the working conditions of the facilities important for dam safety are checked
- trees and other plants are checked for dam safety
- it is verified that the hazard risk assessment is up to date
- it is checked that the assessment of the impact on health and the environment is up-to-date
- the site plan and the associated action plans are inspected and the viability of the arrangements required by them is evaluated and the site plan revised if necessary
- it is established whether changes have taken place in conditions or in the type of impounded substance affecting the dam class
- the dam classification is verified, which may imply that the existing hazard risk assessment has to be supplemented or, if the situation so requires, the authority (regional environment centre) may have to order a hazard risk assessment to be made for the dam
- the dam safety monitoring programme is checked and necessary changes, if any, recorded
- it is verified that the safety file is up-to-date in every other respect and that any amendments needed are recorded
- if necessary, decisions are made on the follow-up measures and investigations.

7.5 *Methods of monitoring*

The surveillance and monitoring of features for evaluating dam safety is done by periodic visual inspections, sample taking and different kinds of measurements. Periodic measurements and data acquisition on site are carried out by operators or inspection members. Non-destructive techniques (NDT) are recommended for existing dams to prevent any harms caused by the measurements which could be the case with the drillings, sample taking, etc.

Monitoring equipment for continuous measuring (which is suited to the importance of the dam) is installed in order to measure the deformations of the dam and its foundation, the seepages, the uplifts, the temperatures, the pore pressures, and, possibly, the phreatic surface in earth fill dams. Automatic acquisition, transmission and processing of significant behavioural data are constructed for generating data alarms, if data exceed allowable limits. Continuously-operating monitoring methods need also periodic calibration and testing of automatic monitoring systems (alarm tests). In addition the periodic test of appurtenant devices (operational equipment like spillway and outlet gates) must be done at least once a year.

Satellite surveillance can be used as a monitoring method for tailings facilities. Satellite images provide an effective method to check such features as pool size and position, wetness of slopes, etc. Evaluation of tailings dams by these images should be made on a regular basis (ICOLD, 1996b).

7.6 *Requirements for inspectors*

In addition to dam owners the personnel participating in the commissioning and regular inspections consist of the head designer and other competent persons. It is also recommended that the dam designer or a person with the corresponding expertise should participate in the annual inspections (at least for the dangerous "P" dams). Personnel undertaking the monitoring proper should be trained at the dam, appropriate attention being paid to the special features of each dam. The persons undertaking monitoring and surveillance should be instructed adequately so that they are aware of the likely dam damage, hazards and their manifestations. They should also be made aware of the measures they have to take if factors endangering the safety of the dam are noted. The dam owner should train his operating and maintenance staff to watch the changes that occur in structures in dam areas and dams and make sure that they know whom they should inform about the changes observed. The member of the dam owner's staff responsible for dam safety decides whether or not the changes observed are significant for dam safety. The surveyor who undertakes measurements should be well informed about the limits of the normal values and if these limits are exceeded (either upwards of downwards) he should immediately inform the person responsible for the dam. A record should be kept of the monitoring and observations (MMM, 1997).

With regard to water storage dams, in the United Kingdom there is a qualified supervising engineer, who is appointed by the dam operator to check the safety level and inform the owner of dangerous events. To this end, one or several inspections are made every year. At least every ten years, a qualified inspecting engineer makes an expert evaluation. The creation of a panel of qualified engineers is required by law in the United Kingdom. The panel is open to any engineer with the requisite qualifications, which are set by the regulatory authority. The inspections of the dams are conducted then by the members of this panel (Bradlow et al, 2002).

8 CONCLUDING REMARKS

Bradlow et al (2002) made a comparative study about the regulatory frameworks for dam safety. World Bank's Operational Policy goes beyond the water storage dams and extends to tailings, slimes and ash impoundment dams. In addition to regulated improvement of dam safety the voluntary initiatives by industry and the dam owners must be taken into account. International standards ISO 9001 and ISO 14001 have already been in use in the mining industry. It appears that in practice dam safety legislation and regulations are applied only in very few countries to both water retention and tailings dams. This unfortunately means that laws do not include clear requirements for safety procedures, inspections etc. for tailings dams.

There are many kinds of authorities for supervision of tailings facilities in different counties. The responsibilities of the authorities can be focused or shared and the jurisdiction itself can be centralized or local. However, the main point is that all authorization procedures are well planned, checked and use the best available expertise concerning tailings dams.

Experience has shown that most critical parameters and factors in tailings dam safety management are involved with inspections, monitoring and surveying the behaviour of tailings dam structures, water/liquid levels and transportation of wastes. The proper planning and design parameters will have little or no meaning at all if the tailings dam is not properly managed, operated, inspected and monitored. It can be stated that most of the recent tailings dam failures could have been prevented had their monitoring been done properly.

Most critical parameters in the monitoring of tailings dams are the level of phreatic surface, water/liquid level in the basin and the amount of the seepage. Dam structures, like pipelines and waste transport systems and materials used in dams (of which the changes in quality are critical for safety issues) can be added to the same list. Extreme weather conditions and changes of seasons must be taken into account by considering extra monitoring of the dams.

Re-evaluating the dam classification periodically during long-term inspections has already be done. But evaluating the initial situation again and checking if the original design parameters are still valid are only under development. The discussion among dam safety experts has aroused concerning the time schedules of 10–15 years to make another "commission inspection" in order to check that the dam is still safe enough in the current circumstances.

REFERENCES

BETCGB (Bureau d'Etude Technique et la Contrôle des Grands Barrages) (2001) *European Club – Dam Legislation,* February, 80p.

Bradlow D D, Palmieri A and Salman S M A (2002) *Regulatory Frameworks for Dam Safety. A Comparative Study,* The World Bank, Washington DC, 159p.

BRGM (2001) *Management of mining, quarrying and ore-processing waste in the European Union,* BRGM/RP-50319-FR. Study made for DG Environment of the European Commission TENDER DC XI E3/ETU/980116, 83p.

Bureau of Mining Regulation and Reclamation (1994) *Guidance Document for Preparation of Operating Plans for Mining Facilities,* 3p.

Canadian Dam Association (1997) *Dam Safety Guidelines.*

Environment Protection Agency (2001) *Integrated Pollution Control Licence,* Licence register Number 516, Tara Mines Limited, 41p.

Glos G H (1999) *Using administrative controls to reduce tailings dam risk,* Mining Engineering, September, pp31–33.

Golder Associates Ltd (2001) *Report to United Nations Environment Programme Giving Guidance for Developing Tailings Regulation,* Report no. 01640072, 9p.

Hamor T (2002) *Legislation on mining waste management in Central and Eastern European Candidate Countries,* Joint Research Centre of the European Commission. Ispra, EUR 20545 EN, 196p.

Hurndall B (1998) *Dam Safety Guidelines,* Proc Workshop on Risk Management and Contingency Planning in the Management of Mine Tailings, Buenos Aires, Argentina, ICME/UNEP, 276p.

ICOLD (1989) *Tailings Dam Safety – Guidelines,* Bulletin 74, Paris.

ICOLD (1994) *Tailings Dams Design of Drainage – Review and Recommendations,* Bulletin 97, Paris.

ICOLD (1995a) *Tailings Dams and Seismicity – Review and Recommendations,* Bulletin 98, Paris.

ICOLD (1995b) *Tailings Dams. Transport Placement and Decantation – Review and Recommendations,* Bulletin 101, Paris.

ICOLD (1996a) *A Guide to Tailings Dams and Impoundments,* Bulletin 106, Paris.

ICOLD (1996b) *Tailings Dams and Environment – Review and Recommendations,* Bulletin 103, Paris.

ICOLD (1996c) Monitoring of Tailings Dams – *Review and Recommendations,* Bulletin 104.

MAC (Mining Association of Canada) (1998) A *Guide to the Management of Tailings Facilities,* Ottawa, Ontario.

Martin T E, Davies M P, Rice S, Higgs T and Lighthall P C (2002) *Stewardship of Tailings Faclities, Mining, Minerals and Sustainable Development,* report no 20, April International Institute for Environment and Development, 35p.

Mine Metallurgical Managers' Association of South Africa (1995) *The Management of Gold Residue Deposits – A Code of Practice,* March.

MMM (1997) *Dam Safety Code of Practice,* Ministry of Agriculture and Forestry.

MMSD (2002) *Mining for the Future, Appendix A: Large Volume Waste Working Paper,* Mining Minerals and Sustainable Development, report no 31, April, International Institute for Environment and Development, 55p.

MODAM (1999) *Tailings Dam HIF Audit,* Mining Operations Division Audit Management System, The Department of Industry and Resources, Western Australia.

NRE (Department of Natural Resources and Environment) (2002). *Discussion Paper: Tailings Storage – Guidelines for Victoria,* State Government Victoria.

Penman A D M (2001) *Tailings Dams: Risks of Dangerous Occurrence,* J. Köngeter (Ed), Mitteilungen des Lehrstuhls und Instituts für Wasserbau und Wasserwirtschaft der Rheinisch-Westfällischen Technischen Hochschule Aachen. Band 124, pp229–247.

Sol V M, Peters W M and H Aiking (1999) *Toxic waste storage sites in EU countries. A preliminary risk inventory,* IVM Report number: E-99/02. 59p.

UNEP (2001) *APELL for Mining – Guidance for the Mining Industry in Raising Awareness and Preparedness for Emergencies at Local Level,* technical Report No 41. United Nations Environment Programme, Division of technology, Industry and Economics, 67p.

Geotechnical and Environmental Aspects of Waste Disposal Sites – Sarsby & Felton (eds)
© *2007 Taylor & Francis Group, London, ISBN 978-0-415-42595-7*

Shear strength of coalmining wastes used as coarse-grained building soils

A. Gruchot, P. Michalski & E. Zawisza
Agricultural University of Krakow, Poland

ABSTRACT: This paper contains results of tests of shear strength of colliery spoils and natural coarse-grained soil carried out with a medium-sized shear box apparatus. The application of coalmining wastes for engineering purposes requires determination of their geotechnical properties and particularly the shearing strength. The waste material coming from coalmines has coarse-grained soil characteristics and due to this fact all the tests should be carried out using large or medium-sized apparatus. The shear strength tests were carried out on the waste material coming from four coal mines and on natural gravel coming from the bed of a mountain river, using direct shear box apparatus having box dimensions of $300 \times 300 \times 200$ mm. All the tests were carried out at three compaction indices and two shearing speeds to determine the influence of these factors on the test results. Results obtained revealed that shear strength of colliery spoils is very high – the values of the angle of internal friction were of 30–40° with cohesion of 30–50 kPa, according to compaction state. These values were very close to the values obtained for the natural soil having similar grading. Generally, it can be concluded that colliery spoils are as useful for earth structures as natural gravel soils having similar grading.

1 INTRODUCTION

Every year about 40 million tonnes of coal mining wastes are deposited on dumping grounds in the Upper Silesian Industrial Region. The problem of utilisation of such great amounts of wastes in the highly urbanised and industrialised region requires a prompt solution since the area for depositing of these materials is very limited and because of the need for environmental protection. The main possible use of these waste materials is in different kinds of earth structures. Coalmining wastes have been used for; construction of road and railway embankments, formation of various kinds of hydraulic structure, filling sinkholes caused by mining subsidence (Zadroga and Olańczuk-Neyman, 2001; Skarżyńska, 1995, 1997).

The usage of coalmining wastes for Civil Engineering purposes requires determination of their geotechnical properties, particularly mechanical ones. The most important parameter is shear strength since it has a basic influence on stability of embankment slopes and bearing capacity of subsoil. The geotechnical properties of co mining wastes are variable depending on the source they are coming from. Hence, in every case the basic geotechnical parameters should be determined. Colliery spoils can be treated as coarse-grained anthropogenic (man-made) building soils having partial properties of cohesive soils (Skarżyńska, 1997). Therefore, laboratory tests should be carried out using large or medium-sized apparatus since the results obtained from standard ones are incorrect, because the material tested does not contain coarse fractions.

2 NATURE AND ORIGINS OF THE MATERIAL TESTED

Coalmining wastes which are the subject of the reported investigations were mixtures of rocks accompanying coal seams and which are separated from pure coal in washery plants. The basic rocks forming the material are claystone, siltstone, sandstone and carbon shales in various proportions. The subject of the tests were colliery spoils coming from three coal mines (Makoszowy, Sośnica and Anna) and from HALDEX-Plant Michał which deals with the recovery of pure coal from coalmining wastes. For comparison purposes tests were carried out also on natural coarse-grained soil (coming from the bed of a mountain river – the Skawa) having similar grading to colliery spoil.

3 PETROGRAPHIC AND GEOTECHNICAL CHARACTERISTICS

Petrographic composition of the materials tested is presented in Table 1. It is seen that claystones predominate in all the colliery spoils whereas sandstone is the main component of the natural soil. Siltstone content is significant only in wastes from HALDEX-Plant and in the natural soil. The sandstone and carbon shale contents are negligibly small in colliery spoils.

The basic physical properties are given in Table 2. The grain size distribution obtained from combined wet sieving and hydrometer analyses shows that in general, grading of all the materials tested is finer than 40 mm with a predominance of gravel fraction (60–84%) and small clay content (1.5–3.1%). Grading of the natural soil can be assumed to be similar to that of colliery spoil. The much greater sandstone content in petrographic composition of natural soil results in greater values of specific gravity and maximum dry density when compared with colliery spoils, whereas optimum moisture contents in all the materials were almost identical.

Table 1. Petrographic composition of the materials tested.

		Colliery spoils from				
		Makoszowy mine	Sośnica mine	Anna mine	HALDEX-Plant Michał	Natural soil
No.	Kind of rocks	Content [%]				
1	Claystones	88	93	81	92	–
2	Siltstones	1	–	14	2	11
3	Sandstones	1	2	2	5	89
4	Carbon shales	–	–	2	1	–
5	Coal	10	5	1	–	–

Table 2. Basic physical properties of materials tested.

		Colliery spoils from				
No.	Property	Makoszowy mine	Sośnica mine	Anna mine	HALDEX-Plant Michał	Natural soil
1	Fraction content [%]:					
	– cobbles >40 mm	5.7	2	–	–	–
	– gravel 40–2 mm	66.8	82.5	81	84	59.5
	– sand 2–0.05 mm	18.4	11.0	13	9	34
	– silt 0.05–0.002 mm	6.0	2.8	4.5	4	4
	– clay <0.002 mm	3.1	1.7	1.5	3	2.5
2	Uniformity Coefficient [-]	158	18.2	72.7	65.2	150.9
3	Specific gravity [$g \cdot cm^{-3}$]	2.15	2.29	2.45	2.44	2.68
4	Maximum dry density [$g \cdot cm^{-3}$]	1.77	1.81	1.855	1.91	2.15
5	Optimum moisture content [%]	7.8	8.1	8.5	7.7	7.9
6	Classification according to PN-B-02480:1986	Gravel	Loamy gravel	Gravel	Loamy gravel	Loamy gravel

4 AIMS AND TEST PROCEDURE

The aim of the investigations was to determine the influence of compaction and shearing speed in a direct shear box apparatus on the values of shear strength obtained from the tests. The additional aim was an assessment of the usability of colliery spoils for earth structures and a subsoil from shearing strength point of view, as a substitute for natural soils having similar grading. All the tests were carried out using a medium-sized direct shear box apparatus having dimensions of 300 × 300 × 200 mm. The boxes were provided with intermediate frames placed between them to form a shearing zone of the thickness of 30 mm to reduce the apparent cohesion resulting from interlock of particles with one another.

All the shearing tests were carried out at moisture contents close to the optimum and at three compaction indices (according to the standard Proctor Test), i.e. 0.90, 0.95 and 1.00. Vertical (normal) stresses applied were in the range 50–500 kPa. In every case five shear tests were made on the separate samples individually prepared after 30 minutes of consolidation under the chosen vertical stress. A shearing speed of 1.0 mm/min was applied for all the materials and a speed of 0.1 mm/min was used for the colliery spoil from Anna mine, HALDEX-Plant and the natural soil.

5 TEST RESULTS AND DISCUSSION

5.1 Angle of internal friction

Table 3 contains of the angle of internal friction ϕ and the influence of compaction on the angle of internal friction is illustrated in graphs given in Figure 1.

As expected, increasing the degree of compaction increases the angle of internal friction (ϕ). The increase of I_S from 0.90 to 1.00 resulted in greater values of ϕ, for the colliery spoils, by an average of 15% for both shearing speeds. In the loamy gravel this relation was the same. The angle of internal friction was greater in loamy gravel than in colliery spoils due to the different petrographic composition – the main rock forming the loamy gravel was sandstone having more sharp-edged and rough grains when compared with clay grains forming the colliery spoils. Waste materials from Anna mine revealed the greatest values of ϕ among colliery spoils because of the significant content of siltstone and the minimal clay fraction content, as well as very high resistance to weathering and mechanical disintegration (Krzyk, 2003).

Generally, the angle of internal friction obtained for the colliery spoils ranged in all cases between 30° and 40°. Maximum differences between ϕ values of the colliery spoils obtained at a shearing speed of 1.0 mm/min were at the compaction index of 0.90 – the difference was 7°. This difference diminishes with the increase of compaction achieving and achieves at $I_s = 1.00$ a value as low value as 3°.

Table 3. Values of angle of internal friction.

| Shearing speed v_s [mm/min] | Compaction index I_s [-] | Angle of internal friction, ϕ [°] | | | | |
| | | Colliery spoils from | | | | |
		Makoszowy mine	Sośnica mine	Anna mine	HALDEX-Plant Michał	Loamy gravel
0.1	0.9	–	–	33.4	30.3	39.5
	0.95	–	–	36.1	31.9	41.3
	1.0	–	–	39.6	34.0	45.5
1.0	0.9	32.8	32.6	37.4	30.4	37.2
	0.95	34.4	–	39.3	32.1	39.6
	1.0	38.7	38.0	39.7	36.6	42.9

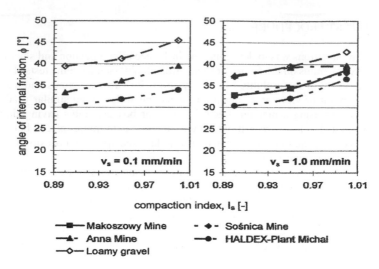

Figure 1.   Influence of compaction on friction angle (two shearing speeds).

Table 4.   Values of cohesion.

Shearing speed v_s [mm/min]	Compaction index I_s [-]	Cohesion, c [kPa]				
		Colliery spoils from				
		Makoszowy mine	Sośnica mine	Anna mine	HALDEX-Plant Michał	Loamy gravel
0.1	0.9	–	–	32.3	35.9	0
	0.95	–	–	34.9	46.4	15.4
	1.0	–	–	38.8	50.9	20.4
1.0	0.9	30.1	18.5	30.4	34.9	10.1
	0.95	43.4	–	32.1	40.3	27.2
	1.0	49.6	46.0	36.6	41.9	44.5

The influence of the shearing speed on the φ values obtained from the tests is visible but very small. Values of φ in colliery spoils obtained at the greater shearing speed (1.0 mm/min) are not more than 1–2° (4%) greater than those obtained at the lower shearing speed. Therefore, φ values obtained for the loamy gravel are smaller by 2–3°.

5.2   Cohesion

Most shear strength tests carried out on coarse-grained soils using both shear box and triaxial apparatuses give values of cohesion. This creates problems with interpretation of test results because, according to some authors (e.g. Wiłun, 2000) coarse-grained soils should not have cohesion as cohesion is a typical property of fine-grained soils. According to the authors of this paper values of the cohesion obtained from the test include two parts, namely; real cohesion of fine fractions (clay + silt fraction content ranges from 4.5% to 9.1%) filling up the soil skeleton formed by coarser fractions, and 'apparent cohesion' of the coarser fractions as a result of interlocking of grains. At this stage it is impossible to determine the proportions of the particular parts within the total value of cohesion obtained from the tests. The total value of cohesion influences the slope stability or bearing capacity, therefore from the practical engineering point of view, this theoretical problem is of no importance.

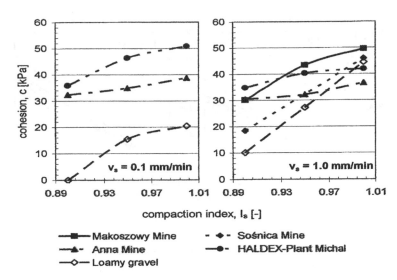

Figure 2.   Influence of compaction on cohesion (at two shearing speeds).

Relationships between values of cohesion and compaction at two shearing speeds are shown in Figure 2. It is evident that greater compaction causes higher values of cohesion in all cases. An average value of cohesion determined for the materials from Anna-mine and HALDEX-Plant was greater at the maximum compaction by 31.5% for the shearing speed of 0.1 mm/min and by 20% obtained at the shearing speed of 1.0 mm/min. The average increase of cohesion of all the colliery spoils determined at the shearing speed of 1.0 mm/min was 53% and this was much greater than the increase of ϕ values caused by the increase of compaction from $I_s = 0.90$ to $I_s = 1.00$. The increase of cohesion determined at the shearing speed of 1.00 mm/min for the natural loamy gravel is much greater (3.4 times) when compared with colliery spoils.

Generally, the average value of cohesion determined for all the colliery spoils obtained at the shearing speed of 1.0 mm/min ranged between 28.5 kPa ($I_s = 0.90$) and 43.5 kPa ($I_s = 1.00$) and in the natural loamy gravel at maximum compaction cohesion was almost the same (44.5 kPa). The results obtained at the lower shearing speed gave higher values of cohesion (by 4 to 12%) in colliery spoils (compared to that obtained at greater shearing speed). In contrast, the natural loamy gravel had much lower values (by a factor of 2). The influence of compaction on the values of cohesion in all the materials tested was greater than the influence on the angle of internal friction. The scatter of results obtained for the cohesion is much greater than for ϕ.

5.3   Shear strength

The total shearing strength of soils depends both on the angle of internal friction and cohesion. Total values of shear strength (τ_f) as a function of vertical stress (σ_v) have been calculated and relevant relationships are presented in Figure 3 for two compaction indices ($I_s = 0.90$ and $I_s = 1.00$).

The presented graphs show that, for low compaction conditions, differences between values of total shearing strength of particular materials are slightly greater than at maximum compaction and they increase with vertical stress (σ_v). However, the fully compacted colliery spoils have almost the same total shearing strength (the maximum differences between them are less than 20 kPa) and this value is slightly smaller than that for the natural loamy gravel. Generally it can be assumed that the total shearing strength of colliery spoils is very close to the shearing strength of natural soil having similar grading.

Shear strength parameters were also obtained from the tests carried out on colliery spoils using triaxial apparatus. The ϕ values obtained from the shear box and triaxial apparatuses were very

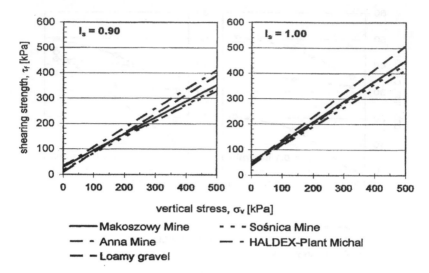

Figure 3.   Total shearing strength of materials for different compaction indices.

similar. However, values of cohesion determined in triaxial apparatus were two times greater (80 to 110 kPa) than those obtained from the shear box apparatus (Gruchot, 2003). So, the values of shearing strength parameters obtained from the tests carried out using shear box apparatus can be assumed to be on the safe side for designing Civil Engineering works.

6   CONCLUSIONS

Specific test results have been obtained:

- Increase of compaction state caused an increase of shear strength parameters for all of the material tested.
- Influence of shearing speed on the shear strength parameters obtained was very small, therefore the higher shearing speed (of 1.0 mm/min) is recommended in all tests for Civil Engineering purposes.
- Comparing results for the direct shear box and the triaxial apparatus; the values of angle of internal friction are almost the same, the triaxial apparatus gives much higher cohesion. Therefore, geotechnical design based on parameters obtained from the shear box apparatus can be assumed to be conservative and safe.

Colliery spoils from a variety of sources have been shown to have relatively high shear strength similar to natural soils having similar grading. Due to this fact they can be used as a good substitute for natural building soils for all types of earthworks.

REFERENCES

Gruchot A T (2003) *Badania parametrów wytrzymałościowych odpadów poęglowych – wpływ metody badan (Shearing strength tests of colliery spoils – influence of test methods)*, PhD thesis, Krakow.
Krzyk P (2003) *Wpływ czynników dezintegrujących na właściwości geotechniczne odpadów górnictwa węgla kamiennego (Influence of disintegrating factors on geotechnical properties of coal mining wastes)*, PhD thesis, Krakow.

Michalski P and Skarżyńska K M (2000) *Zasady wykorzystania odpadów pogórniczych do wznoszenia nasypów hydrotechnicznych (Rules of using of colliery spoils for hydraulic embankments erection)*, Proc. Symp. "Hydrotechnika III", Agricultural University, Krakow, pp 243–257.

Skarżyńska K M (1995) *Reuse of coal mining wastes in Civil Engineering. Part 1. Properties of minestone*, Waste management, 15.1, pp 3–42.

Skarżyńska K M (1995) *Reuse of coal mining wastes in Civil Engineering. Part 2. Applications of minestone*, Waste Management, 15.2, pp 83–126.

Wiłun Z (2000) *Zarys geotechniki*, Publ. Komunikacji i Łączności, Warsaw.

Zadroga B and Olańczuk-Nejman K (2001) *Ochrona i rekultywacja podłoża gruntowego. Aspekty geotechniczno-budowlane (Protection and reclamation of subsoil. Geotechnical and building aspects)*, Publ. Gdansk Technical University.

Zawisza E (2001) *Geotechniczne i środowiskowe aspekty uszczelniania grubookruchowych odpadów powęglowych popiołami lotnymi (Geotechnical and environmental aspects of sealing coarse-grained colliery spoils with fuel ashes)*, Zeszyty Naukowe of Agricultural University, Krakow, 280.

Geotechnical and Environmental Aspects of Waste Disposal Sites – Sarsby & Felton (eds)
© 2007 Taylor & Francis Group, London, ISBN 978-0-415-42595-7

Novel solutions in tailings management

T. Meggyes
Federal Institute for Materials Research and Testing (BAM), Berlin, Germany

A. Debreczeni
University of Miskolc, Department of Mining and Geotechnics, Miskolc, Hungary

ABSTRACT: Tailings are fine-grained residue of the milling process in which the desired raw materials are extracted from the mined rock and appear, due to mixing with water during this process, as slurries. The deposits of these residues in ponds or lagoons, usually confined by man-made dams, can present a serious threat, especially where there is improper handling and management. The existing technologies which use the conventional upstream, centreline and downstream methods have left the legacy of many unstable tailings facilities with potentially liquefiable zones and steep slopes which are prone to erosion. The technology of producing high-density, very low-moisture thickened tailings, i.e. Paste Technology (PT) has made extremely rapid progress from 1995 onwards and offers significant economic incentives and environmental benefits. No particle segregation takes place during the thickening process, the paste material exhibits a much greater stability than conventional tailings, there is no pond on top of the deposit, the paste forms a gently sloping surface after placement which promotes the runoff of rain water and the overall costs are lower than for conventional methods.

1 INTRODUCTION

The basic differences between conventional tailings disposal and PT are:

- In the conventional tailings disposal system the tailings properties are fixed by the processing plant and all confining dykes and 'control structures' have to be engineered to withstand the stresses imposed on them by nature.
- In the PT system the tailings properties are 'engineered' to suit the unstructured natural topography of the disposal area – a much safer and friendly approach to the environment.

By thickening the tailings to a heavy slurry prior to disposal it is possible to create a self-supporting deposit of tailings and to eliminate the typical superimposed settling pond (Robinsky, 1999). The reason why the failure of a conventional tailings disposal dam is so environmentally disastrous is not the dam itself, but the fact that the dam retains a mass of very loose unconsolidated tailings and a great deal of process water. If the dam fails, the contents liquefy completely as they flow through the breach. In this liquid state, they can flow many kilometres downstream, as it happened in the Stava, Italy catastrophe in 1985, where 250,000 m³ of liquefied tailings rushed down the Stava Valley following a dam breach, achieving a peak velocity of 90 km/h and burying two villages and claiming 268 lives, making it the worst tailings facility disaster ever to have occurred in Europe (Chandler and Tosatti, 1995).

2 TECHNOLOGY BASICS

The reason the material is so loose in the first place is that it has been discharged into a pond filled with water. The tailings particles drift down through the water, their weight reduced by 50% due to buoyancy, to form a loose structure, like a 'house of cards' or lake-bottom mud. It does not take much of a shock to collapse the material into a liquefied state.

2.1 Paste technology

The aim of PT is to create a self-supporting ridge or hill of tailings and thus to minimize the requirement for confining dams. To achieve this the tailings must be strengthened which can be done by the removal of most of the process water. This is attained by passing the tailings that include process water through high density thickeners (large circular tanks) wherein the tailings particles settle to the bottom and are extracted as underflow for release into the disposal area. Most of the process water is taken off the top and recycled back into the plant. The thickening process must be sufficient to change the tailings and process water from a mixture to a non-segregating, but still pumpable slurry. For some tailings, it may be necessary to filter part of the underflow and recombine the filtered solids with the remaining underflow to attain the desired consistency.

When the tailings are released, due to their heavy consistency and thus high viscosity, they will flow great distances without segregation. Eventually the flow stops at a gentle slope. The slope is controlled by the degree of thickening. The aim is to attain a slope of 2% to 6% (1.1° to 3.4°) in moderate climates. Such slopes are sufficiently gentle to avoid erosion, yet provide good drainage for vegetation. In very dry conditions even steeper slopes may be contemplated. The non-segregating property of thickened tailings is also responsible for bonding the tailings particles and thus reducing both erosion potential and dusting.

2.2 Tailings disposal

Tailings disposal areas may consist of valleys or flat terrain somewhere in the vicinity of the processing plant. To form a sloping tailings deposit in a valley, the thickened tailings would be discharged at the head of the valley or along one of the side hills. The heavy slurry will flow down the valley until it encounters a slope flatter than its own, or alternatively, until it is stopped by a small dam. On flat terrain, thickened tailings would be discharged from an artificial ramp or tower, resulting in a ridge or cone of tailings (Figure 1).

Only a low perimeter dyke is required to direct precipitation and a small amount of extruded process water to a pond, ideally located beyond the limits of the tailings deposit, for recycling. Such a pond will receive, due to thickening, only one-third of the amount of process water that flows to conventional ponds.

It is the homogeneity of the slurry (the result of thickening) and the gently sloping surface of the tailings that provide the major advantages of the PT system over conventional flat, wet disposal areas. During the active life of the disposal area, as well as after closure, the surface of the sloped tailings drains rapidly, and allows drying to take place. One of the aims of the system is to provide sufficient surface area during deposition to allow drying of discharged tailings, thus strengthening them considerably.

The elimination of the conventional pond on top of the deposit also provides a major environmental advantage: the hydrostatic head that causes seepage of process and rainwater to occur throughout conventionally deposited tailings is eliminated. Another very important environmental advantage is that no confining dam(s) are needed or, at least, reduced substantially in height.

Finally, the adoption of the system may permit progressive reclamation in some topographical settings – a feature that permits the close-out of one end of the disposal area even as the discharge is moving progressively towards the other end. This advantage may result in a much smaller active disposal area. If a mine is abandoned for any reason, most of its disposal area would have already been reclaimed. Progressive reclamation may reduce the assurance bond demanded in some jurisdictions to guarantee eventual reclamation of the area after closure. A test on 68 tailings and processing plants from all over the word proved that all would satisfy the requirements for PT. The materials tested included tailings from gold, silver, copper, zinc, bauxite, phosphate, tar sand, diamond etc. Conversion from conventional to thickened tailings disposal can be realistically

Figure 1. Paste disposal from an artificial tower – note the gently sloping surface (Newman, 2003; Engels, 2003).

achieved for any existing tailings operation. Discharging a sloping tailings cap on an existing conventional flat tailings disposal pond will facilitate reclamation. The alumina industry fighting the problems of red mud disposal is probably the most forward looking industry by adopting PT to a large number of tailings facilities.

2.3 Direct benefits

The capital cost of thickening for a PT system should be weighed against the cost of constructing and operating a conventional disposal system, including the final reclamation costs (Robinsky, 1999). PT also inhibits acid drainage. The best way of doing this is to keep sulphides in a saturated condition thus preventing them from coming into contact with oxygen. The homogeneous slurry of thickened tailings due to the fines promotes high matric or capillary suction that raises water near the surfaces and keeps the tailings saturated.

2.4 Physical properties

Paste is a high-density mixture of water and fine solid particles. It has a relatively low water content (10–25%) such that the mixture has a consistency as measured by the ASTM slump cone test from slightly greater than zero up to nearly 305 mm (12 in.). Particles of different size classes will not segregate or settle when the paste is not being agitated or when it is stationary in a pipeline, i.e. the paste has no critical velocity. Cement may be a component of the paste. Larger particles of aggregate can be added to a paste without greatly changing the pipeline transport characteristics (Brackebusch, 1994).

The moisture content or density of a paste for a given slump consistency depends on the size distribution of the particles. The finer the particles, the more surface area must be wetted. This yields higher moistures and lower densities for a given consistency. With larger particles, the surface area is smaller and this results in lower moistures and higher density.

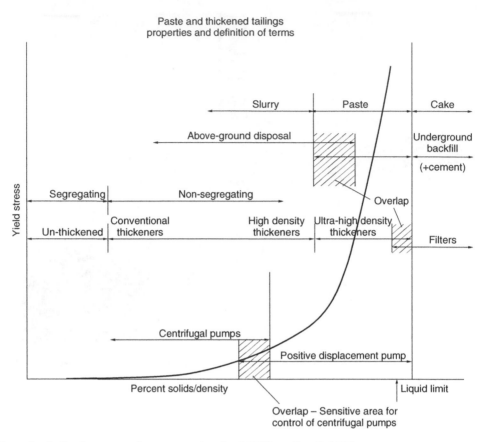

Figure 2.   Indicative ranges of paste properties after P Williams (Jewell, 2002).

Having a sufficient amount of fine particles is the most important requirement to produce a paste. In most cases, pastes must contain at least 15% by weight of particles less than 20 μm in diameter (625 mesh) (Tenbergen, 2000). Mineralogy and particle shape affect the amount of fine particles necessary (Slottee and Schreiber, 1999).

As experience has shown that it is difficult to scale paste flow characteristics from small-scale tests to full-scale pipeline conditions, pilot-scale pumping tests are usually necessary. PLC-control is essential because only slight changes in moisture content cause wide variations in viscosity and pipe friction.

Figure 2 shows a classification system for solid-liquid mixtures (Jewell, 2002). With increasing solids concentration, slurry is followed by paste and cake identifying various grades of consistence. A yield stress range of 200 ± 25 Pa is proposed as marking the transition between slurry and paste. A transition from slurry to paste may occur as the mixture flows down a slope from the discharge point and releases excess 'bleed' water. This happens with tailings derived from hard rock ores where the clay sized fraction (passing 2 μm) is primarily comprised of granular shaped particles. These tailings would cease flowing once the self-weight forces on the solidifying tailings stop the forward momentum.

3   PREPARATION OF PASTE MIXTURES

Tailings from a milling operation are usually discharged as a dilute slurry. Excess water may be recovered for recycling in the milling operation by a tailings thickener. Therefore, dewatering of the tailings slurry is usually the first step. Fine particles, 'slimes' must not be lost during the dewatering operation. A conventional gravity thickener becomes the equipment of choice for the first stage of dewatering. If more than a sufficient amount of fine particles are present in the tailings stream, part of the stream can be processed with a hydrocyclone and the overflow discarded, thus removing some of the water and increasing the thickening and filtration rates. This is called partial classification (Brackebusch, 1994).

3.1 Key factors

Key factors affecting thickening and clarification of a feed stream (Bedell et al, 2002) are:

- Solid-liquid weight ratio in feed
- Size and shape of the particulate solids
- Specific gravity differences between solids and liquid
- Presence of flocculant
- Viscosity
- Temperature of the feed stream
- Method of flocculant application
- Particle wetting characteristics
- Feed method/arrangement
- Rake time and speed
- Wind and evaporation.

3.2 Thickening components

The most frequently used reagents in the thickening process are (Bedell et al, 2002):

- Flocculants: high molecular weight synthetic or natural polymers that aid in enhancing the settling rates of most suspended solids. They are generally polyacrilamide-based compounds manufactured in anionic, non-ionic and cationic forms.
- Coagulants: natural minerals such as alum, lime, ferric salts etc. which are effective for colloid suspensions. They are less effective than flocculants.

Flocculant selection aspects (Bedell et al, 2002):

- Design of the actual dewatering equipment to be used
- Minerals present in the slurry: surface chemistry, concentration, particle size
- Chemical make up of the slurry liquid: ionic strength, pH
- Type of floc structure required: weak, strong or small.

Common components of thickeners (Bedell et al, 2002):

- Tank
- Feedwell
- Overflow and underflow withdrawal system
- Rake
- Drive
- Support structure.

3.3 Tailings dewatering

Cyclones cannot generally be used solely as the first stage of dewatering because slimes are lost in the overflow. However, cyclone overflow can be dewatered in a conventional thickener and remixed with dewatered cyclone underflow or an alluvial material to form a paste. For a common milling operation with quartz, carbonates and feldspars as predominant minerals, the underflow from the thickener should be 65 to 70% by weight.

The underflow from a conventional thickener should be a stable slurry. A stable slurry does not exhibit segregation of particle sizes or rapid settling of large particles. A stable slurry can be easily pumped with centrifugal pumps and pipeline velocity is not so critical as with dilute slurries. Many different types of dewatering filters can be used including disk and drum vacuum filters, horizontal belt vacuum filters, belt filter presses and hyperbaric disc filters. The product of the final dewatering step is a moist filter cake that can be handled by belt conveyors. Some storage of filter cake is usually necessary to level out surges in the paste preparation process. The filtration step can be avoided in the preparation of paste by mixing the thickener underflow slurry directly with a suitable dry alluvial material to produce a paste mixture.

3.4 High-density thickeners

Using high-density thickeners to produce a high slump paste directly from a dilute tailings slurry to prepare pastes is becoming the preferred method. High-density thickeners have a deeper sludge bed than conventional thickeners and the rake is submerged in the sludge bed. Consequently, rake torque requirements are high. The aluminium industry has pioneered special high density thickeners for processing red mud so that it can be deposited in a stack or mound. The Baker Process Deep Cone Paste Thickener System is designed specifically to produce pastes with relatively high viscosity and a yield stress. The thickener maintains a deep blanket of settled solids, maximizing gravity compression, and achieving discharge solids that can approach the limits of flowability. The solids content of the pastes will be greater than for any traditional thickener operation. The target is to control the rheological properties short of consolidation where fluid flow stops. The control techniques are encompassed in the geometry of the rake and underflow removal methods. The geometry of the thickener tank is designed to handle high viscosities and yield stresses. Effective flocculation is important and this is accomplished using a specially designed feed dilution and flocculant mixing system as part of the feedwell (Slottee and Schreiber, 1999).

The components of a paste mixture, including filter cake, cement, aggregate, and water, must be mixed thoroughly to produce a homogeneous paste for pipeline transportation. A paste mixing plant is similar to a concrete batch plant. Components must be accurately weighed and rapidly supplied to the mixing process. A batch mixing process is easier to control and is usually preferable to a continuous mixing process. High-intensity mixers developed by the concrete industry are suitable for mixing tailings paste.

The historic development of thickeners is schematically illustrated in Figure 3 (Bedell et al, 2002; Slottee, 2003). The high rate thickeners were developed by the two pioneers EIMCO Process Equipment Company (Baker Process) and Enviroclear in the 1980's. Understanding the 'flux and concentration' relationship for flocculants resulted in a special deep feedwell and special dilution features in thickener design. Key to the process is the compressive yield stress and permeability of the thickened material (Scales and Usher, 2003).

The routine dewatering for most tailings is a combination of thickening followed by vacuum filtration. Thickening efficiencies have improved over the last decades as a variety of flocculants and reliable dry flocculent mixing and addition systems became available. Thickener capacities have improved from the routine historical rate of 0.45 t/m^2/h to some conditions that allow 2.7 t/m^2/h to be used.

The ultra high-rate and ultrahigh-density thickeners (Figure 4) evolved in the 1990's. They encompass a deep thickener tank (Green, 1995) and use specially designed dewatering centre cones. This feature plus the cylinders ringing the outside of the feedwell produces a good clear overflow. The rapid removal of water from the feed collapses the hindered settling zone of the thickener taking solids from the free settling zone into the compaction zone. The deep tank plus the 60° cone provides a deep compression zone that results in high-density underflow solids (Figure 8). In a case of application two 12 m diameter E-CAT™Clarifier Thickeners replaced two 100 m diameter conventional thickeners while producing a consistent underflow of >60% solids and overflow clarity of less than 50 ppm. These thickeners require a high degree of instrumentation and automatic control plus variable speed controlled pumps (Bedell et al, 2002).

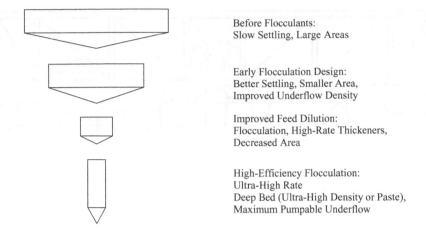

Before Flocculants:
Slow Settling, Large Areas

Early Flocculation Design:
Better Settling, Smaller Area,
Improved Underflow Density

Improved Feed Dilution:
Flocculation, High-Rate Thickeners,
Decreased Area

High-Efficiency Flocculation:
Ultra-High Rate
Deep Bed (Ultra-High Density or Paste),
Maximum Pumpable Underflow

Figure 3.   Thickener evolution (Bedell et al, 2002).

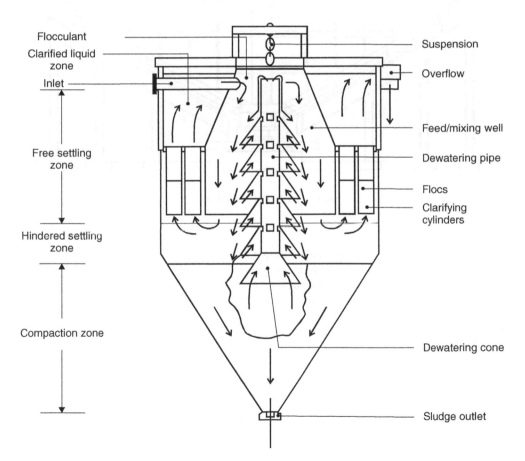

Figure 4.   E-Cat ultra-high rate 'rakeless' thickener, EIMCO (Bedell et al, 2002).

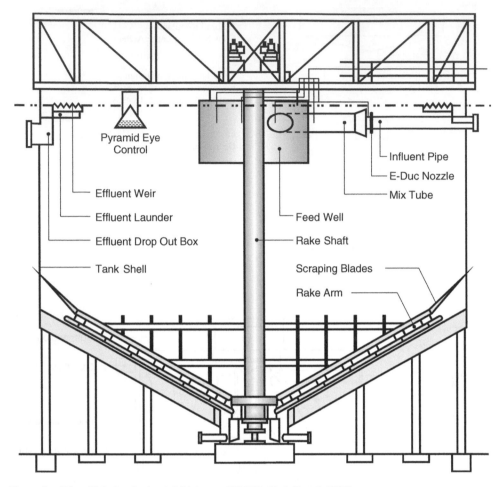

Figure 5.   Ultra-high density 'paste' thickener, EIMCO (Bedell et al, 2002).

A large variety of flocculants are now available. Most suppliers optimise their reagents as to ionic specificity and the molecular weight of the polymer. Reagent screening is followed by graduate cylinder testing of the preferred polymer to allow settling rates to be determined at several addition levels, as well as determine the terminal density of the particular tailings. Once the polymer has been mixed with the slurry feed stream at the right density in the feed well, gravity has to do the rest. Sometimes extreme dilution may be necessary before effective flocculation and settling occurs. Ultra High Density "Paste" Thickeners produce underflows of paste consistency by (Figures 5 and 6):

- Maximising flocculant flux efficiency
- Using feed dilution systems (E-Duc® System or AUTODIL®)
- Using a deep tank for compression
- Using a 30–45° cone
- Using a specially designed raking system
- Using shear thinning principles
- Using a high degree of instrumentation.

Figure 6. Typical underflow from ultra-high density paste thickener (Bedell et al, 2002).

Table 1. Paste suitability tests at pilot plant.

Thickening at plant site
Filtration at plant site
Slump vs. mixing power requirement
Full scale pumping loop test

Table 2. Paste suitability tests in laboratory.

Grain size distribution by laser method	Cycloning test
Mineralogy and density	Paste mixing – visual observation
Grain shapes	Pipe column flow test
Compaction curves and optimum density	Yield stress
Liquid and paste limits	Shear strength
Porosity	Process water quality
Permeability	Viscosity
Abrasiveness	Liquefaction
Bin flowability	Consolidation & desiccation
Slump vs. water content	Water retention curve
Thickening test	Hydraulic conductivity
Filtration test	

4 TESTING

4.1 *Material properties*

Numerous tests are needed to determine suitability for paste transport (Brackebusch, 1994; Zou, 1997; Ouellet et al, 1998; Fourie et al, 2002) as outlined in Tables 1 and 2.

277

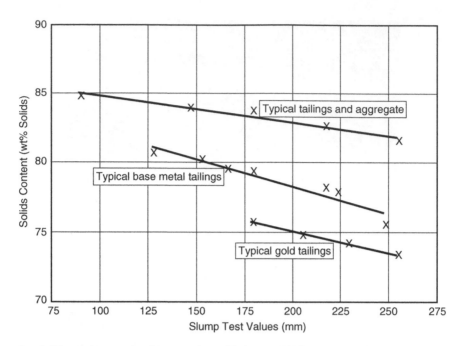

Figure 7. Solids contents as related to paste slump (Tenbergen, 2000).

The rheology of a paste is controlled by a number of different variables such as particle size distribution, particle surface chemistry, liquid viscosity, flocculant quantity and characteristics, concentration of particles and the amount of energy put into the mixture. A common means to measure rheological properties of a paste is the ASTM slump cone test, which uses a truncated cone 305 mm (12 inches) in height. The cone is filled with paste and then removed allowing the contents to assume a pile shape with a natural slope. The distance of the top of the pile from 305 mm is the slump. A slump measurement reflects the yield stress of a paste. Dense paste would have low slump of 50–200 mm (2–8 inches). Materials at the transition between a slurry and paste would have high slump, up to a maximum of 305 mm (12 inches). The slump value determines the suitability of the paste for disposal – for most applications a slump of 175–250 mm is ideal (Slottee; Schreiber, 1999 and Fourie et al, 2003; Newman, 2003).

4.2 *Paste plant design*

The design procedure for a paste plant was summed up by Tenbergen (2000) as:

- Laboratory rheological testing of tailings to review the suitability of the tailings as a paste for underground or surface disposal
- Uniaxial cylinder strength testing to define the paste recipe
- Pump loop testing to assess paste pipeline friction losses for design purposes
- Paste plant flowsheet development. At this point major design criteria will be assessed and agreed upon, such as:
  - The tailings dewatering method
  - Requirements for binder and aggregate addition
  - Requirements for pumping for surface or underground disposal
  - Batch mixing or continuous mixing of the paste recipe components.

Figure 7 shows typical solids by weight versus slump relationships for relatively coarse gold tailings, base metal tailings and blended gold tailings with aggregate (Tenbergen, 2000).

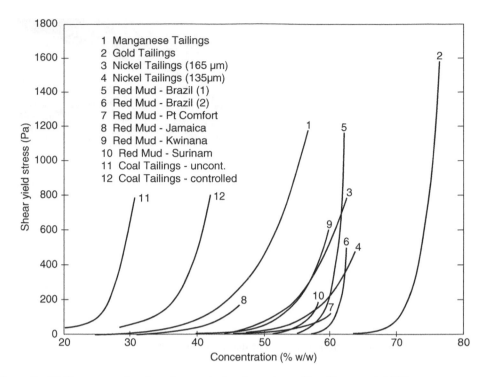

Figure 8.   Yield stress as a function of concentration for various tailing (Boger et al, 2002).

5   FLUID MECHANICS OF PASTE MIXTURES

5.1   *Paste characteristics*

Paste mixtures are non-Newtonian fluids, usually with Bingham properties (Boger, 2003). Many pastes exhibit pseudoplastic features, which means that viscosity decreases with greater pumping rates – a beneficial property for pipeline transportation (Brackebusch, 1994). Yield stress being one of the key transport parameters for pastes is shown as a function of concentration for various tailings in Figure 8 (Boger et al, 2002).

5.2   *Transportation aspects*

Horizontal paste transportation distances may be in excess of 1 km. Pipeline diameters range from 100 to 200 mm and flow velocity is less than 1 m/s while typical dilute slurry velocities are around 3 m/s. This reduces friction losses and energy consumption by nearly one order of magnitude due to a proportionality to the square of velocity. Figures 9 and 6 show the underflow of ultra-high thickened tailings and paste consistencies.

In the design of a paste plant and delivery system it is necessary to assess pipeline friction losses for different slump pastes. Friction losses will decrease as the slump of the paste increases. Friction losses however, can vary widely between pastes from different mines even though the slumps may be similar. Figure 10 shows friction losses for a particular gold tailings paste, at various paste slumps, with and without aggregate addition. Friction losses varied by a factor of five in pump loop tests between pastes varying in slump between 178 and 254 mm. Surface disposal of paste would be possible with a slump of 254 mm (Tenbergen, 2000; Pullum, 2003). The high friction loss experienced with the low slump paste limits the horizontal distance to which paste will flow

Figure 9.   Typical underflow from an ultra-high rate thickener (Bedell et al, 2002).

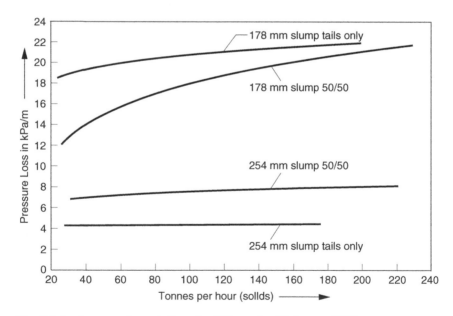

Figure 10.   Friction losses vs. flow rate through a 150 mm pipe (Tenbergen, 2000).

under gravitational forces, even though a lower slump is desirable from the paste strength point of view. The addition of aggregate to the gold tailings lowers the friction losses significantly and allows a wider distribution. As a basic design step friction loss data from pump loop tests are used (Tenbergen, 2000).

5.3 *Tailings pumps*

Both centrifugal and positive displacement pumps can be considered for slurry and paste transport. Slurry handling centrifugal pumps are compact, relatively cheap, robust and versatile machines and are almost universally used in low head slurry transport systems. Lining and special alloys increase wear resistance. To obtain the heads required for long distance pipelines, centrifugal pumps are connected in series. To increase capacity, centrifugal pumps are arranged in a parallel configuration (Paterson et al, 2002; Paterson, 2003). Positive displacement pumps provide high discharge pressures, capacities range from $30 \, m^3/h$ to $800 \, m^3/h$ at pressures of up to 30 MPa. Disadvantage is the high capital cost which is often offset by their low operating cost and high reliability.

6 ADVANTAGES OF PASTE TECHNOLOGY

The main advantages of paste technology are:

- Reduced capital cost (no large dam is required)
- Increased safety
- Water conservation
- Decreased area of tailings disposal site
- Reduced soil and groundwater pollution
- Inhibition of acid mine drainage (AMD)
- Reduced liability for the mine
- Possible co-disposal of other mine waste
- Improved conditions for reclamation
- No segregation of particles
- Easy rainwater drainage due to sloping tailings surface
- Potential use as a foundation material (Crowder et al, 2000; Debreczeni and Debreczeni, 2001).

A major factor contributing to increased safety is that even if the (small) dam required by PT failed, the material behind the dam, being well consolidated and without water can slump but cannot flow. The conventional slimes pond has been eliminated, and without water to wash out the tailings, a dam collapse is a local problem and cannot trigger an ecological disaster.

Environmental benefit of paste technology is that a true 'walk-away' facility can be created. The paste can be mixed with soil, seed and fertiliser and, in the case of the mound disposal method, by the time the mine closes the majority of the structure will already be reclaimed land.

Saving water is another environmental plus point for paste technology. Water is reclaimed at the tailings paste plant and returned to the mine or mill for use. This is in particular attractive in dry areas as compared to slurry technology, when large quantities of water are evaporated by the tailings pond.

Conventional tailings dams are particularly vulnerable in areas of seismic activity. Leakage can also occur under the structure which is why drainage is of utmost importance, particularly as they may impound toxic substances. Surface paste disposal will obviously be much safer as the liquid cannot escape. Tailings paste requires stabilization and issues surrounding seismic stability are still important. The minimal amount of free water in the paste also limits or eliminates the potential for surface/groundwater contamination.

7 ECONOMICS

Dams are notoriously expensive structures. Surface paste technology significantly reduces the cost of embankment construction, engineering, monitoring, closure and reclamation. However, the high technology of the paste plant is reflected in its price of US$3–10M (for daily paste production of

3000 to 10,000 tpd). Newman (2003) reaches the conclusion in his cost analysis for a ten-year, one million tonnes per year mining operation in the Barrick Gold mine in Bulyanhulu, Tanzania that paste technology ensures lower overall mine-life costs (US$15.3 million) compared to conventional tailings technology costs (US$16.2 million). The operating costs are greater than for tailings pond system, reflecting the added processing cost of paste manufacture, the capital costs are similar and reclamation costs are significantly lower.

REFERENCES

Bedell D, Slottee S, Parker K and Henderson L (2002) *Thickening process,* In Jewell, R.J., Fourie, A.B., Lord, E.R. (eds) (2002): Paste and thickened tailings – A guide. The Australian Centre of Geotechnics. The University of Western Australia. Nedlands, Western Australia. 49–79.

Boger D V (2003) *Exploiting the mineralogy of mineral tailings,* 2003 International Seminar on Paste and Thickened Tailings. 14–16 May 2003. Melbourne, Victoria, Australia. Section 6.1.

Boger D, Scales P J and Sofra F (2002) *Rheological concepts,* In Jewell, R.J., Fourie, A.B., Lord, E.R. (eds) (2002): Paste and thickened tailings – A guide. The Australian Centre of Geotechnics. The University of Western Australia. Nedlands, Western Australia. 23–34.

Brackebusch F W (1994) *Basics of paste backfill systems,* Mining Engineering. October, 1175–1178.

Chandler R J and Tosatti G (1995) *The Stava tailings dams failure, Italy, July 1985,* Proceedings of the ICE, Geotechnical Engineering, 113, 67–79.

Crowder J J, Grabinsky M W F, Welch D E and Ollila H S (2000) *Tailings as a construction and foundation material,* Tailings and Mine Waste '00. Balkema. Rotterdam. 131–140.

Debreczeni E and Debreczeni Á (2001) *Environmentally friendly transport and landfilling of dense fly ash/slag slurries,* In Sarsby, R.W., Meggyes, T. (eds): Proceedings Green 3. 3rd International Symposium on Geotechnics Related to the European Environment. June 21–23, 2000. BAM, Berlin. Thomas Telford. 367–374.

Engels J (2003) *Existing tailings placement and lagoon and dam design and formation practices,* In Meggyes, T. (ed): TAILSAFE WP1, Definitions: technologies, parameters. Review Report. www.tailsafe.com

Fourie A B, Davies M P, Fahey M and Lowson R (2002) *Material characterisation,* In Jewell, R.J., Fourie, A.B., Lord, E.R. (eds) (2002): Paste and thickened tailings – A guide. The Australian Centre of Geotechnics. The University of Western Australia. Nedlands, Western Australia. 35–47.

Green D (1995) *High compression thickeners are gaining wider acceptance in minerals processing,* Filtration & Separation. November/December. 947.

ICOLD Bulletin No. 106 (1996) *A Guide to Tailings Dams and Impoundments.*

Jewell R J (2002) *Introduction,* In Jewell, R.J., Fourie, A.B., Lord, E.R. (eds) (2002): Paste and thickened tailings – A guide. The Australian Centre of Geotechnics. The University of Western Australia. Nedlands, Western Australia. 1–8.

Jewell R J (2003) *An introduction to thickened tailings applications,* 2003 International Seminar on Paste and Thickened Tailings. 14–16 May 2003. Melbourne, Victoria, Australia. Section 1. 1–5.

Jewell R J, Fourie A B and Lord E R (eds) (2002) *Paste and thickened tailings – A guide,* The Australian Centre of Geotechnics. The University of Western Australia. Nedlands, Western Australia.

Newman P (2003) *Paste, the answer to dam problems,* Materials World. Jan. 24–26.

Ouellet J, Benzaazoua M and Servant S (1998) *Mechanical, mineralogical and chemical characterization of a paste backfill,* Tailings and Mine Waste '00. Balkema. Rotterdam. 139–145.

Paterson A (2003) *The hydraulic design of paste transport systems,* 2003 International Seminar on Paste and Thickened Tailings. 14–16 May 2003. Melbourne, Victoria, Australia. Section 7. 1–11.

Paterson A, Johnson G and Cooke R (2002) *Transport (pumps and pipelines),* In Jewell, R.J., Fourie, A.B., Lord, E.R. (eds) (2002): Paste and thickened tailings – A guide. The Australian Centre of Geotechnics. The University of Western Australia. Nedlands, Western Australia. 81–94.

Pullum L (2003) *Pipeline performance,* 2003 International Seminar on Paste and Thickened Tailings. 14–16 May 2003. Melbourne, Victoria, Australia. Section 8. 1–13.

Robinsky E I (1999) *Tailings dam failures need not be disasters – The thickened tailings disposal (P&TT) system,* CIM Bulletin. Vol. 92, No 1028. 140–142.

Scales P and Usher S (2003) *The thickening process (compression),* 2003 International Seminar on Paste and Thickened Tailings. 14–16 May 2003. Melbourne, Victoria, Australia. Section 13. 1–5.

Shou G and Martínez M (2003) *Feasibility of using centrifugal pumps to transport thickened tailings,* 2003 International Seminar on Paste and Thickened Tailings. 14–16 May 2003. Melbourne, Victoria, Australia. Section 10. 1–10.

Slottee S (2003) *Recent developments in paste thickeners,* 2003 International Seminar on Paste and Thickened Tailings. 14–16 May 2003. Melbourne, Victoria, Australia. Section 14. 1–5.

Slottee S J and Schreiber H (1999) *Paste backfill using deep paste thickener systems Solid/Liquid Separation including Hydrometallurgy and the Environment,* In Harris, G.B., Omelon S.J. (eds): Proceedings of the 29th Annual Hydrometallurgical Meeting. Hatch, Montreal, Quebec.

Tenbergen R A (2000) *Paste dewatering techniques and paste plant circuit design,* Tailings and Mine Waste 2000. Balkema. Rotterdam. 75–84.

Zou D H S (1997) *An innovative technology for tailings treatment,* Tailings and Mine Waste '97. Balkema. Rotterdam. 633–642.

Geotechnical and Environmental Aspects of Waste Disposal Sites – Sarsby & Felton (eds)
© 2007 Taylor & Francis Group, London, ISBN 978-0-415-42595-7

Usability of sludges from coal mining industry for sealing Civil Engineering structures

P. Michalski, E. Zawisza & E. Kozielska-Sroka
Agricultural University of Kraków, Poland

ABSTRACT: The paper presents results of investigations carried out on post-flotation sludges produced by the Polish coal mining industry and deposited in sedimentation ponds. The material tested came from three plants recovering black coal from coal mining wastes deposited on tips and from the washing plant of one coal mine. The investigations were aimed at determining the usability of these materials for sealing engineering 'structures', e.g. as sealing layers under industrial waste dumping grounds or sedimentation ponds, sealing elements of hydraulic embankments or as admixtures to reduce the permeability of coarse-grained soils to be used for hydraulic embankment construction. Investigations comprised geotechnical properties, both physical and mechanical, i.e. grading, Atterberg limits, compaction parameters, moisture content, permeability, shearing strength and compressibility. The combustible matter content was also determined. Results obtained were compared to relevant specification values. Permeability of mixtures of coarse-grained colliery spoils with sludges was also determined for various proportions of sludge admixtures. The results of the investigations revealed that the sludges from coal mining industry are good material for reducing permeability of coarse-grained soils and for many kinds of sealing element within Civil Engineering structures.

1 INTRODUCTION

Polish coal mining industry concentrated in the Upper Silesian Industrial Region produces large quantities of mining wastes, among which are post-flotation sludges deposited in sedimentation lagoons. The post-flotation sludges have, in a geotechnical sense, the character of fine-grained soils. The sludges can be considered as a useful material for sealing and separation layers beneath sedimentation ponds or dumping grounds for industrial and municipal wastes as well as for sealing material for the reduction of permeability of hydraulic embankments. In the Upper Silesian Industrial Region, there is a great demand for large amounts of soil materials suitable for hydraulic embankment construction. Because of mining subsidence, there is a need for comprehensive regulation of surface waters and this requires erection of numerous river embankments. The demand for soil materials cannot be met by natural soils due to a serious deficit of this material in the area. Therefore, the use of colliery spoils for this purpose is a technical and economic necessity. Unfortunately these materials have too high permeability, greater than relevant specifications require, i.e. 10^{-6} m/s, so reduction of permeability of the material to be used for hydraulic embankment construction is needed. The application of sludges for this purpose requires determination of their geotechnical parameters as well as those of mixtures of sludge and soil to be used for embankment construction. The aim of this work was estimation of the usability of post-flotation sludges coming from three sources for sealing layers and as sealing admixtures for earth hydraulic structures.

2 INVESTIGATIONS

Post-flotation sludges taken from sedimentation lagoons of two plants dealing with the recovery of coal from coal tips (Haldex-Michał, Gwarex-Ryan) and from a coal mine with a washing plant

(Krupiński mine) were the subject of the first step of the investigations (Kawalec and Kawalec, 1995). In the next step colliery spoils from Haldex-Michał plant and the mixtures of this material with post-flotation sludges were tested. The basic physical and mechanical parameters of post-flotation sludges were determined using standard apparatus and procedures. Compaction parameters were determined for Proctor compaction energy, grain size distribution was obtained using sedimentation method and the coefficient of permeability was determined in oedometers having a ring diameter of 75 mm and a height of 19 mm. Shear strength was measured in a shear box apparatus with box sizes of 6×6 cm and compressibility was measured in standard oedometers.

Geotechnical parameters of the colliery spoils from Haldex-Michał plant and mixtures of the colliery spoil with sludges were determined using medium-sized equipment, i.e. the permeameter diameter and sample height were 36 cm, the direct shear box had dimensions of 30×30 cm, the oedometer used for compressibility measurements had working height and diameter of 37 cm. The proportions of sludge in mixtures were in the range 5 to 20%.

3 TEST RESULTS AND ANALYSIS

The results of determination of the geotechnical properties of post-flotation sludges are displayed in Table 1. It is clearly visible that the grading of sludges coming from plants dealing with recovery of coal from mining waste tips (Haldex-Michał and Gwarex-Ryan plants) is very similar and

Table 1. Average properties of post-flotation sludges (from various sources).

Property	Symbol	Unit	Haldex-Michał	Gwarex-Ryan	Krupiński Coal Mine
Physical properties – Grain size distribution					
Fraction content	–	%			
– sand			28	29	–
– silt			48	43	–
– clay			24	28	–
Natural moisture cont.	w_n	%	27	35	–
Bulk density	ρ	Mg/m^3	–	1.56	–
Specific gravity	ρ_S	Mg/m^3	2.26	2.27	2.38
Max. dry density	ρ_{dmax}	Mg/m^3	1.455	1.42	1.54
Optimum moisture cont.	w_{opt}	%	21.0	21.6	17.2
Liquid limit	w_L	%	40.90	43.25	–
Plastic limit	w_P	%	21.34	22.95	–
Plasticity index	I_P	%	19.56	20.3	26
Ignition loss	I_0	%	28.5	27	10.4
Coefficient of permeability	k_{10}	m/s			
at R.C. = 90%			1.32×10^{-7}	–	–
95%			5.55×10^{-7}	–	5×10^{-10} 2.4×10^{-9}
100%			3.42×10^{-8}	–	–
Mechanical properties – shear strength					
			R.C. = 95% $w_n \approx w_{opt}$	R.C. = 81–82% $w = 41$–42%	
Angle of internal friction	ϕ	°	30.9	5.5–8.5	23.0
Cohesion	c	kPa	59.5	10–4	44
			R.C.–95% $w \approx w_{opt}$		
Compressibility modulus 0–400 kPa	M_0	MPa	5.99	–	–

this results in similar geotechnical properties. The materials have high fine-fraction content and in geotechnical classification they correspond to highly plastic soils. According to Garbulewski (2000) these materials satisfy the required specification values of geotechnical parameters within the category of suitable materials for sealing purposes. The liquid limits ($w_L = 40.9$–43.25) are less than 90%, the plasticity index ($I_P = 19.6$–20.3) is less than 65 but greater than 12, the silt content of 43 to 48% is greater than 30% and the clay content is 24 to 28% and thus greater than 20%. Maximum grain size was in the range 2 mm to 30 mm. Shearing strength parameters at moisture content close to the optimum and 95% relative compaction were relatively high.

Due to the fact that natural moisture content is very variable (11–48%, according to the part of sedimentation pond from where samples were taken), before use for engineering applications this material should be thoroughly intermixed and, if necessary, dried to a moisture content close to the optimum. Oedometric modulus determined at the same moisture content and relative compaction as the shear strength was of the order of 6 MPa, so compressibility of this material is very high.

Mining waste materials, coming from Haldex-Michał plan were used to prepare test mixtures with sludges. The grain size distribution of this material was:

• gravel content – 84%, sand content – 9%, silt + clay content – 7%.

The coefficient of permeability (k_{10}) of the spoil was 10^{-4} m/s which is greater than 10^{-6} m/s. The results of permeability tests of the mixture of colliery spoils with sludges are displayed in Table 2

Table 2. Sludge admixture ratio and permeability of colliery spoils (Haldex-Michał).

Sludge admixture ratio (%)	Relative compaction R.C. (%)	Coefficient of permeability k_{10} (m/s)
0	90	9.57×10^{-4}
	100	1.15×10^{-4}
5	90	4.78×10^{-4}
	95	1.86×10^{-4}
10	90	1.24×10^{-5}
	95	7.68×10^{-6}
20	90	5.48×10^{-7}
	95	8.48×10^{-7}

Table 3. Influence of sludge admixture ratio on compactibility parameters and mechanical properties of colliery spoils from Haldex-Michał plant.

				Shear strength		Compressibility
Sludge admixture ratio (%)	Max. dry density ρ_{dmax} (Mg/m^3)	Optimum moisture content w_{opt} (%)	Relative compaction R.C. (%)	Angle of internal friction ϕ (°)	Cohesion c (kPa)	Oedometric modulus M_0 (0–400 kPa) (MPa)
0	1.91	6.4	90	30	35	–
			95	–	–	19.5
			100	36	41	–
10	1.87	9.05	90	34.4	37.5	–
			95	–	–	18.2
			100	35.0	56.2	–
15	–	–	95	33.4	49.4	–
20	1.81	10.5	90	32.0	31.7	–
			95	–	–	14.9
			100	33.7	72.3	–

(Kozielska-Sroka and Michalski, 2002). The results obtained show that mixtures more than 10% sludge give reduction of coefficient of permeability to the specification required level.

As the admixing of sludge can change other geotechnical parameters of mixtures of colliery spoils with sludges, determination of other basic geotechnical parameters was carried out. The results are presented in Table 3 (Gorzynik, 2003; Kandefer, 2003). As expected, maximum dry density decreases and optimum moisture content increases when the proportion of sludge in the admixture is increased. The angle of internal friction was approximately unchanged but cohesion increased, so the greater the proportion of sludge in the mixture the greater is the total shear strength of the mixture and therefore stability of embankment slopes will not be affected. However, oedometric modulus of compressibility diminishes when the admixture ratio is essentially greater than 10%, but an oedometric modulus of $M_0 = 15$ MPa can be accepted in materials to be used for hydraulic embankments.

4 CONCLUSIONS

Post-flotation sludges coming from various sources have similar geotechnical properties.

Sludges can be used as a suitable material for sealing layer construction beneath sedimentation lagoons or dumping grounds of industrial wastes as a substitute for natural mineral soils when the moisture content is uniform and approximately equal to the optimum moisture content and when compaction is not less than 95% of maximum, i.e. Relative Compaction equals 95%.

Sludges are good materials for reduction of the permeability of coarse-grained colliery wastes for use in hydraulic embankment construction when comprising 15 to 20% of the total material.

REFERENCES

Garbulewski K (2000) *Dobór i badania gruntowych uszczelnień składowisk komunalnych (Choice and tests of soil sealings of municipal dumping grounds),* SGGW, Warsaw.
Gorzynik R (2003) *Wpływ uszczelnienia odpadów powęglowych odpadem poflotacyjnym i popiołem elektrownianym na wytrzymałość na ścinanie (Influence of sealing of colliery spoils with sludge waste and fly ash on shearing strength),* MSc thesis, University of Kraków.
Kandefer K (2003) *Wpływ uszczelnienia odpadów powęglowych na moduły ściśliwości w aspekcie ich wykorzystania w budownictwie drogowym (Influence of sealing of colliery spoils on compressibility moduli in the aspect of their use for road constructions),* MSc thesis, University of Kraków.
Kawalec B and Kawalec J (1995) *Odpady kopalniane jako warstwa uszczelniająca składowisk odpadów przemysłowych (Colliery spoils used as a sealing layer of dumping grounds of industrial wastes),*Zesz. Nauk Pol Śl, seria Budownictwo, 81, Gliwice, pp447–458.
Kozielska-Sroka E and Michalski P (2002) *Wpływ uszczelniania karbońskich łupków ilastych odpadami poflotacyjnymi na zmianę ich wodoprzepuszczalności (Influence of sealing of carbon shales with post-flotation sludges on their permeability),* Czasopismo Techniczne PK, z. 7-B, Kraków, pp63–70.
Zawisza E, Setmajer J and Skalicz F (1993) *Ekspertyza w zakresie możliwości zazwałowania osadników szlamu z Zakładu Odzysku Węgla w Trachach odpadami powęglowymi (Assessment of the possibility of depositing colliery spoils in sludge sedimentation ponds of Coal Recovery Plant in Trachy),* Internal Report.

Geotechnical and Environmental Aspects of Waste Disposal Sites – Sarsby & Felton (eds)
© 2007 Taylor & Francis Group, London, ISBN 978-0-415-42595-7

Monitoring of spontaneous vegetation dynamics on post coal mining waste sites in Upper Silesia, Poland

G. Woźniak
University of Silesia, Katowice, Poland

E.V.J. Cohn
School of Applied Sciences, University of Wolverhampton, UK

ABSTRACT: As with many other anthropogenic habitats, coal mine sedimentation lagoons are often refuges for rare and protected plants whose natural habitats have been diminished by industrial, urban and agricultural activities (Adamowski 1998; Shaw and Halton 1998; Tokarska-Guzik 1991, 1996). The distinctive nature of the habitats, their history and their biological communities mean that the pools have intrinsic value, making a unique contribution to the ecological, cultural and industrial heritage of their region. This paper aims to show the spontaneous development of vegetation on coal mine water sedimentation lagoons, how it changes with successional processes over time and with variations in the physical characteristics of the pool.

1 INTRODUCTION

Upper Silesia is one of the largest urban-industrial areas in Central Europe. Its industrial history has resulted in an anthropogenic landscape where the legacy of the extractive industries particularly, is almost always apparent. The nature of the industrial processes, the types of extractive industry and the underlying geology have combined to create an unusual and often unique set of chemical and physical conditions for the establishment of colonising organisms at each site.

Much scientific interest in the past has often been concerned with overcoming problems, such as soil toxicity, and developing techniques for reclamation and afforestation, but there has also been a recognition of the role of natural succession in ameliorating post-industrial sites – Jochimsen (1991) and Jochimsen et al. (1995). The importance of such anthropogenic habitats in the process of establishing and maintaining biodiversity and also as refuges for protected and rare plants has also been considered by many authors – Buszman et al. (1993), Trzcińska-Tacik (1966), Greenwood and Gemmell (1978), Tokarska-Guzik and Rostański (1996) and Box (1993).

2 RECLAMATION OF POST-INDUSTRIAL SITES

Investigations of the spontaneous development of vegetation, often of high wildlife value, through natural succession has been examined in relation to many types of post-industrial site such as:

- Glasswork solvay process spoil tips (Cohn et al. 2001; Tokarska-Guzik 1996).
- Iron and non-ferrous metal spoil heaps (Tokarska-Guzik 1991).
- Calamine land (Dobrzańska 1955).
- Coal mining spoil heaps (Kuczyńska et al. 1984; Jochimsen et al. 1995).
- Brown coal mining sedimentation lagoons (Balcerkiewicz et al. 1984).
- Sand pits (Szwedo et al. 1996).
- Coal mine sedimentation lagoons (Woźniak 1998).

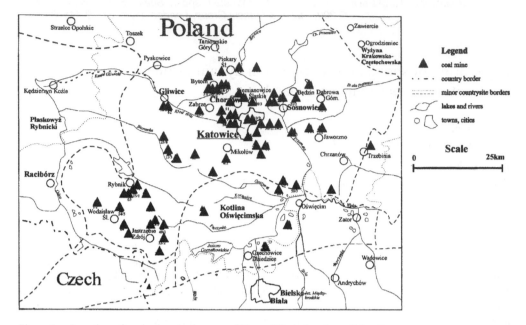

Figure 1.   Location of coal mines in the area of Upper Silesia (southern Poland).

With several hundred coal mines in Upper Silesia (Figure 1), each with at least one sedimentation lagoon, investigation of the features of coal mining landscapes have formed the basis of a detailed and extensive investigation, partly reported in Woźniak (1998) and further reported on here.

Sedimentation lagoons are produced when underground water, sometimes from as deep as 1000 m, is pumped out of the mine together with suspended coal dust and mineral components. After a period of settling, excess water is pumped into rivers and the sedimentation lagoons are left to dry out. As they dry, a variety of habitats develop within the different areas which may be flooded, muddy or dry and plants begin to colonise. The physical characteristics of the sediment vary with the geology of the region. In the north-east, the groundwater is not far beneath the ground, but in the south west and central areas, the Carboniferous sediments are entirely covered with Tertiary layer clayey sediments. Here, the groundwater may be several hundred metres below ground where its mineral content tends to increase with depth (Woźniak 1998).

3   INVESTIGATION METHODS

This study involved examination of 24 coal mine sedimentation lagoons from across Upper Silesia as shown in Figure 1. These lagoons represent the range of geological variation in the region. The vegetation of the lagoons was recorded twice a year, in spring and autumn, for a period of four years. In the first year, transects were laid out from the outer edge of the sediment in the direction of the centre and samples recorded contiguously using a 1 m^2 quadrat. Each transect was the length of the extent of the vegetation, so each year further quadrats were added as the vegetation extended further towards the centre of the pool. The abundance of all vascular plant species present in every quadrat along the transects was recorded as percentage cover.

The foregoing sampling strategy was designed to allow an examination of several aspects of vegetation change. Each quadrat was in effect a permanent quadrat so that vegetation change over time in any one spot (succession) could be observed. The extension of the transect each year showed vegetation zonation associated with local variation in the sediment, particularly in relation to drying out. It also showed the rate of spread of the vegetation across the lagoon.

4 RESULTS

There were 31 transects at the 24 lagoons with a final total of 477 quadrats. Altogether, 1984 samples were recorded over the 4 years of the study.

4.1 *Vegetation development*

The number of samples along each transect increased each year on most of the lagoons (Figure 2) indicating that the vegetation was spreading towards the centre as the pools dried out. The rate of spread varied from one site to another – a rate of 2 to 3 metres per year was typical, but at some sites it was over 10 metres some years. At a few sites, there was little change and even some loss of vegetation.

A total of 109 species was recorded on the transects. Figure 3 shows that there were typically 7 or 8 species per transect, with a few transects having 2–3 times this number.

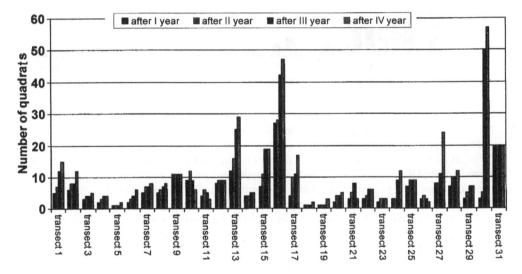

Figure 2.   Change in number of quadrats on each transect over four years.

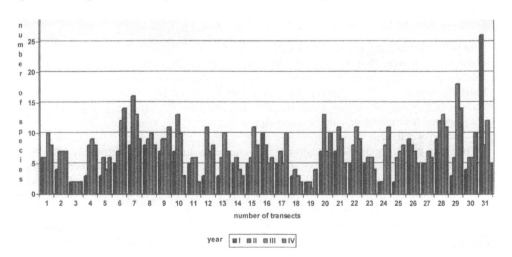

Figure 3.   Number of species recorded in each transect each year.

291

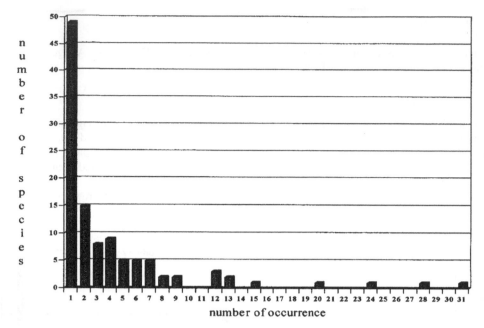

Figure 4.   The frequency of species occurrence.

The typical trend was for species number to increase to a maximum in year 2 or 3 and then to decline slightly to a final average of 7.3 species per transect in the fourth year. At the same time, vegetation cover increased as ruderal dominants such as Calamagrostis epigejos and Cirsium arvense increased their cover. This provided less opportunity for the short-lived early colonisers to expand as competition played an increasingly important part in the vegetation dynamics.

Figure 4 shows that a large number of species occurred only once in the transects, while only a few species were recorded in all 31 transects. Amongst the species recorded were some protected species such as Epipactis atrorubens (Dark-red Helleborine) and Epipactis palustris (Marsh Helleborine) as well as some which are regionally rare, such as Equisetum variegatum (Variegated Horsetail).

4.2   *Vegetation composition*

An examination of the actual species composition of the vegetation and the patterns of species replacement over the four years, showed that the transects fell into broadly 5 groups:

- *Group 1* – ruderal/ruderal dominant species
  These transects were characterised by a low number of quadrats on the transect and only 2–7 species in total. Vegetation development was slow, with little change in species composition from the youngest to the oldest quadrats. For example, Tussilago farfara (Coltsfoot), Cirsium arvense (Creeping Thistle) and Calamagrostis epigejos (Wood small-reed), were both typical pioneers and later dominants on these transects.
- *Group 2* – ruderal – halophilous species
  On transects in this group, the vegetation of the youngest quadrats comprised mainly salt-tolerant species such as Puccinella distans (Reflexed Saltmarsh-grass), Spergularia salina (Lesser Sea-spurrey) and Atriplex prostrata (Spear-leaved Orache). These species were replaced in the older quadrats, where salt has been washed out deeper into the soil, by a range of ruderals and ruderal dominants such as those present in Group 1.
- *Group 3* – halophilous species
  Only halophilous species occurred on these transects, suggesting higher salt levels. The younger quadrats were less species rich than in Group 2 and the older ones had a similar range of species, but with higher cover.

- *Group 4* – flooded transects
There was standing water in these transects and the reedswamp vegetation which developed suggests that salt was less concentrated here than in Group 3. Typical species here were Phragmites australis (common reed), Typha angustifolia and Typha latifolia (Bulrush), and Bolboschoenus maritimus (Sea Club-rush).
- *Group 5* – sandy transect
There was only one transect in this group. It was situated in a sandy surrounding area, so that the lagoon surface was covered with a thin layer of sand. The main plant in both the young and the older quadrats was Corynephorus canescens (Grey Hair-grass).

5 CONCLUSIONS

The analysis has shown the great variability of the sedimentation lagoons by demonstrating the development of a number of distinctive plant assemblages. Salinity and waterlogging, mediated by substrate particle size, were the main site characteristics influencing vegetation development, while neighbouring sources of colonisers and rainfall were amongst the principal extrinsic factors. Many of the colonisers arc specialists, tolerant of the unusual conditions for plant growth associated with the lagoons such as high salinity, mineral nutrient imbalances and hydrological extremes and fluctuations.

While some patterns and trends in the vegetation of the lagoons have been identified, the course and rate of vegetation development, species replacement and successional processes varies from lagoon to lagoon. Long-term studies on the development of vegetation in these distinctive anthropogenic habitats is therefore important to inform their management from both an engineering and restoration perspective, and also from the perspective of their importance for nature conservation. Unless underpinned by a sound understanding of associated ecological processes, attempts at reclamation of the lagoons by artificially attempting to accelerate the development of vegetation cover can result in very expensive failures. Allowing spontaneous colonisation and following nature brings sound economic, as well as ecological, benefits. Thus, it is important that no reclamation work should be attempted without a full assessment of the nature conservation value of the lagoons and an understanding of the consequences of intervention.

REFERENCES

Adamowski W (1998) *Colonisation success of orchids in disturbed habitats,* Plant population biology, Falińska K. (ed.) W. Szafer Institute of Botany, Polish Academy of Sciences, Kraków, pp167–174.
Balcerkiewicz S, Brzeg A and Pawlak G (1984) *Rośliny naczyniowe zwałowiska zewnętrznego Pątnów-Jóźwin,* Badania Fizjograficzne nad Polską Zachodnią 35, Ser. B, pp35–52.
Box J (1993) *Conservation or greening? The challenge of post-industrial landscapes,* British Wildlife, 4(5), pp273–279.
Cohn E V J, Rostański A, Tokarska-Guzik B, Trueman I C and Woźniak G (2001) *The flora and vegetation of an old Solvay Process tip in Jaworzno (Upper Silesia, Poland),* Acta Societatis Botanicorum Poloniae 70 (1), pp47–60.
Buszman B, Parusel J B and Świerad J (1993) *Przyrodnicze wartości leśnych stawów w Tychach Czułowie przeznaczonych na zwałowisko odpadów kopalń węgla kamiennego [Natural value of forest ponds in Tychy-Czułów town appropriated for coal mine dump],* Kształtowanie Środowiska Geograficznego I Zurbanizowanych, UŚ, Katowice-Sosnowiec 8, pp9–15.
Greenwood E F and Gemmell R P (1978) *Derelict industrial land as a habitat for rare plants in S. Lancs. (v.c. 59) and W. Lancs (v.c.60),* Watsonia, 12, pp33–40.
Jochimsen M (1991) *Bergehalden des Steinkohlebergbaus,* Vieweg Wiesbaden, pp155–162.
Jochimsen M, Hartung J and Fischer I (1995) *Spontane und Künsthiche Begrünung der Abraumhalden des Stein – und Braunlcohlenbergbaus,* Berichte der Reinh Tüxen – Ges 7, p88.
Kuczyńska I, Pender K and Ryszka-Jarosz A (1984) *Roślinność wybranych hałd koplani węgla kamiennego Victoria w Wałbrzychu,* Acta Universitatis Wratislaviensis, Pr Bot 27 (553), pp35–60.

Shaw P J A and Halton W (1998) *Nob End, Bolton,* British Wildlife 10(1), pp13–17.

Szwedo J, Woźniak G, Kubajak A, Wyparło H and Rak W (1996) *Ścieżki dydaktyczne po terenach rekultywowanych kopalni piasku "Szczakowa",* Wyd Planta 5, pp34–71.

Tokarska-Guzik B (1991) *Hałda huty szkła w Jaworznie-Szczakowef jako ostoja zanikających gatunków w obrębie miasta [Glass works heap in Jaworzno-Szczakowa (district of Katowice) as the mainstay of the disappearing plant species in the town],* Kształtowanie Środowiska Geograficznego i Ochrona Przyrody na Obsza-rach Uprzemysłowionych i Zurbanizowanych, UŚ, Katowice-Sosnowiec, pp39–42.

Tokarska-Guzik B (1996) *Rola hałd zasadowych w utrzymaniu lokalnej bioróżnorodności [The importance of lime waste heaps in maintenance of local biodiversity],* Przegląd Przyrodniczy 7(3–4), pp261–266.

Tokarska-Guzik B and Rostański A (1996) *Rola zatopisk (zalewisk) pogórniczych w renaturalizacji prze-mysłowego krajobrazu Górnego Śląska [The significance of colliery flood reservoirs in the renaturalization of industrial landscapes of Upper Silesia],* Przegląd Przyrodniczy 7(3–4), pp267–272.

Trzcińska-Tacik H (1966) *Flora i roślinność zwałów Krakowskich Zakładów Sodowych [Flora and vegetation of the Solvay dump of the Krakow Soda Factory],* Fragm Flor Et Geobot 12(3), pp243–318.

Woźniak G (1998) *Primary succession on the sedimentation pools of coal mines,* Phytocoenosis 9(10), pp189–198.

Geotechnical and Environmental Aspects of Waste Disposal Sites – Sarsby & Felton (eds)
© *2007 Taylor & Francis Group, London, ISBN 978-0-415-42595-7*

Fly ash compaction

K. Zabielska-Adamska
Faculty of Civil and Environmental Engineering, Bialystok Technical University, Poland

ABSTRACT: Use of power engineering waste as a material for earth structures is connected with determination of its compactability. It has been found that fly ash/slag mixture samples that had been compacted several times in Proctor moulds did not give representative results. Relationships between dry density and water content for re-used waste samples were determined. The effect of recompaction on grain size distribution, density of solid particles and specific surface of tested wastes was investigated.

1 INTRODUCTION

Soil compactability is the ability to obtain maximum possible dry density (ρ_d) of solid particles as a function of compaction energy and its application, as well as the kind of soil and its moisture content. It is measured by means of degree of compaction (I_S), which is determined from the formula:

$$I_s = \frac{\rho_d}{\rho_{dmax}} \tag{1}$$

Where ρ_d is the dry density determined for soil compacted in an embankment and ρ_{dmax} is the maximum dry density determined in the laboratory (for the same material).

Laboratory testing of soil compactability consists of compaction of soil in a standardized way, at various moisture contents, and plotting the relationship between unit dry weight (or dry density) and moisture content. The moisture content at which the compacted soil attains its maximum dry density is called optimum water content (w_{opt}). Compaction curves (ρ_d against w), determined for various values of compaction energy approach asymptotically the line of maximum compaction (called the zero air voids line), which is calculated assuming that soil pores are completely filled with water as well (Degree of Saturation, S_r, equal to unity).

Various methods of compactability assessment are used in road and geotechnical laboratories. These tests can be classified according to the kind of compaction energy applied. Compaction methods are classified into main types of; dynamic or impact, static (kneading), vibratory. Each of these methods produces different results, both in structure of compacted soil and in the material properties. Laboratory compaction methods are supposed to reproduce field conditions but they are not sufficiently reliable. The most common methods which are applied for determining compaction parameters of fine-grained soils are dynamic methods. Values of ρ_{dmax} and w_{opt} are obtained by two methods, namely; Proctor's method (the Standard Proctor test) with compaction energy corresponding to field compaction conditions fusing lightweight soil compactors, the modified AASHTO method (also called the Modified Proctor test) with energy corresponding to field compaction by heavy compactors. It has been found that higher values of maximum dry density are obtained for well-graded soils – this is the base of classification of soil for embankments in U.S.A. (Liu and Evett, 1984).

2 LABORATORY COMPACTION OF POLISH FLY ASHES

One of the first investigations of Polish power industry waste was conducted on waste from burning brown coal. Waluk (1964) tested, for a wide range of moisture contents, the compactability of fly ash, fly ash and slag mixtures, as well as slag alone. Compaction was by means of manual rammer, static load (equal to 294.2 kPa) and vibration. Waluk stated that, in comparison to waste in a loose state, the best compaction effect was obtained by the dynamic method. The compaction achieved in the Standard Proctor test was nearly twice that obtained by the vibration method.

Similar research was done for wastes from hard coal (Pieczyrak and Sekowski, 1985), when fly ash and slag mixes were compacted by Standard Proctor method and by vibration. In both cases, the same value of optimum water content was obtained, but vibration gave lower porosity.

Analysis of compaction achieved made by means of static loads reaching up to 100 MPa, for three fly ashes from different electric power plants, was done by Dragowski and Pininska (1978). Observations were conducted for several load stages. During each stage, microscopic observations were made and grading, unit weight and porosity, as well as mechanical properties were determined. It was found that during the compaction process weak fly ash grains were first crushed, then closed pores were destroyed and then unburnt coal pieces were crushed and the grains approached one another. As a result of the load the silt-sized fraction was increased, the fly ash density increased 3 to 4 times, and porosity changed significantly.

The relationship between optimum water content and grading of power industry waste has been clearly seen. For example, $w_{opt} = 33\%$ was obtained for fly ash/slag mix with a grading corresponding to a gravel and sand mix (Pieczyrak and Sekowski, 1985), whilst $w_{opt} = 45.8\%$ was obtained for sandy silt (Kawalec et al, 1981). The dry densities in the foregoing cases effectively did not depend on grading, and they equalled 1.01 and 1.06 g/cm^3, respectively. Many researchers have pointed out the flat shape of the compaction curve for ash waste. It can be stated that the generally accepted methods of determining fly ash compaction properties are dynamic methods, i.e. the Standard Proctor and Modified Proctor tests.

3 THE LABORATORY STUDY

Compactability tests of power industry waste were conducted on samples of fly ash and slag mixture formed from hard coal burning in Bialystok Thermal-Electric Power Plant and stored in a dry condition (Zabielska-Adamska, 2002). The slag content is up to 10% of the stored waste – because of the small content of slag in the mixture it has been designated 'fly ash' in the paper. Grain size distribution of all tested fly ash corresponds to sandy silt. In the laboratory grading was estimated by means of the uniformity and curvature coefficients. According to the criterion the tested fly ash qualified as a poor material for compaction.

3.1 Effect of re-compaction of the same 'fly ash' sample

At first fly ash tests to determine optimum water content (w_{opt}) and maximum dry density ($\rho_{d\,max}$) were conducted according to Polish standard PN-88/B-04481 which involves repeat ramming (five times) of the same sample of soil. Compaction curves obtained for one batch of fly ash tested by Standard Proctor method are shown in Figure 1. Values w_{opt} and $\rho_{d\,max}$ obtained for the same batch of fly ash were in the range; $w_{opt} = 36.0$ to 37.5%, $\rho_{d\,max} = 1.144$ to 1.164 g/cm^3. Because of the scatter of the obtained values and the completely different shapes of compaction curves, it was decided that re-compacted specimens could not be used.

Accurate determination of compaction parameters is needed because of the relationship between fly ash mechanical properties and compaction moisture content, as in the case of cohesive soils, and particularly the considerable loss of fly ash bearing capacity (based on California Bearing Ratio) when it is compacted dry of optimum (Zabielska-Adamska, 1997).

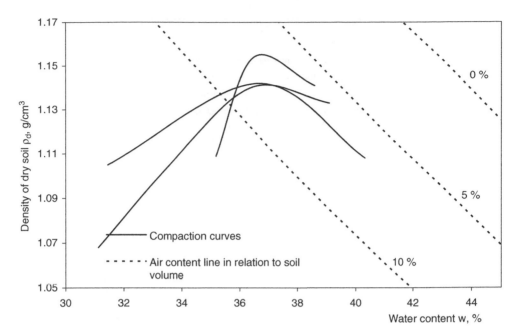

Figure 1.   Standard Proctor compaction curves for the first batch of fly ash.

3.2   *Proposed test method*

Six different fly ash batches were compacted by means of Standard and Modified Proctor methods. It was established that each fly ash sample could be compacted in the Proctor mould only once – each point on the compaction curve was determined for a separately prepared specimen. Samples were wetted to give an increase of moisture content in successive samples of about 2% and then they were stored for 24 hours in closed tins. Compactability tests were done on virgin samples and then the same samples were used several times in order to determine the influence of repeated ramming on fly ash compaction. Compaction curves obtained by means of both Standard and Modified Proctor methods for two fly ash batches, when the same fly ash sample was rammed only once and several times, are shown in Figures 2 and 3.

It should be noted that Figures 1 and 2 contain curves determined for the same batch of fly ash but obtained by different methods, i.e. when re-compaction was permitted and when virgin samples only were used. In Figure 2 the two independent compaction curves of fly ash (first batch) which were only compacted once by Standard Proctor method are very close to each other. Compaction curve shape, representing the relationship between ρ_d and w, depends distinctly on compaction energy and the number of ramming repetitions. Curves obtained for fly ash only compacted once are flat for both compaction methods (and are flatter for Standard than the Modified method).

3.3   *Test results*

It has to be stated that fly ash samples which were compacted many times could not be considered as representative. The values of maximum dry density increase with the number of repeat compactions, whilst optimum water content decreases in comparison with samples compacted only once under the same conditions.

It should be emphasized that laboratory determination of the compaction parameters ρ_{dmax} and w_{opt} according to many national standards leads to incorrect evaluation of compaction effects. The compaction parameters determined on the basis of curves obtained for the same fly ash batch

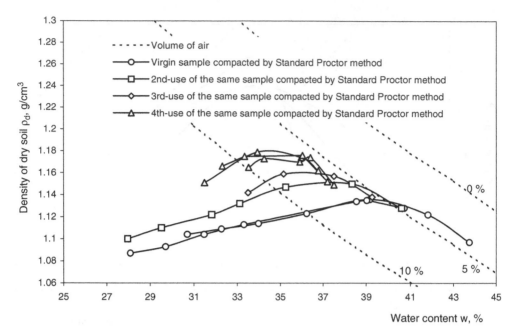

Figure 2.   Proctor compaction curves for a sample compacted only once or repeatedly (first fly ash batch).

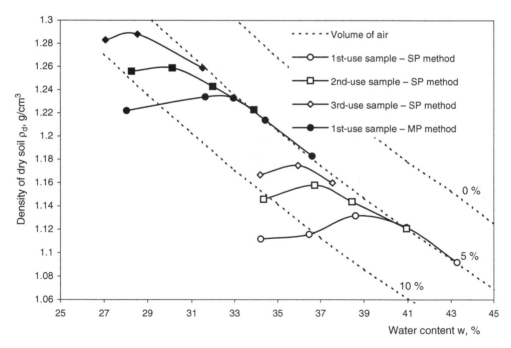

Figure 3.   Standard (SP) and Modified Proctor (MP) compaction curves (second fly ash batch).

obtained by compaction methods which allow re-use of the same sample or not (Figures 1 and 2) are completely different. The phenomenon of obtaining greater dry densities after second ramming of the same fly ash sample is also known for mineral soils. A simply explanation of this effect (Rodrigues et al, 1988) is apparent plastic volumetric-strain, which is caused by successive fly

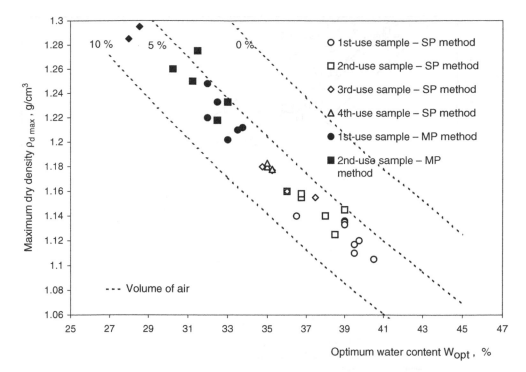

Figure 4. Proctor optimum compaction data points for six batches of fly ash.

ash compaction. The differences between fly ash compaction curves, when the same material was used for determination of other data points and when the same material was tested only once, were observed by Raymond and Smith (1966), Leonards and Bailey (1982) and, in Poland, by Garlikowski (1992). In all these papers researchers stated that crushing of dynamically-rammed fly ash grains contributed to better waste compaction. However, Garlikowski (1992) explained the increase of dry density by increased packing of the solid particles – this was not observed during the research reported herein.

The values of $\rho_{d\,max}$ and w_{opt}, for six fly ash batchs tested by means of two methods (compacted once and many times), are shown in Figure 4. The points obtained lie along a line parallel to the Saturation Degree (S_r) line. On the basis of all test results a linear regression relationship between maximum dry density and moisture content, of the form $\rho_{d\,max} = f\,(w_{opt})$, was established. The relationship is described by the formula:

$$\frac{1}{\rho_{d\,max}} = 0.458 + 0.011 w_{opt} \tag{2}$$

The correlation coefficient (r) calculated for the foregoing relationship was equal to 0.978, which shows that the established regression equation represents over 95.6% of the variations of optimum water content and maximum dry density.

Comparison of the regression line (equation 2) with the zero air void line and lines of Saturation Degree S_r is presented in Figure 5. It was found out that the regression line for the fly ash corresponded approximately to a Degree of Saturation of 0.865 so that the relationship became:

$$\rho_{d\,max} = \frac{86.5 \rho_s}{w_{opt} \cdot \rho_s + 86.5} \tag{3}$$

This equation can be used to determine the reference values $\rho_{d\,max}$ and w_{opt}.

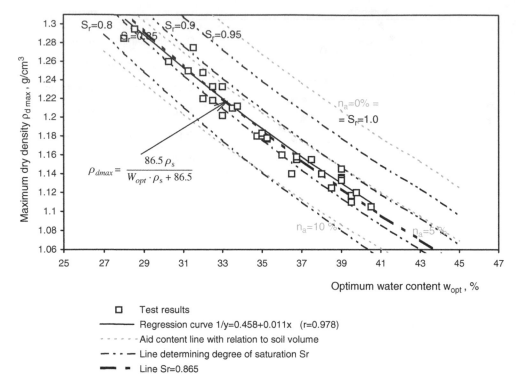

Figure 5. Relationship $\rho_{d\,max} = f(w_{opt})$ for all fly ash test results.

The 'fly ash' from the dry storage yard of Bialystok Thermal-Electric Power Plant is a material which can be easily compacted. The 'fly ash' optimum water contents lie between air voids of 5 to 9% (the most often are 6 to 7%) for the Standard Proctor method and between 5 to 8% for the Modified method. The porosity of 'fly ash' compacted at optimum water content is in the range from 0.501 to 0.516 (average value 0.508) for Standard Proctor and from 0.454 to 0.474 (average value 0.466) for the Modified test. The high values of minimum porosity were found when the densest state of the fly ash was determined by vibration. The range of porosities obtained by this means was from 0.524 to 0.548.

3.4 Influence of compaction on selected fly ash physical properties

To determine the influence of compaction on the physical properties of fly ash the grain size distribution, particle density, sand equivalent and total specific surface of fly ash in a natural state and after compaction were measured. As a result of the measurements grain size distribution, which were done for fly ash in its natural state and after repeated proctor compaction (five-times), it can be concluded that grading of the studied wastes changed by and insignificant amount due to compaction. The coarse fraction (grains larger than 0.071 mm) were equal to;

- 45.86% in the natural state,
- 44.61% after the same sample had been compacted three times,
- 37.05% after the same sample had been compacted five times.

There was no evidence that compaction affected particle density (ρ_s). Average values of ρ_s determined for one sample of fly ash, in its natural state and after being subjected to Proctor compaction five-times, were equal to each other. Thus, experimental results do not confirm the conclusion reached by Garlikowski (1992) that the increased maximum dry density ($\rho_{d\,max}$), obtained after multiple ramming of the same fly ash sample, were due to elimination of closed voids within fly ash grains so that there was an increase in the density (ρ_s) of the solid particles.

The influence of compaction on sand equivalent and total specific surface values has be observed. The percentage volumetric content of sand and gravel fractions (expressed as a sand equivalent) was equal to 35.1 for a typical ash batch in its natural state and decreased to 17.8 after Proctor compaction (five-times) and to 12.4 when compacted by the Modified method. The specific surface increased from $2.60 \, \text{m}^2/\text{g}$ in the natural state to $3.52 \, \text{m}^2/\text{g}$ after compaction – this value was independent of applied compaction energy.

The phenomenon of increase in maximum dry density ($\rho_{d\,max}$) at decreasing optimum water content (w_{opt}) for repeatedly compacted fly ash is accompanied by reduction of the size of the fly ash grains and increase in their specific surface. This is completely the reverse of the compaction process for mineral soil and fly ash in its natural state. For these materials, greater values of $\rho_{d\,max}$ and lower values of w_{opt} were obtained for coarse-grained material and with lower specific surface. Dynamic ramming of fly ash causes its degradation through partial crushing of weak grains. This contributes to better fly ash compaction due to better grading. Grading coefficients C_U and C_C, determined on the basis on grain size-distribution curves for fly ash in its natural state and after repeated Proctor compaction, show the possibility of improved compaction of re-compacted fly ash. The uniformity coefficient C_U is lower for the grading curve of fly ash repeatedly compacted – the value is equal to 4.33 while for fly ash in its natural state it equals 5.33. The curvature coefficient C_C exceeds the threshold value and is equal to 1.12 (for fly ash in its natural state it was 0.91). Considering compaction curve shapes for mineral soils which are well or poorly grained (Rodrigues et al, 1988), it may be stated that compactibility curves obtained for fly ash repeatedly compacted have a more pronounced peak than those for fly ash compacted only once. Furthermore, greater values of $\rho_{d\,max}$ and lower values of w_{opt} prove the improvement of the grading of the compacted fly ash.

4 CONCLUSIONS

Re-compacted fly ash samples cannot be considered as a representative samples. Laboratory-determination of compaction parameters using repeated compaction of the same fly ash sample (which is the usual practice in many laboratories) leads to the incorrect estimation of fly ash compactibility effects. When working with re-compacted ash samples the resultant maximum dry densities were higher than those achieved with virgin samples whilst the optimum moisture contents were lower for the same compaction conditions, in both cases.

The methodology for testing to obtain compaction parameters should be standardized. Each point on the compaction curve (the moisture-density relationship) should be determined using a separate fly ash sample.

Re-compaction of fly ash causes partial degradation of dynamically-rammed grains and increases their specific surface, which improves ash compactibility but does not have an effect on the density of the solid particles. Fly ash compactibility depends not only on grain size distribution but on the structure of individual grains as well.

ACKNOWLEDGMENTS

The investigation was carried out at Bialystok Technical University in Poland. The State Committee for Scientific Research supported this work under project number 5 T07E 05124.

REFERENCES

Dragowski A and Pininska J (1978) *Fly Ash Compaction Process Analysis and Changes its Structure,* Proc Fifth Polish National Geotechnical Conference, Katowice, pp201–210 (in Polish).

Garlikowski D (1992) *The Frost Influence on Fly Ash from Hard Coal Assigned to Structure Embankments,* PhD Thesis, Agricultural Academy in Wroclaw, (in Polish).

Kawalec B, Kopka Z and Soczawa A (1981) *The Usability of Fly Ash and Slag Mixtures from Power Plant Laziska to Road Embankments,* Scientific Paper of Silesian Technical University, Building Engineering No 55, Gliwice, pp25–41 (in Polish).

Leonards G A and Bailey B (1982) *Pulverized Coal Ash as Structural Fill,* Journal of the Geotechnical Engineering Division, Proc American Society of Civil Engineers, Vol.108, No4, pp517–531.

Liu C and Evett J B (1984) *Soil Properties. Testing, Measurement and Evaluation,* Prentice-Hall, New Jersey.

Pieczyrak J and Sekowski J (1985) *Power Engineering Waste from Stalowa Wola in Geotechnical Opinion,* Scientific Paper of Silesian Technical University, Building Engineering No 9, Gliwice, pp107–121 (in Polish).

Raymond S and Smith P H (1966) *Shear Strength, Settlement and Compaction Characteristic of Pulverised Fuel Ash,* Civil Engineering and Public Works Review, No8, pp1107–1113.

Rodrigues A R, Castillo H and Sowers G.F (1988) *Soil Mechanics in Highway Engineering,* Trans Tech Publication, Clausthal-Zellerfeld.

Waluk J (1964) *Physical and Mechanical Properties of Fly Ash from Brown Coal in Application to Transport and Dumping,* PhD Thesis, Technical University in Wroclaw, (in Polish).

Zabielska-Adamska K (1997) *Geotechnical Properties of Power Engineering Wastes from Białystok Thermal-Electric Power Plant in Aspect their Usability to Embankment,* PhD Thesis, Bialystok Technical University, (in Polish).

Zabielska-Adamska K (2002) *Compactibility of Power Plant Coal Wastes,* Inżynieria Morska i Geotechnika, Vol. 23, No1, pp26–31 (in Polish).

Environmental management

Geotechnical and Environmental Aspects of Waste Disposal Sites – Sarsby & Felton (eds)
© 2007 Taylor & Francis Group, London, ISBN 978-0-415-42595-7

Environmental insurance – addressing waste management liabilities

D.A. Brierley
Bridge Risk Management Ltd, Manchester, UK

ABSTRACT: Businesses operating in the Waste Management industry or producing contaminated waste are exposed to potentially ruinous environmental liabilities. Recent legislation and a greater willingness to litigate have aggravated pollution exposures, which can, fortunately, be minimised by transfer to insurance. Since 1992, there has been a specialist market for environmental policies in the UK, including specific products for the Waste Management industry.

1   INTRODUCTION

The Landfill Directive has brought new concerns for waste managers and for those hoping to develop brownfield sites. Process remediation techniques are being developed to deal with contaminated sites but the answer may be in centralised ex-situ soil recycling facilities. The potential for disputes between the facility and its clients is obvious but insurance can help to address this through joint policy arrangements. For over a decade, insurance has been an important part of the Waste Management Industry's past and present. It is ready and waiting to help it to face the future.

2   THE RISK FRAMEWORK

Risk affects all our lives and it needs effective management if it is not to rule our lives. The UK Environment Agency defined risk in 2003 as 'a combination of the chance of particular event, with the impact that the event would cause if it occurred.' It is useful to explore the ways in which pollution risks arise, with particular reference to the Waste Management industry. For convenience, these are divided into risks directly arising from contamination and those resulting from external factors.

2.1   *Ongoing operations*

The ongoing operation of any waste management site is fraught with risk – methane gas can accumulate within domestic waste sites to the point where there is a possibility of explosion, whilst leachates can migrate in such a way as to contaminate land or water.

2.2   *By-products of brownfield development*

The disposal of hazardous substances, often in the form of contaminated material from development sites, presents a high risk to the environment and to people, not least because of the danger of migration into controlled waters, notably groundwater, where the impact may not be noticed for some time.

2.3 Effectiveness of process remediation innovations

Recent years have seen the growth of a host of remediation techniques as alternatives to landfilling of contaminated waste, e.g. soil stabilisation, monitored natural attenuation, thermal desorption. But developers have been slow to embrace them. The Landfill Directive is likely to change this because of factors such as insurance, provided the construction industry is prepared to entertain effective techniques which may take longer to work than 'dig and dump'. Uncertainties about the effectiveness of alternative remediation techniques and the time they take to work meant that in 2001, Civil Engineering techniques, i.e. excavation and removal or capping, were still used in 90% of cases. Bioremediation, a high profile technology, was only used at 5% of sites.

2.4 The isolation issue

The decline of the UK's industrial base has presented a new risk to the businesses that have survived, i.e. isolation. A good example would be the foundry industry, where the past 35 years have seen a reduction in the number of ferrous facilities in the whole country to the number operating in the West Midlands alone in 1968. Poor planning and high housing demand have led to industrial sites becoming islands in a sea of residential sprawl. Inevitably, householders notice the noise and the smell and the twin spotlights of regulation and litigation threaten the existence of an isolated manufacturer, which may be attributable in part to high legal defence costs and stigma damage. Fortunately, these risks are insurable if they are addressed in time. Furthermore, an industrial concern may decide to follow the trend by decommissioning and selling its site to a developer, who will be wary of inherited environmental liabilities. It is hard to imagine a more cost-effective way of achieving a clean exit (and of turning a liability into an asset) than an insurance policy in the names of both parties.

2.5 Legal and regulatory risks

There is a large volume of environmental legislation that is currently operative in the UK but it is useful to concentrate on four key examples:

- *Part IIa of the Environmental Protection Act 1990* created a statutory definition of contaminated land in Great Britain and empowered local authorities to identify sites within their boundaries that fall within the definition and therefore need to be remediated. If the polluter cannot be traced, the owner or occupier will be liable for the cost of clean-up and any accompanying legal costs. This may be enough to put them out of business.
- *The IPPC (Integrated Pollution Prevention and Control) regime*, which affects a number of industries including Waste Management, makes the licensing of operations conditional upon the use of Best Available Techniques (BAT) and the production of a baseline report, which describes the land condition at the time of licensing. Any deterioration noted at the time of licence surrender must be made good by the operator, which could adversely affect future plans including retirement. Insurance is available to address this and the use by operators of appropriate insurance is endorsed at a high level (OECD, 2003).
- *The EU Environmental Liability Directive* is ultimately likely to result in the imposition of compulsory financial provision against serious environmental damage, quite possibly in the form of insurance cover. The intention, as with Part IIa, is that the polluter pays.
- *The Landfill Directive*, which became effective in the UK in July 2004, represents a revolution in the handling of waste by effectively sounding the death knell for the unsustainable 'dig and dump' approach. Co-disposal of hazardous and non-hazardous material is now illegal and the number of landfill sites accepting hazardous waste has been reduced from 277 to 12. Given the widespread perception that process techniques are slow and unreliable the directive could result in widespread fly tipping of dangerous material and pose a direct threat to brownfield development. This point is illustrated by the fact that 120,000 tons of waste was fly tipped in London alone in 2003 (prior to the directive) and 60% of landfilled hazardous waste comes from

the construction and demolition industries. There is a question as to how derelict sites can be reclaimed if there is no safe and proven method of processing or disposing of contaminated soil.

2.6 Dangerous misconceptions

Complacency over environmental vulnerability can result from ignorance or strongly held beliefs with no grounding in fact, for examples:

- *Public Liability* policies only provide pollution cover in respect of sudden and unforeseen damage or injury, which may in any case be difficult to prove. Gradual contamination of land or water is not insured unless specialist environmental cover is put in place.
- *Environmental Consultants' and Contractors' Warranties* will, quite rightly, be insisted upon by those who commission reports and contract works. If something goes wrong which is attributable to a consultant or contractor's negligence, the warranty may be invoked and a subsequent court settlement would normally be funded by the negligent party's Professional Indemnity insurance. If a pollution incident has arisen from rising water tables or changes in the law, which may mean that a previously acceptable site is deemed to be contaminated, the warranty cannot be invoked. Any business that is vulnerable to pollution liabilities needs to identify and manage them rather than believing that warranties cover all risks.
- *Capital funds* can work as an element within a blended risk transfer programme but amount to 'dead money' when they stand alone. It makes more sense to use capital for investment in the business and pay a small fraction of the amount of the perceived liability as a premium which will finance access to Insurers' reserves in the event of a claim. In addition, FRS (Financial Reporting Standard) 12 stipulates that all contingent liabilities must be stated on company accounts. How much better for shareholder confidence if they have been offset by insurance and thus effectively removed from the balance sheet.

3 THE INSURANCE ANSWER

3.1 Background

Having examined the various forms of environmental risk it is necessary to explore the ways in which these risks can be controlled by financial means. Whilst specialist insurance products have only been available in the UK since 1992, they now attract an annual premium spend estimated at £70 million. They are often driven by lenders and investors who are concerned that they may become mortgagees in possession of a contaminated site in the event of a customer's insolvency.

Following the Local Government Association's report ('Something Old, Something New') authorities are now aware that environmental insurance can enhance the market value of their surplus sites at the same time as creating developer confidence to unlock brownfield sites (LGA, 2002). The message to the public sector was reinforced in a British Urban Regeneration Association (BURA) memorandum to the parliamentary urban affairs sub-committee on the effectiveness of government regeneration initiatives. This went as far as to say that in principle, the liabilities that deter purchasers and their lenders from being part of the drive for brownfield development are insurable, (BURA 2002).

3.2 Policy types

- *Environmental Impairment Liability policies* provide on-site and off-site cover for remediation costs imposed by regulators and protection against third party claims for bodily injury and property damage. Legal expenses are also covered and policies may be extended to include Business Interruption and Loss of Rental income arising from pollution. Cover periods may be from one to ten years and the standard policy excess is £25,000.

- *Clean-up Cost Cap* creates confidence in remediation budgets by providing insurance against cost overruns.
- *Contractor's Pollution Liability* bridges the gap between the sudden and unforeseen pollution damage cover provided by a Public Liability policy and the Professional Indemnity insurance which addresses negligence claims, by focussing on gradual pollution incidents caused by contractors' activities on site.
- The Waste Management industry has its own specialist products in addition to the above, which include *Emergency Works policies and bonds* to provide the financial provision that waste managers need to agree with the Environment Agency as a licensing condition. Unlike conventional bonds, these do not affect credit arrangements.
- *Alternative Risk Transfers* such as blended finite programmes may be applied to uninsurable risks or to the very largest sites, combining low cost insurance with capital investment at a guaranteed rate of growth.

3.3 *Practical aspects*

Mainstream insurance brokers are naturally able to assist with general insurances for the Waste Management industry but it is prudent to appoint specialists to deal with Environmental and Professional Indemnity covers. Premiums and scope of cover depend to some extent on the relationship between the broker and the underwriter and, by implication, the underwriter's perception of the broker's competence and knowledge. As with any insurance, the underwriters' willingness to accept a risk is dependent on the quality of information provided to them. In the case of environmental insurance, this would consist of site investigation reports and, as appropriate, correspondence with regulators, licensing documentation, details of Environmental Management Systems and sale/purchase agreements. Premium costs vary hugely from one site to the next depending on factors such as the proximity of controlled waters and the sensitivity of adjacent land uses.

4 THE CLUSTER PROJECT

4.1 *The concept*

In practical terms, excavation and removal to landfill of contaminated soil is no longer an option for developers of polluted sites, a situation which is bound to have implications for Waste Managers who are now unable to accept hazardous material. At the same time, a combination of ignorance and uncertainties over the efficacy of process remediation techniques suggests that the growth of the industry will accelerate but not dramatically. In addition, owners of smaller development plots would not have room on site for windrows or thermal desorption plants. Fortunately, investigations have taken place into the feasibility of centralised soil recycling facilities, notably the Cluster Project (www.entecuk.com). The kernel of the idea is that a recycling hub would take material from a number of source sites, exchanging this initially for clean fill and ultimately for recycled soil processed as appropriate at the hub. The concept combines the sustainability of process techniques with the time advantage of dig and dump, thus providing a practical future for the reclamation of brownfield sites and, potentially, a diversification opportunity for the Waste Management industry.

4.2 *The risks and their solutions*

Classification of Waste remains an unresolved issue. The report from the Cluster project recommends that site specific 'Remediation Permits' should replace Waste Management Licences where material recycled at the hub is to be used at a receiving site. The permit would become effective from the point that the processed material leaves the hub site. Regulators need to accept that 'treated products' are no longer classified as waste but instead are recognised as economically useful and crucial to the future of sustainable development. It is believed that common sense will prevail as it is difficult to see another way forward.

Contractual relationships are necessarily complex under the Cluster framework. A source site owner may employ a remediation contractor to excavate material from the site following site investigation by an environmental consultant. This will then be taken to the hub for processing and recycled soil will be transported to the site and deposited. There is fertile ground for dispute and uncertainty here in the event that a pollution incident occurs in the future or the site is defined as contaminated under Part IIa. There is then a whole range of questions relating to how the problem originated, i.e.

- Did the site investigation miss a 'hot spot'?
- If a 'hot spot' was missed, was this due to negligence?
- Was the excavation sufficiently thorough?
- If the excavation work was satisfactory then is there an alternative explanation for residues having risen to the surface, e.g. as a result of rising water tables?
- Has the hub process been effective in removing pollutants or was the soil returned to site in a contaminated state?

The foregoing questions involve at least four parties, i.e. consultant, contractor, site owner and hub operator, any of whom might eventually be faced with a potentially ruinous bill for remediation costs and legal expenses depending on the judge's decision. For this reason, insurance is crucial to the success of the concept as it is only by tying up all of the potential liabilities in a scheme of cover to include all parties that it can be made safe or desirable for any of them to contribute to it. Instead of initiating a protracted and costly legal action, Party A can submit a claim under the policy, whilst party B, being named as a Joint Insured would be ineligible for recovery by the Insurers.

5 CONCLUDING REMARKS

This paper's intention has been to outline the environmental risks faced by the Waste Management industry, not least in its role in the land reclamation process, and the ways in which insurance can be used as a secure and cost effective means of transferring these risks elsewhere. The uncertainties presented by the Landfill Directive demand a revolution in terms of attitudes to remediation techniques and the very definition of waste itself. Insurance is there to create an atmosphere of confidence for this change and to support the practical steps, such as centralised soil recycling, that will be needed for a sustainable future.

REFERENCES

Anon (2004) *Enviro-Tech New (UK)*, Issue No. 65, April.
Anon (2004) *CLUSTER Project: Soil Treatment Hubs for Off-site Treatment, Recovery and Re-use of Contaminated Soils – a viable alternative to landfill*, Entec Ltd, April.
Brierley D A (2003) *Environmental Insurance – from Contamination to Confidence*, in *Land Reclamation, Extending the Boundaries*, Proc 7th International Conference of the International Affiliation of Land Reclamationists, Balkema Publishers, ISBN 90 5809 562 2 pp135–137.
Brocklehurst M (2004) *'No Time to Waste'*, Environment Business, May.
BURA (2002) *Memorandum by BURA Steering and Development Forum to the Urban Affairs Sub-Committtee into the Effectiveness of Government Regeneration Initiatives* (unpublished report), British Urban Regeneration Association.
ENDS (2004) *Report 351*, April, pp 4–5 and 30–33.
LGA (2002) *Something Old, Something New*, Local Government Association, ISBN 1 84049 294 5.
McBarron M, Nathanail C P, Nathanail J and Morley J G (2004) *Risk-based Management of Land Contamination at Foundry Sites*, Castings Technology International , ISBN 0 953309 061.
OECD (2003) *Environmental Risks and Insurance*, Organisation for Economic Co-operation and Development, ISBN 92-64-10550-6.

Geotechnical and Environmental Aspects of Waste Disposal Sites – Sarsby & Felton (eds)
© 2007 Taylor & Francis Group, London, ISBN 978-0-415-42595-7

Design for recyclability: Product function, failure, repairability, recyclability and disposability

A.J. Felton & E. Bird
University of Wolverhampton, UK

ABSTRACT: The early 1960's saw the growth of the consumer revolution and emergence of 'lifestyle' products and 'designer' label. With consumerism came waste driven by the design and changes in design thinking from product longevity and repairability of the immediate post war era to the short term and throw away mentality of the 1990's. Currently professional design thinking is undergoing change, with the life cycle of the product from design to disposability becoming a major concern. Using past and present case studies this paper interrogates the product life cycle issues of function, failure, repairability, recyclability and disposability. Case studies are drawn from household, consumer and automotive product areas, including packaging.

1 THE PAST PERSPECTIVE

The consumer society as we know it today grew out of the post second world war austerity of the 1940's and 1950's when the industrial infrastructure of both the United Kingdom and mainland Europe was recovering from six years of the second world war. Rebuilding the industrial base to supply peacetime as opposed to wartime demands was slow, hampered by shortages of materials, energy and skilled labour, it was very much a time of 'make do and mend' with the rationing of food, raw materials and consumer goods.

It was not until the late 1960's that we started to see the growth of the consumer society and the emergence of 'lifestyle' products and the 'designer' label. Consumerism was to become rampant over the next thirty years and as consumer societies are wasteful societies by the new millennium domestic waste was endemic and the throw away society had arrived.

1.1 The throw-away society

In 2001–2, according to government statistics (Anon, 2004), 25.6 million tonnes of household waste was disposed of in landfill sites within the UK. According to 'Waste Not, Want Not' published by the UK Government Strategy Unit (Anon, 2002) the predicted volume of waste is set to double by 2020 and the cost of its disposal will increase annually by £1.6 billion a year. This figure refers to all waste, which includes industrial, agricultural, constructional and household. Against this background of projected increase a EU directive requires the United Kingdom to radically reduce its landfill over the same period. The report goes on to outline a number of possible actions, one of which is tackling waste through education. A growing part of the domestic waste is discarded consumer products, 998 out of every 1000 of which have not been designed with recyclability or disposal in mind. Being involved in the education of tomorrow's product designers who will be responsible for the design of future generation of products we believe sustainable and conservation issues need to be high on the design agenda.

In order to help prevent household waste spiralling out of control we need to effect a paradigm shift in thinking among tomorrow's product designers, who not only need to design products but product life cycles as well. The life cycle needs to become an integral part of design thinking for every product whether it is small or large. Designers need to consider a product's life cycle from it's birth through to it's eventual retirement and disposal. We need to establish a culture among tomorrow's product designers of design for longevity, repairability and recyclability.

The post 1960's throw away culture of fashion-led mass consumption has been supported by the constant restyling of products so that the status conscious 'must have' consumer can fulfil their desire and keep up with fashion by possessing the latest model. This process now means that the consumer is spoilt for choice with a global market saturated with so-called 'lifestyle' products and 'designer' labels. Consumption is now a disease, possession is status, shopping is now classed as a leisure activity and waste endemic. We need to effect change. One product design area where there has been a significant cultural shift is the automobile industry. Car design during the 1960's and 1970's started to move towards the throw away concept, with limited life expectancy and design for obsolescence. In order to stimulate sales the car achieved fashion status with regular new models and upgrading to encourage the consumer to trade in every two years. Throw away components, less repairability, body shells prone to deterioration and metal fatigue and constant restyling became the norm. Next to buying a home the car is possibly the next most important financial investment. Gradually the world car market changed. Consumers expected not only reliability but longevity so that today a car is considered to be a long life product backed by at least 5 year anti-corrosion warranty on the body shell and at least 3 year unlimited mileage guarantees on its mechanical components.

1.2 *Legislation driven design*

Legislation particularly through the European Union, ensures that car design not only meets safety, emission and fuel consumption standards, but also now meets recycling and disposal requirements when the product is finally retired.

We need to expand the achievement in the car industry into other product areas. The car possibly ranks as the premier consumer product but few others have legislative requirements for their disposal after retirement unless they contain what is deemed to be hazardous or harmful materials. One such product is the domestic fridge or freezer because of the CFC used as a coolant.

Initially CFC's in fridges were not considered harmful but recent research has acknowledged that their release into the atmosphere contributes to global warming. Modern manufacturing technology used in the construction of fridges (Figure 1) has not helped. In many cases it is difficult to remove the parts containing CFC's without damage and subsequent release into the atmosphere. It may

Figure 1.   Fridges waiting processing at Civic Amenity Site.

312

be a case of having to look back to older construction technologies where all components can be easily removed and the dangerous parts separated from the cabinet.

1.3   *Design for repairability and recyclability*

Up to 1955 the UK was in the grip of post war austerity with rationing in one form or another. It was a time of make do and mend; repairability and recyclability were the order of the day. Products were only disposed of when they had completely reached the ends of their lives and could no longer be repaired. Product construction was simpler and based on a bolt on culture, which meant that parts could be removed and replaced easily. Reconditioning was the norm, motors and generators were rewound while engines were factory rebuilt. Many manufacturers offered a reconditioning service on their components with part-exchange allowances for trading in worn out parts for a reconditioned replacement.

Many manufacturers offered a range of reconditioned products alongside their new ranges, particularly products such as vacuum cleaners, washing machines, radios, TV's and even smaller items such as irons and toasters. In a time when there was a scarcity of new products a thriving second-hand and reconditioned market existed. Contrast this with today's products where few are repairable and reconditioning is non-existent. The consumer has little alternative but to replace by buying new. We need to evaluate the design of individual products and where replacement is economically viable promote it as a design feature.

One phenomenon of the UK's current credit crisis with personal debt reaching a record one trillion pounds, is that 80% of new products being bought are replacements for existing products that are still usable and have plenty of life left in them. We are upgrading manly for status, to posses the newest model while the items we are replacing find their way into landfill.

We need to develop a culture of product thrift and create classic items that do not age and are perceived as forever modern. Perhaps we should learn from the current trend for nostalgia, where manufacturers have found it lucrative to re introduce old products as design classics. We now have a whole range of domestic consumer products such as toasters, food mixers, fridges and kettles that have been reintroduced. It is not a bad thing to learn from the past; a constant search for the new is not always the answer particularly if older solutions mean less waste.

Many new products are developed using the established practices of design for assembly and design for manufacture; however, future design needs to focus more on design for disassembly, maintenance and recyclability.

2   THE PRESENT PERSPECTIVE

2.1   *Educating product designers for the future*

This section is conceived from an educational context and the ways in which product designs can be developed with undergraduate product design students for recyclability and disposability.

Historically many proposed design process models cover the main areas of product development from conception to retirement. Dieter (1986) discusses the final phase of the Morphology of Design as 'Planning for Retirement of the Product'. This is the disposal of the product at the end of its useful life cycle, usually because of one of the following associated factors:

- The product can no longer function correctly through wear and deterioration.
- Technical advancement, when new products out perform older designs.
- Fashion or trend changes.

It is this latter stage of Planning for Retirement of the Product that is becoming a major concern at the University of Wolverhampton it has become an important factor within the curriculum for teaching Computer Aided Product Design to undergraduate students. The goal is to design products that satisfy the customer but minimise environmental impact over the life cycle, often referred

Figure 2.   Product life cycle retirement – many products end up in landfill.

to as Sustainable Design, thus trying to alleviate the many products that often end up in landfill (Figure 2).

With reference to 'environwise' (2004), wherein practical environmental advice is given for business, the term 'Cleaner Product Design' is used. This Cleaner Product Design cycle promotes continual improvement of both new and existing products via;

(1) Product research
(2) Identifying cleaner design priorities
(3) Designing the cleaner product to develop its form and function and
(4) Design Review.

In developing Cleaner Product Designs 'environwise' (2004) also gives a number of key considerations to reduce a product's environmental impact through:

• Reduced raw material
• Reduced use of energy
• Less pollution and waste
• Elimination of hazardous materials
• Increased service life
• Greater potential for recycling.

2.2   Sustainable design projects

Recycling now plays an integral part of the design brief for student projects. Figure 3 and Figure 4 shows first year design and make projects. The design brief was to design a chair made from cardboard for a summer Classical or Rock concert. The chair would be given on entry to the concert as a flat pack and then assembled using only mechanical joints for fixing. At the end of the concert the chair could be taken away or recycled.

Figure 5 shows a student's design solution for tool storage. When buying tools for DIY most of the packaging the tools come in is discarded. In this particular case the packaging becomes the tool storage medium and when sold as individual items the packaging is simply locked together.

Figure 6 gives a student's design solution for Re-Usable postage boxes in place of the protective postage bags that are often used once then discarded. The Re-Usable postage boxes have an

314

Figure 3.   Recyclable cardboard seat for outdoor summer concerts.

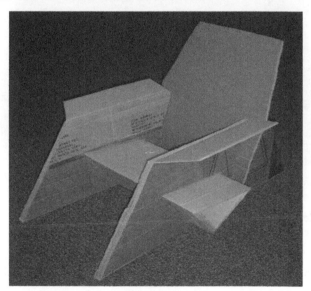

Figure 4.   Recyclable seat produced as flat pack and assembled on entry to concert.

international theme and offer features such that they can be manufactured with the appropriate country's postal logo, bar coded for automatic sorting, high impact protection for sensitive goods such as electronic components, CD's etc, a tagging device for security and sold via the internet.

2.3   *Design competitions*

The title of the Product Brief for 2003–2004 was 'Sustainable Design' and was based around a national design competition sponsored by The Design Council. The design brief asked the student to select and appraise an object and redesign it to;

 (i) Achieve both a decrease in the environmental impact from the creation, use and disposal of the product and,
(ii) An increase in the business potential of the product.

315

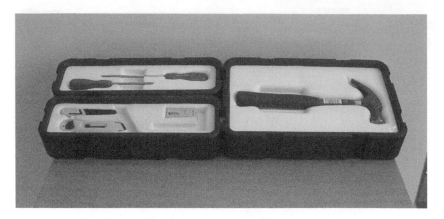

Figure 5. Tool storage device, packaging provides the storage medium.

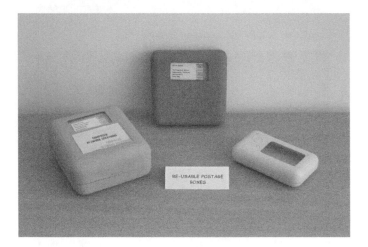

Figure 6. Re-usable postage boxes.

Some of the design solutions presented by students for sustainable designs included:

- ECO Building Block (designed using miscanthus grass as the primary material)
- Oil box (engine oil packaged in an oil bag)
- Integral Shower and bath compartment (compact shower/bath that uses less water/energy)
- Reduced energy cooking pot (cooking vessel that uses reduced energy to cook)
- VegTable (a product for growing vegetables in confined housing space)
- Tree Cuirass (a protection device for young saplings/trees)
- Automatic light switch (switches off interior mains lights in the home when no personnel are present).

Although all of the student's designs had a sustainable design theme the first two products the ECO Building Block (Figure 7) and the Oil Box (Figure 8) paid particular attention to the end of the product life cycle and how they could be recycled with reduced impact on land-fill. The ECO Building Block weighs in at just 4 kg compared with 14 kg for a concrete block with possibly incineration or mulching at the end of its life cycle, whereas the design of the oil

316

Figure 7.   ECO block.

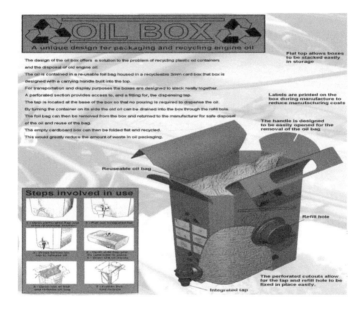

Figure 8.   Oil box.

box offers a solution to the problem of recycling plastic oil containers and the disposal of old engine oil.

3   CONCLUSIONS

Sustainable and conservation issues need to be high on the design agenda for the education of tomorrow's product designers who will be responsible for the design of future generation of products.

In order to help prevent household and consumer waste spiralling out of control we need to effect a paradigm shift in thinking among tomorrow's product designers, who not only need to design products but product life cycles as well from conception to retirement.

Many new products are developed using the established practices of design for assembly and design for manufacture; however, it is necessary to take into account design for disassembly and establish a culture amongst product designers of design for longevity, repairability and recyclability.

REFERENCES

Anon (2002) *'Waste Not, Want Not'*, UK Government Strategy Unit, November.

Anon (2004) *'UK 2004'* The Official Yearbook of the United Kingdom 55ed pp.269.

Dieter G. E. (1986) Engineering *Design; A Materials and Processing Approach*. 1st ed., USA: McGraw-Hill. pp.37–39.

'environwise' (2004) *Cleaner product design: an introduction for industry* (GG294), http://www.environwise. gov.uk

Alternative ways of treating hazardous waste

N. Gaurina-Medimurec & G. Durn
University of Zagreb, Croatia

H. Fröschl
Seibersdorf Research, Austria

ABSTRACT: Technological waste (tank bottom residues, separator sludges, oily soils etc) is generated during oil and gas production. Part of this waste is considered hazardous. Central oilfield pits (COJ) are used for temporary disposal, pretreatment, treatment and final disposal of these wastes. After treatment to change the hazardous waste to non-hazardous material it is deposited in a pit area assigned to final disposal of solidified material. Because of the accumulation of inert material inside COJs and their limited volume, subsequent disposal becomes problem, and the possibilities for reusing the treated waste have to be considered. The paper contains results of laboratory investigation of new ways of treatment of technological waste from the petroleum industry. The new techniques of treatment are being developed for materials which contain organic or both organic and inorganic pollutants. Samples of technological waste from a central oilfield pit were treated under laboratory conditions with single materials (cement, bentonite, organophilic clay, calcined Moler clay) or different material combinations. A sample of technological waste treated with lime was used for comparison of results. The results clearly show that hazardous technological waste can be treated more effectively by dosing with more suitable materials, e.g. organophilic clay, which significantly reduce pollutant discharge from such treated waste into the environment.

1 INTRODUCTION

In petroleum engineering there are two main technological processes that may have an influence on the environment, i.e. drilling and production. Different materials are used in both technological processes and technological waste is generated. Some of the waste can have significant potential for harmful effect on the environment so it is necessary to manage created waste responsibly (Gaurina-Medimurec, 2000; Gaurina-Medimurec, 2002). In virtually all cases, the harmful impact can be minimized or eliminated through the implementation of proper waste management.

Waste is generated from a variety of activities associated with petroleum production. Technological waste from the processes of drilling and production falls into the general categories of drilling waste, produced water, and associated waste. The most general classification considers the origin and volume of generated waste (Table 1). Primary waste is created in large quantities, but it is mostly waste with low toxicity in comparison with associated waste (Gaurina-Medimurec et al, 2001). Associated waste represents mainly the materials that are used and thrown away in production operations. Its small quantity and high toxicity are the characteristics of associated waste.

Items that have the potential to cause environmental damage are heavy metals, salt compounds, organic wastes (including hydrocarbons), acids or bases and total suspended solids (TSS). They can have detrimental effects on human health, animals, fish and other aquatic organisms and on plant growth. The level of impact that discharge makes on the environment depends on the type of material contained in the waste and the environment into which they are discharged.

Table 1. Primary and secondary waste from drilling and production operations.

Wastes	Primary waste	Associated waste
Drilling waste	Drilling muds Drilling cuttings	• Rigwash, • Service company wastes: empty drums, spilled chemicals, workover and completion fluids, spend acids, etc.
Production waste	Produced water	• Oily waste: tank bottoms, separator sludges, • Used lubrication or hydraulic oils, • Oily debris, filter media and contaminated soils, • Untreatable emulsions, • Produced sand, • Dehydration and sweetening wastes, • Workover, swabbing, unloading, completion fluids and spent acids, • Used solvents and cleaners, etc.

There are several options for treatment and disposal of waste created in petroleum engineering. The different treatment methods vary significantly in efficiency and cost. Most waste treatment processes include separation of waste in its particular component, i.e. water-solids-hydrocarbons (Gaurina-Medimurec, 2000; Gaurina-Medimurec et al, 2001).

In recent years, the technology that enables permanent disposal of waste (in the form of waste slurries) by injecting it into the subsurface, without separating solid and liquid phases, has been developed. Selection of the method for managing technological waste depends on law regulations, the eco-system of the location where the operations take place and how economical is the selected procedure (Gaurina-Medimurec, 2002; Gaurina-Medimurec et al, 1998).

2 WASTE MANAGEMENT IN CROATIA

According to the current waste regulations (Anon, 1995), oil field waste in Croatia is treated as a technological waste. Regulation of this waste is undertaken at state level (Anon, 1996; Anon, 1997).

Different wastes are managed using different approaches. The common practice in Croatia is pit disposal of solid wastes (on-site treatment with lime and burial) and injection of wastes in a liquid phase (Anon, 1998). Mud pits are used for collecting and temporarily storing drilling mud, drilling cuttings and sometimes associated waste generated during drilling and well completion. Central oilfield pits are mainly used for collecting associated wastes. Previously-applied strict environmental regulations forced the petroleum industry to close abandoned mud pits and to return the sites to their original condition. In the nineties, a mud pit closure programme was initiated (Anon, 1990). Most existing mud pits were closed by the solidification method and now there are six active central oilfield pits. The central oilfield pits (COJs) are given the names of the places where they are located. Production activities take place in three different districts named Podravina, Posavina and Slavonija. Each of them consists of one or more production units (fields) where waste is generated (Durn et al, 2003).

The estimated annual volume of secondary waste that should be treated and the central reserve pits in which each production unit disposes its waste are presented in Table 2.

2.1 *Overview of central oilfield pits (COJs)*

Table 3 contains data about the construction of central oilfield pits, their nominal capacity to accept generated solid technological waste, as well as current rate of filling of these pits (Durn et al, 2003)

Table 2. Estimated annual volume of solid waste that should be treated.

District	Production unit (Field)	COJ	Volume (m³/year)
Podravina	Šandrovac, Molve, Koprivnica	Šandrovac	1500–2000
Posavina	Ivanić grad, Žutica, Etan, Dugo Selo, Šumećani	Žutica	2500–3000
	PSP Okoli, Stružec	Stružec	500–1000
Slavonija	Lipovljani	Lipovljani	<500
	Beničanci	Beničanci	1000–1500
	Vinkovci	Vinkovci	<500

Table 3. Central oilfield pits (COJ).

COJ	Construction (isolation)	Capacity (m³)	Filling up (%)
Oilfield pit as a disposal places of waste			
Lipovljani	Earth pit (plastic lined)	1000	80
Šandrovac	Earth pit (plastic lined)	5000	60
Beničanci	Earth pit (plastic lined)	10000	60
Žutica	Earth pit (compacted clay)	10000	75
Oilfield pit as a treatment plant			
Stružec	Earth pit (compacted clay)	1000	–
	Concrete pit (plastic lined)	300	–
	Disposal area (compacted clay)	500	100
Vinkovci (1997)	Earth pit (compacted clay)	1000	20
	Concrete pit (plastic lined)	300	–
	Disposal area (compacted clay)	2000	20
Molve	In progress		

With regard to the pits identified in Table 3:

- The central oilfield pits at Lipovljani, Šandrovac, Beničanci and Žutica were constructed by excavating the soil and they were isolated from the surrounding ground by placing a special plastic liner or a compacted clay layer. They only serve as places for disposal of waste without possibility of it being treated.
- Two central oilfield pits at Stružec and Vinkovci were transformed into treatment plants, so they serve as places for treatment and permanent disposal of waste.
- Central reserve pit at Molve is also going to be constructed as a treatment plant.

According to the data presented in Table 3, the filling up of existing pits (COJs) requires the reconstruction of these existing pits and exploration of the possibilities for reusing waste material which has been adequately treated so that it is classed as inert.

2.2 Vinkovci central oilfield pit (COJ-V)

The form and operational mode of central oilfield pits is illustrated by using the pit at Vinkovci as a typical example. COJ-V was constructed in 1988, and started to receive waste the following year. During the next eight years the pit was completely filled with solid waste so disposal of further quantities of waste became impossible. To enable further usage of the pit some reconstruction of the existing pit was undertaken in 1997 and additional quantities of technological waste were dumped (Durn and Gaurina-Medimurec, 2002). The components of the pit are shown in Figures 1 and 2.

The size of the pit at ground level is 47.20 × 51.20 m and its depth is 4.50 m. The volume of COJ-V is 3000 m³. The pit bottom and walls were built to be impermeable to ensure that disposed

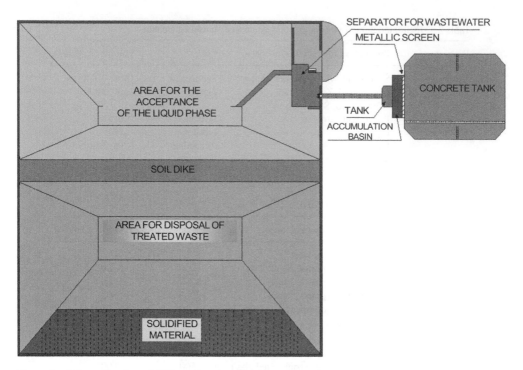

Figure 1. Schematic view of COJ-Vinkovci.

Figure 2. The concrete tank at the oilfield pit COJ-Vinkovci.

waste stays in the pit without negative impact on the environment (Durn and Gaurina-Medimurec, 2002). The central area is divided into two parts by a soil dike/embankment – this provides an area for acceptance of the liquids (mostly water) from the waste water separator and an area for disposal of treated waste. One sidewall of the tank (Figures 1 and 2) contains an accumulation basin. It serves

Figure 3. Part of the COJ-V oilfield pit designed for the acceptance of the liquid phase.

to collect the drained oily liquid phase from the waste and the water due to rainfall as well. The metallic screen on the entrance to the accumulation basin stops coarse particles of waste entering. During waste disposal the collected oil flows from the accumulation basin towards the separator for wastewater. The separator is made up of a deep concrete basin with a partition-wall inside it that permits overflow of fluids. The role of the separator is to divide oil and water. Separated oil is transported by pipeline through the delivery station to the transport system for oil. Separated waste water is drained into a part of the oilfield pit designed to accept the liquid phase (Figure 3). Subsequently, waste water is transported into the pumping unit for brine, where it is mixed with brine and finally is pumped into the injection well.

Associated technological waste (sand, silt, and oil after maintenance operations on production wells, tank bottom sediment after reservoir cleaning, oily soil from production well sites, etc) is continuously deposited in the concrete tank. When a sufficient volume of waste has collected (about 200 m^3) solidification with lime is undertaken. When treating waste with a high hydrocarbon content, clayey soil is added to the waste for better and more successful treatment. A system of paddles mixes the waste with soil and lime in the concrete tank. After the process of solidification, the treated technological waste (which is a solidified material) is transported from the concrete tank to the dedicated disposal area the oilfield pit (Figure 4). The concrete tank is then empty and ready to receive further volumes of waste. However, the free space within the dedicated disposal area is continuously decreasing and requires construction of new places for disposal and/or investigation of the possibilities for reusing the solidified waste material.

3 LABORATORY INVESTIGATION OF NEW WASTE TREATMENTS

3.1 *Tests*

Laboratory investigations have shown that solidification with lime should not be used as a universal method for the treatment of drilling and production waste (Durn et al, 2002; Durn and Gaurina-Medimurec, 2002; Durn et al, 2003). The new approach to this problem is based on the development of new methods that include treatment of waste by materials that can tie organic pollutants prior to the process of stabilization and/or solidification, thereby improving the actual

Figure 4.   The area for disposal of the solidified material (COJ-V).

process of stabilization/solidification (Durn et al, 2002). Organophilic clay, zeolite, and calcined siliceous earth are the most commonly used absorbents.

A waste sample from the Vinkovci central oilfield pit (COJ-V) was used for the investigation of alternative ways of waste treatment. The composite waste sample was treated with different materials or with combinations of several materials (Table 4). For comparison of results a sample of waste treated with lime (VS-1) was also used. In this way samples of waste treated by alternative methods, but containing the same quantity of waste (700 g) and treatment material (300 g), were made to enable comparison of the behaviour of the samples. For treated samples and their leachates the following parameters were analysed:

- Total oil content and mineral oils only – using IR spectrometry,
- Polycyclic aromatic hydrocarbons (PAHs) – using Gas chromatography and Mass spectrometry,
- Heavy metals and potentially toxic elements – using Inductively coupled plasma-mass spectrophotometry and Atomic absorption spectrophotometry.

The leachate test was carried out according to the standard method DIN 38414 Part 4.

3.2 *Results*

To determine the efficiency of the applied method it is necessary to compare the content of measured parameters for the treated sample with the same parameters in the leachate of the treated sample. For that purpose, as a measure of treatment efficiency, two factors were used:

- *Ratio (A/B)* to represent how many times higher is the concentration of the measured parameter in the total sample than in the leachate of the sample
- *Factor of Treatment Efficiency* represents the value obtained by dividing ratio A/B for a parameter in the treated sample with ratio A/B for the same parameter in the lime treated sample. For example, if the Factor of Treatment Efficiency for an analysed parameter of a treated waste is 2,

Table 4. Alternative treatment of the laboratory waste sample.

Sample Code	Total sample weight (g)	Portion of each component in sample (g)					
		Waste	Organophilic clay	Bentonite	Calcined Moler clay	Lime	Cement
VS-1	1000	700				300	
VS-2	1000	700					300
VS-3	1000	700	300				
VS-4	1000	700			300		
VS-5	1000	700	150	50		100	
VS-6	1000	700	150	50			100
VS-7	1000	700			150	150	
VS-8	1000	700			150		150

Table 5. Concentration of total hydrocarbons and mineral oils in samples and leachates.

Sample	Total oils and mineral oils, mg/kg				Mineral oils, mg/kg			
	Sample (A)	Leachate (B)	(A/B)	Factor	Sample (A)	Leachate (B)	(A/B)	Factor
VS-1	140363	2062.6	68.1	1.00	102247	1220	83.8	1.00
VS-2	89339	981.8	91.0	1.34	74332	373.3	199.1	2.38
VS-3	296610	69.1	4292.5	63.03	163421	29.1	5615.8	67.01
VS-4	124748	1350.5	92.4	1.36	84738	401.5	211.1	2.52
VS-5	100264	7157	14.0	0.21	68400	4010.8	17.1	0.20
VS-6	125961	672.1	187.4	2.75	86676	318.5	272.1	3.25
VS-7	100402	2176	46.1	0.68	65658	1220	63.7	0.76
VS-8	119648	6064	19.7	0.29	84727	373.3	25.5	0.30

it means that for the treated waste water leaches out only half the quantity of the chosen parameter that it leaches from the waste sample treated with lime.

Results for the total petroleum hydrocarbons, mineral oils (Table 5), naphthalene and lead (Table 6), are presented whilst for other analyzed parameters only the most important results will be mentioned.

According to data presented in Table 5 it can be concluded that organophilic clay is the most successful treatment material (VS-3) for total oil content and mineral oils only. In second place, according to ratio A/B and the Factor of Treatment Efficiency, is the sample treated with a combination of organophilic clay, bentonite and cement (VS-6).

According to data presented in Table 6 it can be concluded that organophilic clay is also the most successful treating material (VS-3) for dealing with naphthalene and lead. In second place, according to ratio A/B and the Factor of Treatment Efficiency, is calcined Moler clay (VS-4).

In the case of inorganic pollutants related to heavy metals and potentially toxic elements, organophilic clay was the most efficient stabilizer for the most part of analyzed heavy metals (Figure 5). For example, the waste sample treated with organophilic clay discharges into water 136.51 times less lead, 7.31 times less nickel and 2.9 times less molybdenum than the waste sample treated with lime. For chromium and copper organophilic clay is in the second place, after calcined Moler clay, according to the Factor of Treatment Efficiency.

The results of laboratory investigations have shown that organophilic clay was the most efficient stabilizer for most part of the organic pollutants studied (Figure 6). It showed a very high Factor of Treatment Efficiency, especially for total oil content, mineral oils only and naphtalene. Calcined

Table 6. Concentration of naphthalene and lead in the samples and their leachates.

Sample	Naphthalene, μg/kg				Lead, mg/kg			
	Sample (A)	Lecheate (B)	(A/B)	Factor	Sample (A)	Lecheate (B)	(A/B)	Factor
VS-1	2200	111	19.8	1.0	103	0.82	126	1.00
VS-2	4800	575	8.3	0.4	95	0.15	633	5.03
VS-3	14700	0.93	15806.5	798.3	172	0.01	17200	136.51
VS-4	2500	0.93	2688.5	135.8	134	0.01	13400	106.35
VS-5	9600	164	58.5	3.0	139	0.24	579	4.60
VS-6	10400	184	56.5	2.9	85	0.05	1700	13.49
VS-7	4800	419	11.5	0.6	111	0.77	144	1.14
VS-8	3500	425	8.2	0.4	122	0.1	1220	9.68

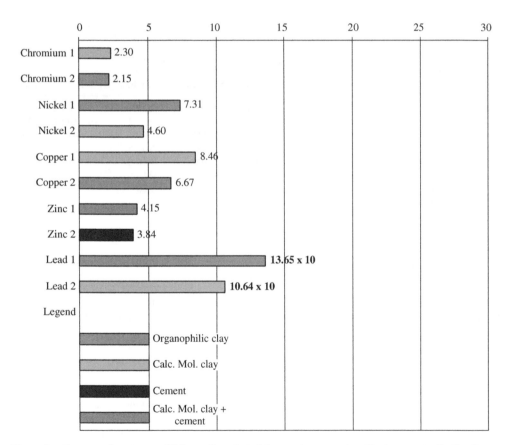

Figure 5. Factors of treatment efficiency for selected inorganic parameters (the two most effective factors are presented).

Moler clay was in second place, according to its Factor of Treatment Efficiency for organic pollutants. In spite of expectations, treatment by a combination of organophilic clay and cement showed very bad results. It is considered that the reason may be the short time period between the treatment with organophilic clay and with cement.

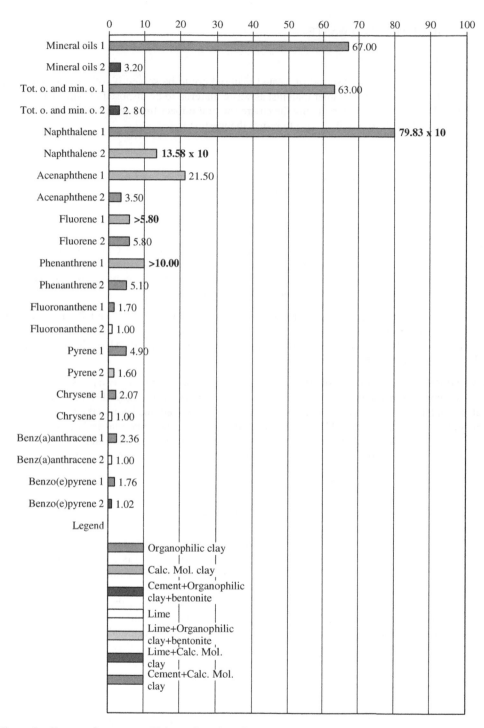

Figure 6. Factors of treatment efficiency for selected organic parameters (the two most effective factors are presented).

4 RE-USE OF TREATED WASTE

4.1 *Options*

The problem of permanent disposal of treated material could be solved if solid wastes could be considered as a raw material, rather than just as waste material to be disposed of. This would offer the possibility of significantly reducing the environmental impact from solid wastes.

There are at least four possible ways to re-use the wastes:

- in cement-producing plants,
- in brick-producing plants,
- for environmental restorations,
- as covering materials for urban waste disposal.

4.2 *Practical examples*

ENI/AGIP experiences can serve as a positive example. In 1999 that Italian company (in accordance with the Italian Ronch law published in 1997) started to apply waste re-using methods in two fields in Italy (Ferrari et al, 2000). In Italy, the wastes can be re-used in cement producing plants and in brick producing plants when the following specifications are met:

- WBM (water based mud) and cuttings: total hydrocarbons <50 kg/t and PAH (polycyclic aromatic hydrocarbons) <10 ppm
- OBM (oil based mud) and cuttings: total hydrocarbons <300 kg/t and PAH (polycyclic aromatic hydrocarbons) <10 ppm

Beside PAH and total hydrocarbon limitations, wastes suitable for the final two re-use possibilities, i.e. environmental restoration and covering material for urban waste disposal, should additionally match regulatory levels from the leaching test. Wastes which meet the hydrocarbon specifications (for total hydrocarbons and PAH) can be re-used directly as raw material in both cement and brick-producing plants without further leaching test requirements (Ferrari et al, 2000).

In Sicily drilling spoil was subject to consolidation and fixation and was transported to a cement producing plant. There, in accordance with cement manufacture composition and process, about 3–5% of drilling cuttings were blended with the raw materials necessary to produce Portland cement (Ferrari et al, 2000).

The expanding waste re-use and successful outcomes are significant positive signs and have encouraged economic benefits. Based on the experiences gained in Italy, ENI/AGIP's pertner in Egypt has taken a leap towards re-utilizing wastes in cement and brick industries. A 2-year contract has already been signed with local cement and brick industries, and it is estimated that a total of 30,000 tonnes of oily cuttings will be furnished to the industries. Reusing these wastes in both these plants costs only 17$ per ton (Ferrari et al, 2000).

Because of pressures due to filling of existing central oilfield pits, the Croatian national oil company is searching for more efficient methods of waste treatment and re-use possibilities rather than constructing new waste treatment plants (Durn et al, 2002; Durn and Gaurina-Medimurec, 2002; Durn et al, 2003).

5 CONCLUSIONS

Each day, the oil industry generates large volumes of waste. Injecting liquid wastes in disposal wells solves permanent disposal problem for fluids, but the problem of permanent disposal of solid hazardous waste still requires further research for adequate solutions.

Stabilization with lime is not always a suitable method for treatment of waste from the petroleum industry, particularly those that are contaminated with both organic and inorganic contaminants. In order to select the most appropriate method of treatment and to avoid the problem of pollutants leaching into the environment, detailed laboratory analyses are needed of the waste material to be processed.

Results of laboratory investigations clearly show that hazardous technological waste can be treated more effectively by application of more suitable materials, e.g. organophilic clay, which reduce pollutant release from such treated waste in the environment. In this way the problem of permanent disposal of treated waste could be solved.

There are new possibilities of reusing treated waste as a raw material in various industries but this is only possible after effective treatment of hazardous waste to transform it into inert, non-hazardous material.

REFERENCES

Anon (1990) *The closing of mud pits in INA – Naftaplin*, The main standard mining project, Zagreb, Croatia.
Anon (1995) *Waste Law*, Official Bulletin No. 34/95, Zagreb, Croatia.
Anon (1996) Regulation of waste types, Official Bulletin No. 27/96, Zagreb, Croatia.
Anon (1997) Regulation of the waste management conditions, Official Bulletin No. 123/97, Zagreb, Croatia.
Anon (1998) Regulation on the hazardous waste management conditions, Official Bulletin No. 32/98, Zagreb, Croatia.
Durn G and Gaurina-Međimurec N (2002) *Study of the efficiency of solidification, selection and efficiency of alternative methods of treatment of technological waste in central oilfield pit Vinkovci*, Faculty of Mining, Geology and Petroleum Engineering, University of Zagreb, Zagreb, Croatia, pp1–168.
Durn G, Gaurina-Međimurec N, Meandžija I, Veronek B and Mesić S (2002) *Geochemical and mineralogical assessment of lime stabilized waste from the petroleum industry in Croatia*, ETCE 2002, Houston, pp1–10.
Durn G, Gaurina-Međimurec N, Veronek B, Mesić S, Fröschl H and Čović M (2003) *Laboratory investigation of new materials for the treatment of hazardous waste from petroleum engineering*, 2nd International Oil and Gas Conference "Oil and Gas Technology Development in the World Globalization Processes", Zadar, Croatia, pp1–10.
Ferrari G, Cecconi F and Xiao L (2000) *Drilling wastes treatment and management practices for reducing impact on HSE: ENI/AGIP experiences*, paper SPE 64635 presented at the SPE International Oil and Gas Conference and Exhibition in China, Beijing, pp1–11.
Gaurina-Međimurec N (2000) *The environmental impact of drilling and production operations*, 4th International Symposium on Power and Process Plants, Dubrovnik, Croatia, pp631–636.
Gaurina-Međimurec N (2002) *Waste disposal methods in petroleum engineering*, Annual meting of the Croatian Academy of Engineering, Croatian Academy of Engineering, Zagreb Croatia, pp35–43.
Gaurina-Međimurec N, Krištafor Z and Matanović D (1998) *The subsurface injection of drilling fluid waste*, 3rd International Symposium on Power and Process Plants, Dubrovnik, Croatia, pp698–701.
Gaurina-Međimurec N, Rauker S, Bratuša Z and Veronek B (2001) *Drilling and environment protection*, 1st International Oil and Gas Conference, Zadar, Croatia, pp1–11.

Geotechnical and Environmental Aspects of Waste Disposal Sites – Sarsby & Felton (eds)
© 2007 Taylor & Francis Group, London, ISBN 978-0-415-42595-7

Construction applications of Foamed Waste Glass (FWG) – use in a rooftop vegetation system

Y. Hara
Nihon-Kensetsu-Gijyutu Company, Japan

K. Onitsuka
Saga University, Japan

M. Hara
Kensetsu-Kankyo-Engineering Company, Japan

ABSTRACT: FWG (Foamed Waste Glass) is a new material made from waste glass, and is lightweight and stiff because of its multi-porous structure. The specific gravity of FWG can be controlled from 0.3 to 1.5 by its production conditions. It is also possible to make two kinds of void structure of FWG. One structure has continuous voids with high water absorption, and the other one has a discontinuous void structure with low water absorption. The FWG of continuous void structure is usable in slope vegetation systems on bedrock, rooftops and outside gardens because it is a water holding material.

1 INTRODUCTION

Society has been taking the form of mass-consumption and mass production due to active industry growth since the last half of the 20th century. At the same time, industrial wastes have also been growing along with the rapid industrial growth. Many industrial wastes have some kind of harmful substance in themselves. They can be one of the causes of environmental damage if preventative measures are not taken. In order to reduce the global environmental disruption there is a need to make the transition to an optimum-production, optimal-consumption, minimum-waste type of society in the new millennium. However, at the present time, secure waste disposal facilities are needed to deal with wastes and recycling of the wastes is desperately needed.

For the purpose of recycling wastes, 'Miracle Sol' (Foamed Waste Glass or FWG), a new material made from waste glass, has been developed. FWG is now utilized in engineering works as a new material. New systems using FWG are divided into two groups according to the characteristics of the FWG – one is an environmental vegetation system and the other is an environmental engineering system.

2 "CLUSTER PLAN" (CONSTRUCTION APPLICATIONS MATRIX) OF FWG

The characteristics of FWG are a lightweight nature combined with stiffness because of its multi-porous structure. Depending on the production conditions the specific gravity of FWG can be from 0.3 to 1.5 and the water absorption ratio of FWG can be controlled as well. Both high water absorption types and low water absorption types of FWG are available.

Figure 1.   Cluster plan of FWG.

Plate 1.   Shapes of FWG.

Plate 2.   Inner structure.

The high water absorption type of FWG is usable for;

• slope vegetation (including over)bedrock,
• rooftop vegetation (as a water holding material),
• water purification (as a filtering material).

The low water absorption type of FWG is usable for;

• lightweight fill,
• lightweight concrete aggregate,
• concrete shotcrete as a lightweight material,
• material of ground improvement,
• weeds control medium.

This new technology using FWG can reduce construction costs and it is safe for the environment because of its lightness and harmless nature (non-polluting and stable). As shown in Figure 1, various construction systems have been proposed and these form 'The Cluster Plan of FWG', i.e. a matrix of construction applications.

This report focuses on 'The rooftop vegetation system' – an environmental greening system which uses the high water absorption form of FWG. The shape and the internal structure of the high water absorption type of FWG are shown in Plates 1 and 2 respectively.

Plate 3.   Cross-section, Miracle sol board.   Plate 4.   Cross-section, Miracle our board.

3   BACKGROUND TO ROOFTOP VEGETATION

In heavily populated urban areas in Japan it is difficult to increase the extent of green areas because of the needs of effective land use. In recent years, the 'island-warming phenomenon' (heating up of the atmosphere in summer) has become a serious problem in large cities due to over-use of air conditioners in summer seasons, and urban floods have often occurred. Moreover, the numbers of small wild creatures (such as Firefly, Dragonfly, Killifish and Water strider) are decreasing due to disturbance of their ecosystem. Thus, something must be done to prevent the destruction of the nature. Rooftop vegetation on the buildings has been proposed as one solution to the foregoing problems. In particular, a new 'thin and light' rooftop vegetation system using new materials made from wastes seems to be very suitable for use on city rooftops. This system could lead to control of global warming.

4   FWG ROOF TOP VEGETATION SYSTEM

As shown in Plates 3 and 4, there are two kinds of vegetation system available;

- the one using 'Miracle sol' with a 0.4 specific gravity is applicable to flat types of rooftop,
- the one using 'Miracle board' is applicable to both flat and sloping types of rooftop.

The advantages of these new vegetation systems are their thinness and lightness.

Though some rooftop vegetation has been installed in urban areas, because to date there were no suitable water-holding materials, a soil foundation has been used as the water-holding material so far. Consequently, a vegetation-foundation thickness of around 1,000 mm has been needed in the usual vegetation systems. Such thick-layer vegetation systems have caused overload problems for some buildings.

In order to solve the foregoing problems a new lightweight rooftop vegetation system using Foamed Waste Glass has been developed. The new system also assists maintenance because frequent watering is unnecessary due to the water holding ability of FWG. Foamed Waste Glass reduces the need to water to a minimum. In addition, the use of a thinner layer of the vegetation-supporting foundation reduces the weight of the soil vegetation system. This new system has been devised on the basis of utilizing the beneficial properties of FWG and has been in development since 1998.

4.1   *Rooftop Vegetation System Using FWG*

In this case particles of 'Miracle sol' were used to provide both the water-holding stratum and the water-draining stratum. Firstly 10 to 50 mm sized particles of 'Miracle sol' were spread on a liner sheet or a waterproof mat (2 mm thick) to form a 50 mm stratum. This stratum was then covered with a 2 mm thick Sandofu synthetic layer (made from waste plastic bottles) to prevent leaching

Plate 5.   Shadow zone before vegetation.

Plate 6.   Area 33 months after vegetation.

out of soil. Finally a 30 mm thick layer of vegetation foundation mat consisting of topsoil with 10% of 'Miracle sol' mixed in was placed on top.

The total thickness of the system is 80 mm (ignoring the thickness of the liner sheet and the Sandofu). The total weight is 34 kg per square metre.

4.2   *Case histories of the rooftop vegetation system using FWG*

Specific examples of the 'Miracle sol' rooftop vegetation system (undertaken in 2001 and 2002) are shown in Plate 6 (office building, about 80 m^2) and Plate 7 (private house, about 100 m^2). Plate 5 shows the shadow zone of the office building before execution of the vegetation works.

One feature of the two examples is the light and thin construction due to use of new material made from wastes. The other point is easy maintenance of soil water regime due to recycling of wastewater that had condensed from external air conditioning equipment in summer (as shown in Plate 8) together with the water-holding ability of the 'Miracle sol'. Even in the summer seasons the vegetation zone only needs watering once a week.

It was determined that the most suitable mixing ratio of 'Miracle sol' (with grain size of 1 to 25 mm) in the vegetation-supporting foundation layer was 10% (by volume) of topsoil. This addition enhanced the water-holding ability of the soil. The example shown in Plate 6 did not have any artificial watering within a period of 33 months after execution of the works. Even though this thin and light rooftop vegetation system was developed for high water-demand plants like grasses, of course, it is also applicable to drought-resistant plants such as Sedamus. However, in the case of drought-resistant plans the volume of water-holding stratum can be reduced.

Plate 7.   Private house verandah (about 100 m²) before and after vegetation.

Plate 8.   Water from tube of air conditioner equipment (outside).

A standard load limit on rooftop vegetation is that it should be below 60 kg per square metre, with the 'Miracle sol' vegetation system the imposed load can be reduced to nearly half of this value.

For the former case the usual situation for combined rooftop vegetation and a rooftop garden would be that more than 1000 mm depth of vegetation stratum was needed for trees and shrubs, but with 'Miracle sol' the vegetation stratum only needed to be between 300 and 500 mm thick due to its water-holding ability. In other words, the new system only needs between one-third and half of the thickness of vegetation-supporting stratum of the former vegetation systems.

In order to avoid wind damage to the plants in a rooftop garden, the drainage and water-holding layer above the liner sheet and the vegetation foundation mat layer are fastened together, to make them a strong unit, by using wire netting and metal fittings.

'Miracle sol' or 'Miracle board' as used in the new rooftop vegetation system are the products of recycling used bottles, different kinds of packaging wastes, and the mat which is used for prevention of leaching out of soil is a 100% recycled product made from plastic bottles as well. Using these wastes in the whole new vegetation system can lead to protection and conservation of the environment.

4.3   Grass growing condition and internal and external temperature change

Growing condition of grass roots when 12 days have passed since it was planted is shown in Plate 9. The average growth 12 days after planting is around 10 mm in length. The roots are around 30 mm in

335

Plate 9.   Grass growing condition 12 days after vegetation.

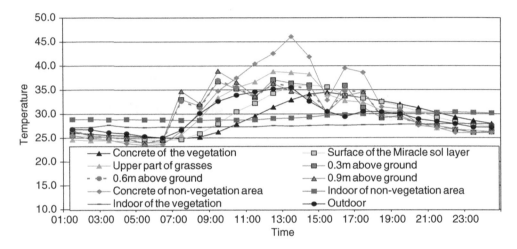

Figure 2.   Temperature change on July14, 2001 (with a 2.5 cm in layer of Miracle sol).

length 170 days after planting. The grass roots have penetrated through the Sandofu (mat which prevents leaching out of soil) and have reached toward the 'Miracle sol' (water-holding material) layer.

Measurements were made of the temperature regime in July 2001 for the building which had the 'Miracle sol' rooftop vegetation system installed in June 2001. The system consisted of a 25 mm stratum of 'Miracle sol' together with a 25 mm stratum of topsoil (total layer thickness of 55 mm) with the imposed weight amounting to 26 kg per square metre. The temperature measurements are shown in Figure 2.

As shown in Figure 2, during the daytime, when the surface temperature of the exterior concrete without vegetation cover had climbed to 46 degrees the temperature at the bottom of the vegetation system was only 32 degrees. In fact, placing the rooftop vegetation system which had a total thickness of 55 mm had generated a 14 degree temperature difference. This result clearly demonstrated the cooling effect of the vegetation system. Moreover, the temperature of the internal ceiling of the part of the rooftop without vegetation went up to 30.4 degrees maximum (30 degrees on average) whilst, on the other hand, the mean temperature of the internal ceiling of the part of the rooftop covered with vegetation was down at 27.6 degrees. The temperature difference between vegetated and non-vegetated internal parts of the rooftop was 2.4 degrees. This clearly proved the cooling effect of the rooftop vegetation system both indoors and outdoors. The temperature difference between vegetated and non-vegetated parts decreased to 1.5 degrees (from 2.4 degrees) overnight (from 1:00 to 11:00am) – the same tendency was observed on other days when temperature measurements were made.

It is generally believed that the cooling effect of rooftop vegetation systems results from watering or rainwater in the soil. In the new, thinner and lighter vegetation system using FWG plants and soil cut off direct sunlight as heat insulating materials and the FWG absorbs and holds the rainwater. Hence, it should be possible to improve thermal-insulation and cooling effects due to vegetation systems. Based on the results obtained, it is expected that further cooling effect can be obtained by increasing the volume of the FWG in the vegetation system. Therefore, installing this new vegetation system using FWG (even on existing buildings) should lead to the environmental improvement due to its cooling effect, particularly in an urban area. Hopefully, by promoting tree planting combined with city planning a great effect can be achieved to lead to environmental improvement.

5 CONCLUSIONS

The flat-roof vegetation system using FWG, which is designed to support only grasses or other small plants, imposes a weight of only 26 to 34 kg per square metre. This system incorporates a 25 to 50 mm thick stratum of FWG within the total 55 to 84 mm thickness of the whole system. The vegetation-supporting system using FWG materials is considerably thinner and lighter than conventional systems and is excellent for holding both water and fertilizer for a rooftop garden. Its low density also means that it reduces the imposed loading on the buildings.

By using a minimum layer of 25 mm of FWG in a total stratum thickness of 55 mm it was possible to obtain a temperature difference of 2.4 degrees between vegetated and non-vegetated parts of a rooftop. If the volume of FWG were increased the cooling effect would become larger because of its water-holding effect. Therefore, use of FWG could become a very effective measurement to combat the global warming.

Using the FWG vegetation system contributes to protection, preservation and creation of both human and global environment by effective utilization of recycled materials made from waste glasses and waste plastic bottles.

REFERENCES

Anon (1999) *Handbook for rooftop and wall vegetation technology*, Urban re-vegetation engineering development organization edition, 27p.
Hara Y. (1999) *Slope Vegetation System Using FWG*, The Japanese Geotechnical Society, Tsuti-to-kiso, Vol.47, No.10, pp35–37.
Hara Y. (2001) *Recycling of waste glass – Rooftop Vegetation System Using FWG*, Japan Society of Waste Management, 12th Workshop of JSWM, pp469–471.
Hara Y. (2001) *Case study on Slope Vegetation with Consideration of Natural Environment – Vegetation System using FWG*, The Japanese Geotechnical Society, Tsuti-to-kiso, Vol 49,No 10, pp13–15.
Hara Y. (2001) *New technology project*, Japan Society of Engineering, 38th Environmental Engineering Forum, special edition, pp46–48.
Hara Y. (2002) *Miracle sol construction method*, Miracle sol association, the 4th revision printed, pp1–37.
Hara Y. (2002) *Case study of rooftop vegetation in considering of the environment*, 37th Research seminar of the Japanese Geotechnical Society, pp639–640.

Use of geosynthetics technology for river embankment protection using a cellular confinement system

E. Korzeniowska-Rejmer
Krakow University of Technology, Institute of Geotechnics, Poland

A. Kessler
WODEKO Water Well Drilling and Environmental Engineering Co, Krakow, Poland

Z. Szczepaniak
Design Office Structum, Lublin, Poland

ABSTRACT: A category of geosynthetic material named the "geocell" has been successfully introduced to the construction industry during the past years. The technology of the Geosynthetic Cellular Confinement System gives a new engineering approach for designing and constructing geotechnical earth structures for slope protection and earth retaining. The Cellular Confinement System (CCS) solves construction problems associated with properties of materials used for road embankments, their confinement, drainage and stability of steep slopes with difficult ground-water conditions.

1 INTRODUCTION

The GEOWEB cellular confinement system (CCS) was developed through a cooperative research effort with the U.S. Army Corps of Engineers during the late 1970s. Advanced research, testing and field evaluation have been the foundation of GEOWEB performance ever since. Today this effort involves a host of facilities, organizations and distributors throughout the world. Cellular confinement technology is based on confinement principles of granular materials in a cellular system. The GEOWEB system is an engineered, expandable, polyethylene, honeycomb-like cellular structure. The system is utilized in the areas of slope protection, channel protection, load support and earth retention,and dramatically improves the performance of infill materials in these applications. The CCS GEOWEB is available in two distinct surface finishes; smooth and textured. The textured finish has diamond shaped indentations molded into the surface of the cell walls. This unique diamond pattern significantly increases the frictional interlock between the surface of the cell and the infill material, enhancing performance for all applications.

2 DESIGN AND CONSTRUCTION

2.1 Requirements

The flood hazard at Sandomierz was identified in July 1997 when extreme water levels occurred in the Vistula river. The level of water almost reached the top of the levee (concrete wall) situated along the road No 777 i.e. along Krakowska street. The area behind the wall (dense development) was partially flooded and a historical part of the city was seriously endangered.

Figure 1. The buttress made of CCS GEOWEB.

The old riverbed, constituting a historic area with many protected species (such as water–lily plants Nuphar luteum), had to be preserved as requested by the Monuments and Nature Conservators. However, the area assigned for construction of a new, high flood bank (between the sandbank of the Vistula river and the road) was extremely narrow.

2.2 *Solution*

It was decided to construct a buttress using CCS GEOWEB to support the new embankment at the side of the deep old riverbed. The buttress would consist of 6 layers of CCS GEOWEB (each 20 cm deep) and the bottom layers of the buttress would be visible. This was considered preferable to the alternative solution of a sealed LARSSEN wall driven in the load-carrying soil layer. The costs of this latter solution were at least twice those of the cellular buttress. The average depth of driving of the LARSSEN wall would be 8–9 m or even 12 m maximum. The buttress made of CCS GEOWEB was an excellent substitute for the LARSSEN wall. Additional sub-base of broken stone, with separation ensured by means of GEOFIBRE layers, would provide a system with elasticity and coupling with the ground layers and the foundation of the Road No 777.

The ground was characterised by variable load-carrying capacity – even plastic and soft plastic organic soils originating from river accumulation were found. Possible construction of a complete river embankment with weak soils being displaced towards the old river bed had been also considered. However, such solution was unacceptable for Monuments and Nature Conservators, because the greater part of the old river bed would be occupied. The buttress made of CCS GEOWEB (Figure 1) ensured uniform distribution of loads perpendicular to the river embankment as well as in the longitudinal direction – the complete length of protection amounts to 600 m. The assumed behaviour has been verified by several flood disasters which caused extreme river embankment loads.

2.3 *Construction*

Construction works associated with the buttress were carried out with insignificantly reduced water level in old river bed in order to maintain proper vegetation conditions for water–lily plants (*Nuphar luteum*) without displacing them out of working area. The plants survived abnormal hydrological conditions and influence of engineering activity area without any problems (Figure 1).

340

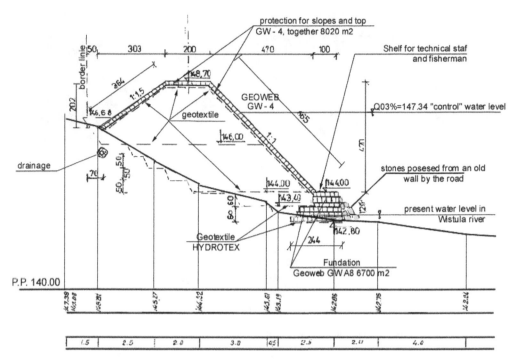

Figure 2.   Cross-section through embankment and buttress.

Figure 2 shows a cross-section of the general solution adopted (it is an abstract from the Construction Design). Apart from the buttress protection both side slopes and embankment top are protected by means of CCS GEOWEB. This latter is backfilled with soil and then grass is sown. The river embankment, which is practically an earth dam, has been provided with a drainage system along its whole length on the road side. The water drains into three embankment sluices which incorporate provisions for water pumping in case of long-lasting high water level in the Vistula river (Figure 2).

The work was initiated without any significantly reduction of the water level in the old riverbed. The first element was the 1.2 m high buttress made of CCS GEOWEB on the sub-base made of broken stone and rock filling. Existing protected plants (water–lily Nuphar luteum) were maintained. Anglers had to leave their normal positions and move to another part of the bank during the construction period. Installation of the buttress and earthworks associated with the river embankment were carried out in a particularly confined area. Plant requiring small working area and optimal logistic arrangements had to be used at the site by the contractor. Working conditions were extremely difficult as a result of river water level increases and flooding of the working area (several times). However, the CCS GEOWEB protection system was resistant to the variable conditions during the construction period and no damage arose during the work period which could be classified as 'flood damage'.

An earth dam was built on the subsoil with the support of the buttress. The maximum practicable embankment compaction parameters were assumed in the project, as would be the case for a dam in Safety Category 1, i.e. compaction indicator $I_{SW} > 0.95$ for cohesive soils and compaction ratio $I_{DW} > 0.70$. The conditions were extremely important owing to non-standard overall dimensions – the top of the levee was only 2 m wide and the river-side slope was 1:1). In performing the earthworks the contractor complied with rigorous controls associated with earthworks technology and for ensuring proper subsoil for bank protection work within the GEOWEB system. It should be noted that the earthworks were performed without stopping traffic on Road No 777.

341

Figure 3. Rear of levee showing filter fabric beneath CCS GEOWEB.

The foundation made of GEOWEB did not undergo any settlement or displacement even when the sub-soil was surcharged with the full height embankment. Protection against wash-out and excessive seepage was provided during the course of the construction. The earth 'dam' has been provided with a drainage system adjacent to the area behind the walls and with filtration fabrics under CCS GEOWEB layers situated on the slopes, over the whole surface. Installation of CCS GEOWEB on the river-side slope was under difficult building conditions. The whole protective structure is sat on fabric supplied by a Polish manufacturer. At the embankment base, on a 'sandbank of the Vistula' a 'shelf' has been created for future use by anglers. The population of protected plants has been not affected by the final phase of work (Figure 3).

CCS GEOWEB installation on the top of the levee and on the slopes (as shown in Figure 4) was followed by immediate backfilling with soil and sowing of grass mixtures. The soil was supplied directly from roadway. This location was characterised by extremely difficult working conditions, in common with other phases of the work (Figure 4).

3 PRESENT STATE OF RIVER EMBANKMENT

At the end of July 2001 a catastrophic flood wave, the greatest one which has ever been recorded by the water–level gauges of the PIMGWA (State Hydro-Meteorological Institute), passed along the valley of the Vistula river. The water level measured in Sandomierz was 60 cm higher than the previous absolute extreme value in 1997. The water–level gauge in Sandomierz was covered by water and a temporary level gauge had to be installed. Figure 5 shows the flood bank along Krakowska Street after completion of the works and passage of the extreme flood wave in July 2001. There were no technical problems with the behaviour of the flood bank. Thus in 2001 flooding of the areas to the right of the road, and also those situated close to historical buildings, was avoided due to the use of CCS GEOWEB modern technology. At the same time period other sectors of Sandomierz, which were situated on the right hand bank of the Vistula, continued to battle with the results of flooding caused by breeching of another flood bank.

Figure 4.   CCS GEOWEB installation on top of levee and on the slopes.

Figure 5.   The catastrophic flood of July 2001.

4 CONCLUSIONS

The use of CCS GEOWEB in flood protection for Sandomierz resulted in some measurable and unquantifiable advantages:

- The flood bank has been fully protected against damage by high water levels in the Vistula river (erosion, wash-out, excessive seepage, run-off pressure phenomena etc). The drainage system of the flood bank was effective. Extended flood bank sluices containing return flaps which were supported on specially-installed CCS GEOWEB and also sealed with bentonite (the VOLCLAY system), performed their functions satisfactorily. There was no damage, deformation, wash-out, movement or surface erosion in the flood bank after passage of the extremely high flood wave.
- The protection of the flood bank has been achieved without using the sealed LARSSEN wall which was the alternative solution. Driving of such a wall would mean that vibrations could be generated which would have a negative effect on adjoining buildings. On economic grounds it has been found that the proposed solution, i.e. the costs of using the buttress made of CCS GEOWEB, were 50% lower, even without considering any compensation for damage caused by vibrations.
- Construction of a Class 1 standard flood bank was practically impossible without any special revetments and protection of the slope in this case. Stability calculations of the flood bank as an earth dam showed that, without CCS GEOWEB use (and on the river-side slope also), the alternative sealing wall would have to be driven to significant depth and installed 1.5 above the level of the old river be. This would result in increased cross-sectional area of the LARSSEN sheets, higher costs associated with the task and it would be impossible to form a 'sand bank on the Vistula river' for use by anglers.
- The technology applied allowed all the demands of environmental protection, as well as the demands of landscape protection, to be met. Rapid completion of the river embankment using this technology protected several buildings and transport infrastructure against flooding.

REFERENCES

Korzeniowska-Rejmer E (2000) *Geotworzywa w konstrukcjach wzmacniających i ochronnych stromych skarp na przykładzie odbudowy kopców krakowskich,* Konferencja Naukowo-Techniczna, Zagospodarowanie gruntów zdegradowanych. Badania, kryteria oceny, rekultywacj,. Mrągowo, Poland, November.

Martin S M, Senf F D and Crowe E (1995) *A New Era In Earth Retention Using The Cellular Confinement System,* Geossinteticos 95, Sao Paulo, Brazil, June.

Martin S M, Senf F D, Kessler A and Jędrzejewski T (1998) *Umacnianie i rekonstrukcja grobli, obwałowań, nasypów oraz ubezpieczanie koryt budowli wodnych systemem Geoweb,* Gospodarka wodna Nr 1.

Szczepaniak Z (1998) *Projekt wykonawczy wału przeciwpowodziowego przy drodze krajowej regionalnej nr 777 Sandomierz-Złota od km. 164+715 do km. 165 +555,* Lublin, Poland, August.

Geotechnical and Environmental Aspects of Waste Disposal Sites – Sarsby & Felton (eds)
© 2007 Taylor & Francis Group, London, ISBN 978-0-415-42595-7

Leakage from the oil pipeline between Struzec pumping station and Sisak refinery

B. Muvrin

Faculty of Mining Geology and Petroleum Engineering, University of Zagreb, Croatia

I.B. Kaličanin

INA- Naftaplin, Zagreb, Croatia

ABSTRACT: Pipelines are the most economical way of transporting crude oil and so are the most common mode of transportation. The pipeline transportation system in Croatia is almost 400 km long with pipe diameters ranging from 0.1 m to 0.5 m. Each section of pipeline is 5 to 80 km long and has a storage capacity of over 7,000 m³. Any kind of damage or breakage in such a pipeline is dangerous for the environment. These pipelines often go through environments which are highly sensitive from an ecological point of view, such as rivers, woods, agricultural regions, populated areas, etc. Even small leakages can cause major ecological accidents, so it is of great importance to locate any defects as soon as possible. After leakages in the pipeline between Struzec and Sisak in 1999 and 2000 inspection of the condition of the pipeline started. A combination of radiographic and ultrasonic measurements of pipe wall thickness was undertaken at various locations. Where the pipeline was badly corroded it was changed (a length of 800 m).

1   THE STRUZEC–SISAK REFINERY OIL PIPELINE

The main oil pipeline between OS Struzec (pumping station) and Sisak Refinery is 0.5 m in diameter and is of vital importance for the petroleum consortium INA-Naftaplin because it is the transportation route for almost all of the consortium's oil production (Figure 1). Crude oil from the oil fields of the Moslavina basin and condensate from the gas-condensate fields of the Podravina basin are transported through this main pipeline between the pumping station at Struzec and the refinery at Sisak.

1.1   Pipeline details

Characteristics of the pipeline between pumping station Struzec and Sisak Refinery are (Anon, 2002):

- Length of pipeline: 13.992 km
- Diameter: 0.5 m
- Capacity: 3,323 m³
- Pipe quality: Conforming to Standard API X-52
- Transported fluid: Crude oil 35, 0–51, 5° API, sulphur content 0.35%, water content less than 1%
- Built: 1980
- Corrosion protection: Cathodic protection

1.2   Pipeline route

The main pipeline is divided into two sections (Figure 2):

(a) From the pumping station at Struzec to the receiving and transfer/cleaning station Cvor1.
(b) From cleaning station Cvor1 to the receiving station at Sisak Refinery.

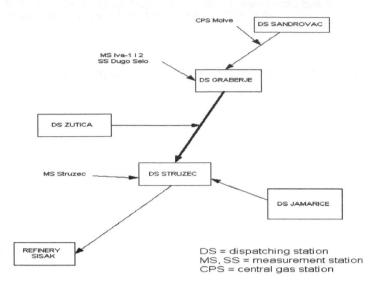

Figure 1.   Main pipelines the Podravina and Posavina regions.

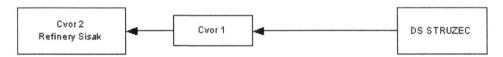

Figure 2.   Main pipeline from pumping station Struzec to Sisak Refinery.

The pipeline is built into the frame of the bridge over the Sava river and it has several dips along its route, where it;

- passes under the Obzev canal,
- crosses under regional roads (several times),
- passes beneath the Lonja river,
- passes through the Brezovica forest (in this region the level of the pipeline is depressed by 12 m),
- crosses under the old riverbed of the Sava river,
- passes over the Sava river.

In addition the Lonjsko Polje nature park is situated along the route of the pipeline and so the integrity of the pipeline is of major importance from an ecological point of view.

2   LEAKAGE HISTORY

The first leakage happened in September of 1999 in a field around 100 m from the pumping station at Struzec. A clamp with oil resistant rubber seal was installed because the leak was from a very small hole. The next leakage happened in November 2001 in the same field. After excavation at the place of the leak two holes were found, about 2 m from one another and both a few metres away from the first leak of 1999. The repair was made by replacing 15 m of pipeline with a new pipe. Parts of the pipeline were sent to be analyzed at two places, namely; the Laboratory of Materials and Tribology at the Faculty of Mechanical Engineering and Naval Architecture, the Laboratory for Power Engineering and Process Technology (TPK).

Table 1. Location of field measurements.

No.	Type	Position
1	Ditch 1	Excavation 220 m along the line close to pumping station Struzec
1	Ditch 2	Excavation 240 m along the line close to pumping station Struzec
1	Ditch 3	Excavation 2 m from the Obzev canal
1	Ditch 4	Excavation in the Brezovica forest
1	Ditch 5	Excavation in Topolovac village
1	Ditch 6	Excavation 200 m after the Lonja river
1	Ditch 7	Excavation 300 m before the Lonja river

Table 2. Measured pipe wall thickness (at Ditch 1).

167	171	172	172	173	175	177	178	179	182
1.5	**4.0**	**4.4**	**6.4**	**1.5**	**5.6**	**6.4**	**3.4**	**6.0**	**6.4**
183	184	184	185	186	186	188	189	190	192
4.8	**2.3**	**5.2**	**2.6**	**6.2**	**4.4**	**5.2**	**6.6**	**5.7**	**4.4**
196	197	199	201	201	202	203	203	205	206
3.8	**6.9**	**6.9**	**3.6**	**5.2**	**6.8**	**4.6**	**6.3**	**1.8**	**5.0**
206	208	209	210						
3.4	**3.6**	**6.7**	**5.3**						

No. radiogram.
Measured thickness (mm).

3 RESULTS OF THE INVESTIGATION

3.1 Laboratory testing

The part of the pipe that was removed showed damage on the bottom of the pipe with pitting corrosion on the inside of the pipe. The first pitting was started by sulfide and chloride (proved by analyses of sediment from inside the pipe). After that the pitting grew inside the material because of the high ratio of area of cathode (undamaged part of the pipe) to anode (pitted part). That kind of loss of material led to breakage of the pipe and leaking of fluid. Since metallurgical testing did not reveal any defect in the material it was concluded that the damage was caused by transport media which remained at the place of corrosion for a longer period of time.

3.2 Field measurements

The measurements were made at the seven locations shown in Table 1. A combination of radiographic and ultrasonic measurements of pipe thickness were used because it was possible to empty a part of the pipeline – at locations No.1 and No.2 and partially at location No. 3. A film (550 × 240 mm in plan) was placed on the bottom of the pipe. Radiography showed darker places that represented cavities on the pipe. After determining the exact locations of the cavities, the ultrasonic measurement was used to determine the precise thickness of the pipe at those places. At other locations, ultrasonic scanning of the bottom of the pipe was used to measure the thickness of the pipe walls.

3.3 Measurements

The radiographic method was used primarily to inspect the condition of the pipeline at ditches 1 and 2. At places where the walls of the pipes were thinneras shown by darker places on the film), ultrasonic measurements were then made to determine the exact thickness of the pipe walls – typical results are shown in Table 2.

347

Table 3. Measured pipe wall thickness (at Ditch 3).

400	560	1250	1700	1800	1950	2000	2300
4.7	5.7	4.7	4.1	4.0	4.6	4.8	4.7

Distance × [mm].
Measured thickness (mm).

At Ditch 3 ultrasonic scanning of the bottom of the pipe was used to measure the thickness of the pipe walls (~10 mm) over the first 2,500 mm from the Obzev canal because of the incoming oil from the depression beneath the canal (Table 3). The places of corrosion are apparent in the table. The radiographic method was used to assess the state of the pipeline. At places where the film was darkened by radiography the actual thickness was obtained by ultrasonic measurements.

Inspection of the bottom of the pipeline by ultrasonic scanning alone was conducted at Ditches 4 to 7 inclusive.

During the measurements, there was a breakage of pipeline in the first ditch but without any ecological consequences. The leakage was solved temporarily with a clamp with oil resistant rubber seal.

4 CONCLUSIONS

After finishing the measurements in the vicinity of Ditches 1, 2 and 3 it was decided to change that part of the pipeline because the results of measurements of thickness of the pipe walls showed they were not satisfactory. A part of the pipeline was changed, 800 m in length, from the pumping station Struzec to past the Obzev canal. After leakage of the pipeline beneath the bed of the Obzev canal a part of the pipeline (loop) under the canal, 200 m in length, was changed.

The method of pipe testing, especially a combination of radiographic and ultrasonic measurements covering 100% of the testing area, gave very good results, but at a high cost. Some 130 m of main pipeline was inspected, partly by a combination of radiographic and ultrasonic measurements and partly by ultrasonic measurements only at a cost of more than 30,000 €. Unfortunately, the inspected section represents just 1% of whole length of pipeline.

Pitting corrosion was found at all the locations inspected, but without the condition of the whole pipeline being known. Further survey work using ultrasonic inspection tools is in progress.

REFERENCES

Anon (2002) *Technical documentation* INA–Naftaplin Co, Zagreb.

Geotechnical and Environmental Aspects of Waste Disposal Sites – Sarsby & Felton (eds)
© *2007 Taylor & Francis Group, London, ISBN 978-0-415-42595-7*

Acid sulphate soil remediation techniques on the Broughton Creek Floodplain, New South Wales, Australia

B. Indraratna & A. Golab
University of Wollongong, NSW, Australia

W. Glamore
University of New South Wales, NSW, Australia

B. Blunden
New South Wales Department of Environment & Conservation, Wollongong, Australia

ABSTRACT: One-way floodgates were commonly installed on flood mitigation drains in coastal areas of Australia during the late 1960s. In acid sulphate soil affected regions, the floodgates create reservoirs of acidic water that discharge at low tide. Several successful remediation techniques have been used in south-eastern NSW, Australia. Groundwater elevation and quality were monitored and modelled using finite element software and fixed level v-notch weirs were installed at three elevations in a drain. The weirs successfully maintained the groundwater elevation above the acid sulphate soil layer and reduced the acid discharge rate to the drain. Following geochemical modelling, modified two-way floodgates that allow tidal ingress were installed. The modified floodgates buffer the drain water pH before discharging into adjacent waterways. Monitoring proved that saline intrusion into the surrounding soil from the drains was not a major concern for the pastureland or other agricultural activities.

1 INTRODUCTION

Disturbed acid sulphate soils cause acidification of coastal waterways and this is a high-profile environmental, social and economic problem in Australia that requires urgent attention (Indraratna and Blunden, 1999). Soils that contain iron sulphides, predominantly pyrite, are commonly referred to as Acid Sulphate Soils (ASS) and these are abundant in coastal Australia (White et al, 1997). Pyrite is chemically inert if left undisturbed and submerged by groundwater, however, when the groundwater falls below the pyritic soil horizon, atmospheric oxygen diffuses through overlying soil layers and the pyrite oxidises to form sulphuric acid. Even greater environmental problems occur when the acid mobilises iron and aluminium ions into the groundwater.

Deep flood mitigation drains cause major problems in regions affected by ASS (Sammut et al, 1994). Channelised, high-density drainage systems greatly increase the rate of lateral water outflow, causing a lowering of the groundwater table. Air (oxygen) rapidly travels through the drained upper soil layers caused by old root channels, desiccation, weathering, and soil disturbance by agricultural activity (Blunden and Indraratna, 2001). One-way floodgates are very common on flood mitigation drains in coastal Australia (Williams and Watford, 1997). The floodgates cause the drawdown of the surrounding groundwater and expose pyrite to oxidising conditions (Figure 1). Acidic drain water is released by the one-way flap valve into the nearby waterway at low tide but brackish water is prevented from entering the drain at high tide, thereby restricting the process of tidal carbonate/bicarbonate buffering. The acidic drain water corrodes steel and concrete infrastructure, blocks waterways with iron flocculates and the high iron and aluminium content of the water kills fish.

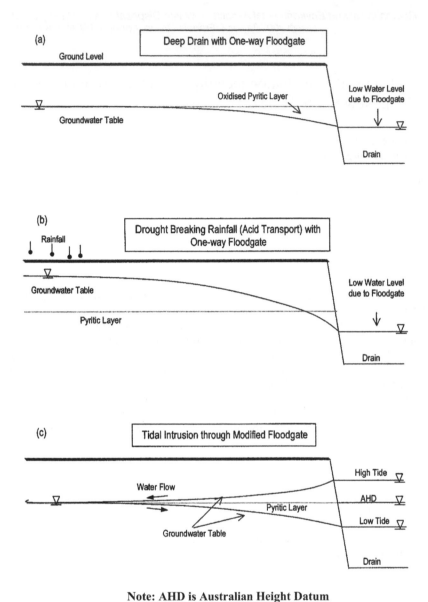

Note: AHD is Australian Height Datum

Figure 1. Groundwater table level as affected by (a) one-way floodgates, (b) heavy rain, (c) tidal intrusion (adapted from Indraratna et al, 2002).

Due to this process, coastal estuaries and floodplains become severely acidified, leading to losses of agricultural and fishery productivity.

During dry periods, evapo-transpiration lowers the groundwater table further, exposing very large volumes of sulphidic soil to oxidising conditions. When rainfall recharges the groundwater, the acidic oxidation products are transported to the drains (Indraratna et al, 2001). However, if the availability of oxygen is reduced then the oxidation of pyrite will decrease. Oxygen diffuses through water about 10^5 times slower than through air. Therefore, to reduce oxygen transport to the pyritic layer the groundwater table must be maintained above the potential acid sulphate soil horizon, for example through the use of weirs (Blunden and Indraratna, 2000).

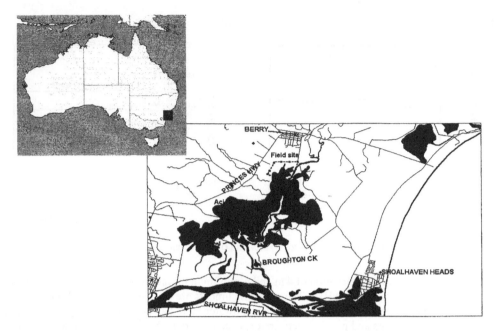

Figure 2. Map of study site, south of Sydney, Australia (areas shown shaded are affected by acid sulphate soils).

Another option for remediating acidic drain water is to utilise carbonate/bicarbonate buffering, or tidal buffering, whereby incoming tides transport acid buffering agents throughout an estuary. Bicarbonate (HCO_3^-) and carbonate (CO_3^{2-}) are the major buffering constituents of seawater. The modification of floodgates to allow tidal flushing into flood mitigation drains may;

(1) decrease the hydraulic gradient between the drain and groundwater,
(2) decrease the 'acid reservoir effect',
(3) increase dissolved oxygen levels,
(4) diminish aluminium flocculation,
(5) combat exotic freshwater weeds,
(6) enhance runoff during wet periods and,
(7) allow fish passage into important breeding grounds.

(Pollard and Hannan, 1994, Portnoy and Giblin, 1997, Dick and Osunkoya, 2000, Glamore and Indraratna, 2001, Indraratna et al, 2002)

A number of factors influence the effectiveness of tidal buffering including; concentration of buffering agents, acid concentration, the hydrodynamics and salinity regime of the estuary. The use of modified floodgates to restore tidal influx improves the geochemical, hydrodynamic and acid transport conditions within acid sulphate soil affected drains (Sutherland et al, 1996, Indraratna et al, 2002, Johnston et al, 2003).

2 STUDY SITE

The study site is a small sub-catchment, approximately 120 ha of coastal lowland near the township of Berry (34°S, 150°E) on the South Coast of New South Wales, Australia (Figure 2). Deep flood mitigation drains (approximately 3.5 m deep × 8 m wide) were installed, in the late 1960s, through the pastureland to minimise the risk of flooding and increase surface runoff. One-way, top-hinged gates are located at the mouth of most drains to prevent tidal intrusion and ensure low drain water elevations. At low tide the drains discharge into Broughton Creek, a tributary of the Shoalhaven River.

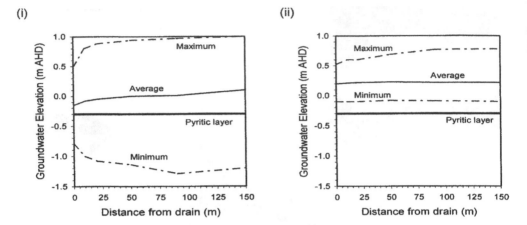

(i)   (ii)

Figure 3.   Average groundwater elevations for (i) pre- and (ii) post-weir installation.

The site has a maximum elevation of 4 m and a minimum elevation of less than 1 m relative to the Australian Height Datum (AHD). The pyritic sediments are at a depth of 1.3 to 1.4 m and are overlain by shallow layers of alluvium. The groundwater table has been lowered by the network of relatively deep drains and pyritic layers are exposed even during mild drought conditions. The primary drains are fed by acidic groundwater and several secondary feeder drains transport highly acidic surface water (pH < 3.0) from the surrounding floodplain.

3   FIRST REMEDIATION APPROACH – WEIRS

The first remediation approach that was considered was fixed level weirs to increase the height of the groundwater table and saturate the ASS layer, thereby preventing further pyrite oxidation. Initially, finite element analysis, using PC-SEEP and FEMWATER (originally developed by Lin et al, 1997) was used to estimate the effect on the groundwater table (Blunden et al, 1997). Additional modelling using various hydrologic and hydraulic procedures showed that the weirs would allow the groundwater table to rise to a specific level without flooding the pastureland. Finally, it was decided to use weirs in the flood mitigation drains.

3.1   Weir performance

Three v-notch weirs were installed in the flood mitigation drains at the study site at 0.1, 0.4, and 0.6 m above the pyritic layer. The groundwater elevation and composition were monitored at 59 locations for a period of 14 months pre-weir installation and 12 months post-weir installation.

During the pre-weir monitoring period, the water table fell below the pyritic layer for over 100 days at most locations. However, for most of the post-weir monitoring period the groundwater table was maintained above the pyritic layer at most locations. An increase of 0.3 m in drain water level was also recorded. The increase in drain water elevation reduced groundwater drawdown close to the drains and caused low hydraulic gradients (Figure 3). At the beginning of the pre-weir monitoring period, the groundwater pH was 2.5–3.5. The weirs reduced the rate of discharge of acidic oxidation products from the groundwater to the drain, but did not substantially improve the quality of the groundwater or soil. The acidity of the groundwater and soil following weir installation indicates that biotic oxidation of pyrite may be still occurring beneath the groundwater table by iron and sulphur oxidising bacteria. The bacteria enhance pyrite oxidation through the reduction of Fe^{3+} at pH < 4, as shown in Equation (1) (Willett et al, 1992).

$$FeS_2 + 14Fe^{3+} + 8H_2O \rightarrow 15Fe^{2+} + 2SO_4^{2-} + 16H^+ \qquad (1)$$

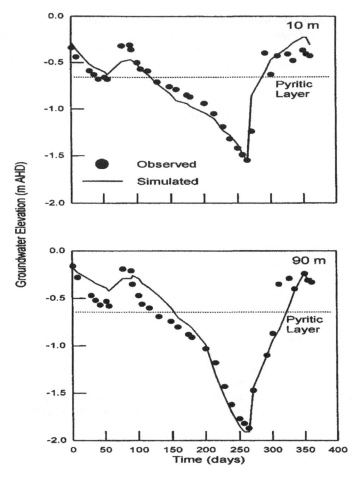

Figure 4. Measured and predicted groundwater levels t 10 m and 90 m from the drain (the pyritic is shown as a dashed line).

3.2 *Numerical modelling of groundwater conditions*

The groundwater conditions were simulated using the FEMWATER finite element software. The computational discretisation used by the software is a three-dimensional mesh. The elements used in the simulation were triangular prisms, each with six nodes, as described by Blunden and Indraratna (2001). The modelling period included a wet period with shallow groundwater, as well as a prolonged dry period that resulted in groundwater falling below the pyritic layer. The FEMWATER-modelled groundwater elevations agree well with the observed groundwater eleva-tion data collected over a 12-month period pre-weir (Figure 4), showing that the model adequately simulates the groundwater hydrology of the study site. The finite element simulation indicated that, in the presence of a weir, the groundwater at 10 and 90 m from the drain would be 0.5 m and 0.4 m higher, respectively, than under the existing drained conditions (Blunden and Indraratna, 2000).

4 SECOND REMEDIATION APPROACH – MODIFIED FLOODGATES

The weirs successfully maintained the groundwater table above the pyritic layer but did not improve the quality of the drain water. A remediation solution was subsequently sought that would improve the drain water quality prior to discharge into Broughton Creek. Therefore, the second remediation approach involved modification of floodgates to allow tidal ingress and hence buffering of the acidic drain water by the carbonate/bicarbonate in seawater. As the drains pass through productive agricultural pastures, any floodgate modifications would involve changing the hydrodynamics of the drain. Several environmental and hydraulic concerns were addressed prior to commencing civil works, including;

• optimising the drain water level with tidal influx without overtopping the levee bank or reducing agricultural productivity,
• predicting the potential change in drain water qualify due to tidal buffering, and
• studying the impact of increased salinity on the sub-soil matrix.

4.1 Drain water quality

To satisfy the foregoing criteria, moderated tidal restoration in a flood mitigation drain was simulated using a digital terrain map, geographic information tools, and measured water levels to predict drain overtopping due to tidal variations (Glamore and Indraratna, 2004). The GIS simulations indicated that full tidal flushing could safely be allowed in the primary drainage network. In the second stage, changes in drain water quality were simulated using an ion-specific program code written within PHREEQC to evaluate the mixing of alkaline creek water of varying ionic strengths with aluminium- and iron-rich acidic drain water (Glamore and Indraratna, 2002). The main inputs to the mixing program were elemental concentrations and electrical conductivity which were determined using samples of brackish river water, rain diluted river water, fresh creek water and acidic drain water. The program was developed to simulate mixing in 10% intervals, until twice as much alkaline water had been added compared to acidic water, i.e. 200%.

The model and laboratory measurements agreed well for mixing of the four types of water samples (Figure 5) for both low and high ionic strength waters. Upon mixing the pH increases and the concentration of sulphate decreases, indicating a marked decrease in acidity caused by tidal buffering.

Both the simulations and the laboratory results indicate that drain water quality improved with mixing, with a decrease in soluble aluminium and iron of 73% and 56% respectively.

4.2 Tidal flushing

Based on the positive numerical simulations, two styles of modified two-way floodgates were installed;

• a manually-operated winch that lifts the floodgate vertically and controls the amount of water permitted upstream of the floodgate, and
• an automated 'Smart Gate' System that allows tidal flushing based on real-time water quality parameters.

Prior to floodgate modifications the drain water was consistently acidic (average pH 4.6) but after the 'Smart Gate' was installed, the drain water pH increased to 6.04 and aluminium and iron concentrations decreased by more than 50%. The modified floodgates also increase the drain water elevation with each high tide but do not prevent the one-way flap gates from discharging the drain water at low tide. Therefore, the modified floodgates do not maintain constant drain or groundwater elevation (Figure 1).

Tidal flushing of the flood mitigation drain allowed saline water to intrude into the soil matrix due to hydrodynamic dispersion and advection. Soil salinity was monitored and increases only occurred close to the drain and did not exceed the ANZECC (1992) guidelines. The plume of saline intrusion occurred as a saline wedge that was limited to 8 m from the drain and 2 m below the ground surface. The small size of the plume and its transient nature show that agricultural productivity on the surface should not be affected by tidal flushing within the drain.

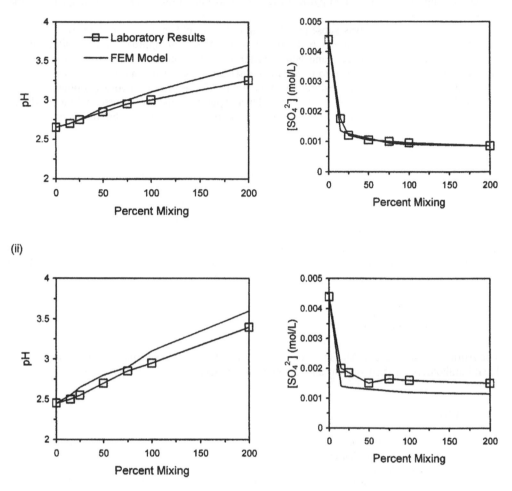

Figure 5.   Comparison of simulations from PHREEQC and laboratory results for mixing with; (i) low ionic strength samples, (ii) high ionic strength samples.

5   CONCLUSIONS

Fixed level weirs in flood mitigation drains are able to manipulate the elevation of the groundwater table. By using weirs, the amount of groundwater drawdown is decreased close to the drains. Measurements of groundwater table elevations at the field site, as well as groundwater simulation results showed that the use of the weirs minimised the volume of exposed pyritic soil. Prior to installation of the weirs, the elevation of the groundwater table was controlled by groundwater drawdown towards the drains, rainfall, and evapo-transpiration. Following significant rainfall, the groundwater table would rise through the pyritic zone, entrain the pyrite oxidation products from the preceding dry period and cause extremely acidic groundwater conditions (pH 2.5–3.5). After the weirs were installed, groundwater levels increased and the production of 'new' acid from the oxidation of pyrite by oxygen was minimised. The maintenance of higher drain water levels also caused lower groundwater hydraulic gradients towards the drain, thereby reducing the rate of acid discharge to the drain. The elevated groundwater levels did not improve long-term groundwater quality. The nature of the 'stored' acid in the soil profile previously caused by pyrite oxidation and continuing biotic oxidation of pyrite prevented a substantial improvement in the quality of the groundwater or soil.

A remediation solution was subsequently sought that would improve the drain water quality prior to discharge into Broughton Creek. The chosen remediation approach was modification of floodgates to allow tidal ingress and hence buffering of the acidic drain water by the carbonate/bicarbonate in seawater. Prior to floodgate modifications, extensive modelling was performed to determine the impact of tidal restoration on the drains. Two innovative two-way floodgates were designed. The first manually lifts the floodgate flap and permits full tidal intrusion within the drain. The second design automatically adjusts the gate to control tidal ingress within the drain based on real-time environmental parameters.

After the redesigned floodgates were installed and full tidal flushing was established within the primary drain, water quality significantly improved. The drain water pH increased by two orders of magnitude and dissolved aluminium and iron decreased by more than 50%. Tidal forcing on the phreatic zone and saline intrusion was limited to within 8 m of the drain.

The two remediation techniques presented in this paper have different purposes and in many cases would work well together. The weirs maintain constant high drain and groundwater elevations above the pyritic layer and thereby reduce the amount of pyrite oxidation. The modified floodgates improve the drain water composition but do not increase the groundwater elevation. By using a combination of the two techniques the volume of acidic discharge into the drain is decreased and the acidity in the drain water is buffered.

ACKNOWLEDGEMENTS

The Authors wish to thank Mr Bob Rowlan, Dr Anand Tularam, Shoalhaven City Council, and local landholders for their support and technical assistance.

REFERENCES

ANZECC (1992) *Australian Water Quality Guidelines for Fresh and Marine Waters*, Australian and New Zealand Environment Conservation Council, Canberra.

Blunden B G and Indraratna B (2000) *Evaluation of surface and groundwater management strategies for drained sulfidic soil using numerical simulation models*, Australian Journal of Soil Research, 38, pp569–590.

Blunden B and Indraratna B (2001) *Pyrite oxidation model for assessing ground-water management strategies in acid sulfate soils*, Journal of Geotechnical and Geoenvironmental Engineering, 127(2), pp146–157.

Blunden B, Indraratna B and Nethery A (1997) *Effect of groundwater table on acid sulphate soil remediation*, In Geoenvironment 97, Bouazza, Kodikara and Parker (Eds), Balkema, Rotterdam, pp549–554.

Dick T M and Osunkoya O O (2000) *Influence of tidal restriction floodgates on decomposition of mangrove litter*, Aquatic Botany, 68(3), pp273–280.

Glamore W and Indraratna B (2001) *The impact of floodgate modifications on water quality in acid sulphate soil terrains*, Proc 15th Australasian Coastal and Ocean Engineering Conference, Gold Coast, Australia, pp265–270.

Glamore W and Indraratna B (2002) *Management of acid sulphate soil drainage via floodgate manipulation*, Proc 5th International Acid Sulfate Soils Conference – Sustainable Management of Acid Sulfate Soils, R. Bush (Ed), Tweed Heads, pp75–76.

Glamor W and Indraratna B (2004) *A two-stage decision support tool for restoring tidal flows to flood mitigation drains affected by acid sulphate soil: Case study of Broughton Creek Floodplain, New South Wales, Australia*, Australian Journal of Soil Science.

Indraratna B and Blunden B (1999) *Nature and properties of acid sulphate soils in drained coastal lowland in NSW*, Australian Geomechanics Journal, 34(1), pp61–78.

Indraratna B, Tularam G A and Blunden B (2001) *Reducing the impact of acid sulphate soils at a site in Shoalhaven Floodplain of New South Wales, Australia*, Quarterly Journal of Engineering Geology and Hydrogeology, 34, pp333–346.

Indraratna B, Glamore W C and Tularam G A (2002) *The effects of tidal buffering on acid sulphate soil environments in coastal areas of New South Wales*, Geotechnical and Geological Engineering, 20, pp181–199.

Johnston S G, Slavich P and Hirst P (2003) *Floodgate and drainage system management: Opportunities and limitations*, Proc Third International Conference on Acid Sulphate Soils, Coolangatta, Queensland, pp79–81.

Lin H, Richards D, Talbot C, Yeh G, Cheng J, Cheng H and Jones N (1997) *FEMWATER: A three dimensional finite element computer model for simulating density dependent flow and transport in variably saturated media*, Technical Report CHL-97-12, US Army.

Pollard D A and Hannan J C (1994) *The ecological effects of structural flood mitigation works on fish habitats and fish communities in the lower Clarence River System of South-Eastern Australia*, Estuaries, 17(2), pp427–461.

Sammut J, White I and Melville M (1994) *Stratification in acidified coastal floodplain drains*, Wetlands (Australia), 13, pp49–64.

Sutherland N M, Scott P A and Morton W K (1996) *Field practicalities of treating acid sulfate soils*, Proc 2nd National Conference of Acid Sulfate Soils, R.J. Smith and H.J. Smith (Eds), Coffs Harbour, pp221–224.

White I, Melville M D, Wilson B P and Sammut J (1997) *Reducing acidic discharges from coastal wetlands in eastern Australia*, Wetlands Ecology and Management, 5(1), pp55–72.

Willett I R, Crockford R H and Milnes A R (1992) *Transformation of iron, manganese and aluminium during oxidation of a sulfidic material from an acid sulfate soil*, Biomineralization Processes of Iron & Manganese: Modern and Ancient Environments, H.C.W. Skinner and R.W. Fitzpatrick (Eds), Catena-Verlag, Cremlingen-Destedt, pp287–302.

Williams R J and Watford F A (1997) *Identification of structures restricting tidal flow in New South Wales, Australia*, Wetlands Ecology and Management, 5(1), pp87–97.

Geotechnical and Environmental Aspects of Waste Disposal Sites – Sarsby & Felton (eds)
© 2007 Taylor & Francis Group, London, ISBN 978-0-415-42595-7

Basic study on development of permeable block pavement with purifying layer for polluted rainwater on road

S. Iwai & X.Z. Xi
C.S.T., Nihon University, Funabashi, Japan

ABSTRACT: It is necessary to purify polluted rainwater that permeate into the ground through porous pavement for preservation of the underground environment. Permeable block pavement is one such porous pavement and to provide a simple maintenance method it was decided to add the purification function to the sand cushion layer. Direct current (DC) was applied to the silica sand used for the sand cushion layer. In order to study the purification function of the silica sand layer under DC electric field, column test and model tests of permeable block pavement were carried out in the laboratory. As a result, the adsorption of acid rain and polluted ion in silica sand under DC electric field was confirmed. It became evident that polluted rainwater can be prevented from infiltrating into the ground by using permeable block pavement with a silica sand layer under DC electric field.

1 INTRODUCTION

Rain falls to the ground capturing suspended particulates and various ions as it passes through the air. According to the third investigation on acid rain protection measures by the Ministry of the Environment in Japan, the annual mean pH of the rainfall in Japan is 4.8 to 4.9 and nearly as much acid rain is observed as in Europe and the United States (Anon, 2002). Phosphorus and nitrogen compounds also fall on the ground, leading to eutrophication of water areas. According to Kunimatsu and Muraoka (1997), phosphorus is precipitated in the form of particulates and nitrogen compounds are for the most part entrained in the rainfall. In the urban area around the campus in Funabashi City runoff water on the road surface during rainfall was intermittently measured with the result that a high concentration of phosphoric ions was detected.

The technology of porous pavement has been introduced to improve the circulation of water and alleviate the 'heat island phenomenon' in urban area. However, infiltration of acid rain, phosphorus and nitrogen compounds into the ground may cause not only ground water pollution but also damage to underground facilities. In this study, the idea of incorporating a purifying function into the sand cushion layer beneath the permeable block pavement was implemented. First, a column test was conducted in which a DC electric field was applied to a sand layer corresponding to the sand cushion layer to confirm the ion adsorptivity under electric field. Next, a permeable block pavement model was tested to investigate the ion adsorptivity of the sand cushion layer under an electric field and confirm the possibility of rainwater purification in the road.

2 PERMEABLE BLOCK PAVEMENT WITH PURIFYING LAYER

It has been shown clearly that silica sand under a DC electric field adsorbs positive ions (Fukuda and Hirata, 1999; Iwai et al, 2001). Therefore, a permeable block pavement, as shown in Figure 1, was devised in which the sand cushion layer formed of silica sand is sandwiched between upper and lower electrodes through which a voltage is applied to the silica sand layer to make it adsorb ions from infiltrating rainwater, thus ensuring rainwater purification.

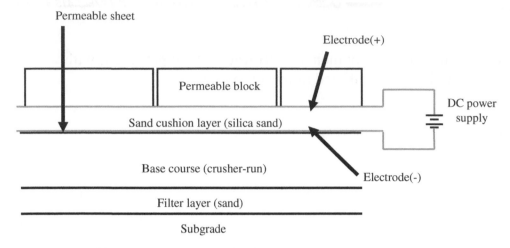

Figure 1. Permeable block pavement with purifying layer.

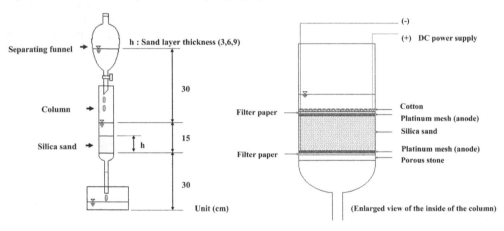

Figure 2. Column test apparatus.

3 COLUMN TEST

A column test apparatus, with a sand layer corresponding to the sand cushion layer, was used to confirm the purifying function of the sand cushion layer under a DC electric field. Silica sand was selected for the sand cushion layer. Pseudo-acid rain for the test was a mixture of dilute nitric acid (3–5 mg/l in nitrate nitrogen concentration) and dilute sulfuric acid (6 mg/l in concentration) adjusted to a pH of 4.1 to 4.3. In addition, polluted water containing total phosphorus (referred to as T.P. polluted water) was prepared by mixing 8 mg of dibasic potassium phosphate with 7 litres of ultrapure water and adjusting to a phosphoric acid concentration of 5.0 to 5.5 mg/l. Tests were performed with silica sand saturated with the test solution and with unsaturated silica sand.

Figure 2 shows the column test apparatus. The column was internally filled with silica sand up to a predetermined height (h) and the resulting silica sand layer was sandwiched between two platinum meshes, upper (anode) and lower (cathode), so as to apply a DC voltage. Tests were conducted with;

- two different column diameters (φ) – 2 and 5 cm,
- three different sand layer thicknesses (h) – 3, 6 and 9 cm,
- two different voids ratios (e) of the sand layer – 0.6 and 0.8,
- four different applied voltages (E) – 20, 40, 80 and 120 V.

360

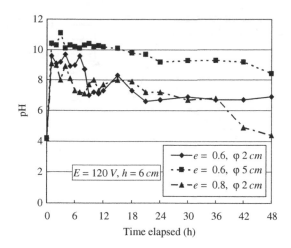

Figure 3.   Variation of pH with time (saturated sand).

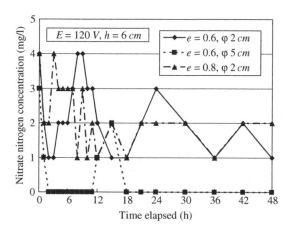

Figure 4.   Nitrate nitrogen concentration versus time (saturated sand).

In the test with saturated silica sand, pseudo-acid rain and T.P. polluted water were made to drop at a constant rate of 80 ml/h. In the test with unsaturated silica sand the precipitation rate was 5 ml/h. The solutions for the test were made to drop through the column, and the pH and ion concentration of the permeating liquid accumulated in the receiving bottle were measured at predetermined intervals of time.

4   COLUMN TEST RESULTS AND DISCUSSION

The results of the column test with saturated silica sand are given in Figures 3, 4 and 5. It is clear that immediately after voltage application the pH of pseudo-acid rain ('the acid rain') varies greatly, indicating a change from acidity to alkalinity. Likewise, nitrate nitrogen and phosphoric ions exhibit an abrupt decrease in concentration.

Figure 6 shows the variation of the pH of the solution with the applied voltage. Irrespective of variations in sand layer thickness, the pH of the solution remains at levels of higher than 9 for an applied voltage of 120 V, indicating high alkalinity. The lowest pH is observed when the sand layer thickness is 3 cm. These results suggest that the adsorption of anions is affected by the contact area and contact time.

361

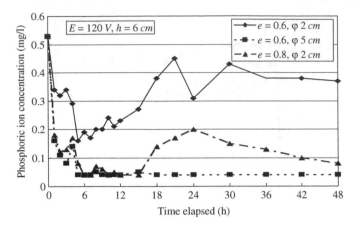

Figure 5.   Variation of phosphoric ion concentration with time (saturated sand).

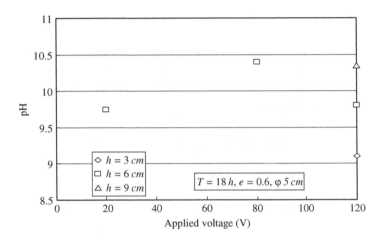

Figure 6.   Variation of pH with applied voltage (saturated sand).

Figure 7 shows how the nitrate nitrogen concentration varies with the applied voltage. There is large adsorption of nitrate nitrogen with applied voltages of 80 V and 120 V. It is obvious that the nitrate nitrogen adsorbing effect of silica sand is affected by the applied voltage. It is also evident that the nitrate nitrogen adsorbing effect was increased by increasing the sand layer thickness (for the applied voltage of 120 V).

Figure 8 shows the change of the phosphoric ion concentration with the applied voltage. It seems that the phosphoric ion adsorbing effect tends to increase with an increase in applied voltage. At an applied voltage of 120 V the phosphoric ion adsorbing effect is greater with the sand layer thickness of 6 cm than with 3 cm and 9 cm.

Figure 9 shows how the pH and the nitrate nitrogen concentration of the acid rain vary with the sand layer thickness. It was found that the pH increases and the nitrate nitrogen concentration decreases with an increase in sand layer thickness.

The variation of the phosphoric ion concentration with the sand layer thickness is shown in Figure 10. It is seen from this figure that the phosphoric ion adsorbing effect is the greatest with the sand layer thickness of 6 cm under conditions of 120 V, 0.6 voids ratio and 5 cm column diameter.

The results of the column test with unsaturated silica sand are shown in Figures 11, 12 and 13. Figure 11 shows how the pH of the acid rain varies with the applied voltage. The acid rain exhibits

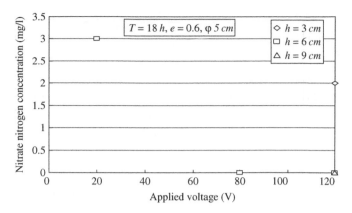

Figure 7.   Nitrate nitrogen concentration versus applied voltage (saturated sand).

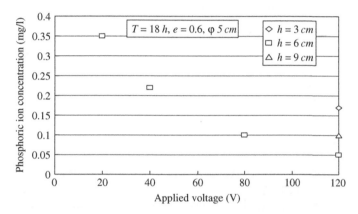

Figure 8.   Variation of phosphoric ion concentration with applied voltage (saturated sand).

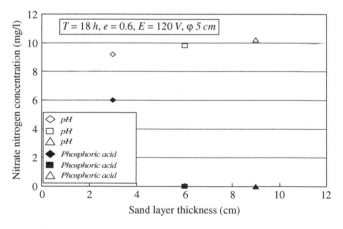

Figure 9.   Variation of pH and nitrate nitrogen concentration with sand layer thickness (saturated sand).

a higher alkalinity than that in the test with saturated silica sand. In addition, the pH of the acid rain tends to increase with an increase in applied voltage.

Figure 12 shows how the pH and nitrate nitrogen concentration vary with the applied voltage. Even at an applied voltage of 40 V the concentration is low (with sand layer thicknesses other

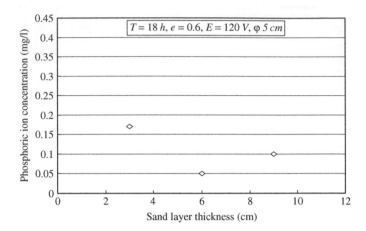

Figure 10.   Variation of phosphoric ion concentration with sand layer thickness (saturated sand).

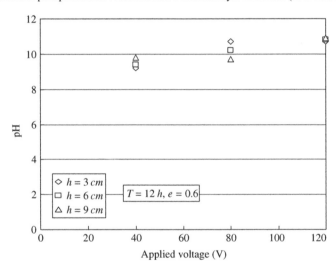

Figure 11.   Variation of pH with applied voltage (unsaturated sand).

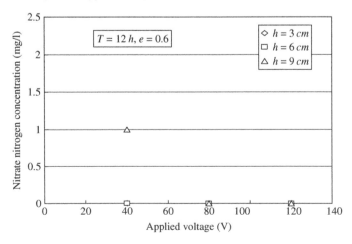

Figure 12.   Variation of nitrate nitrogen concentration with applied voltage (unsaturated sand).

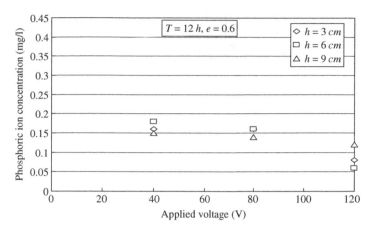

Figure 13. Variation of phosphoric ion concentration with applied voltage (unsaturated sand).

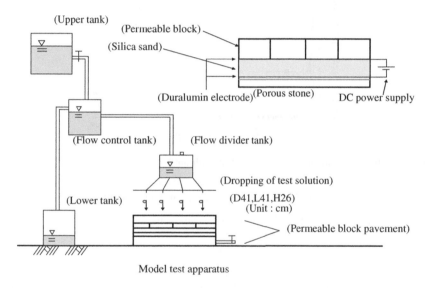

Model test apparatus

Figure 14. Model test apparatus.

than 9 cm) compared to that in the test with saturated silica sand, proving the effectiveness of ion adsorption.

Figure 13 shows the changes of the phosphoric ion concentration with applied voltage. It seems that the phosphoric ion adsorbing effect tends to increase with an increase in applied voltage. It is clear that even unsaturated silica sand has the effect of adsorbing phosphoric ions.

5 MODEL TEST OF PERMEABLE BLOCK PAVEMENT

In order to investigate the ion adsorptivity that a sand cushion layer laid beneath a permeable block pavement exhibits under an electric field a model of a permeable block pavement was made (Figure 14). Pseudo-acid rain and T.P. polluted water were dropped onto the pavement surface. The model is composed of permeable blocks, duralumin electrodes, silica sand, porous stone. The electrodes are set on both sides of the silica sand layer laid on porous stone and permeable blocks are arranged on the sand layer.

365

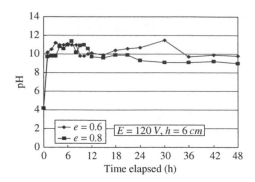

Figure 15. Variation of pH with time.

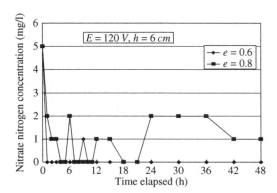

Figure 16. Variation of nitrate nitrogen concentration with time.

As in the case of the column tests experiments were conducted with;

- two different silica sand layer thicknesses (h) – 3 and 6 cm,
- two different void ratios (e) – 0.6 and 0.8,
- four different applied DC voltages (E) – 20, 40, 80 and 120 V.

Pseudo-acid rain and T.P. polluted water were dropped at a constant rate of 500 ml/h through 16 syringe needles. The solution passing through the silica sand layer was collected at the bottom of the model and used for measurement of pH, etc.

6 MODEL TEST RESULTS AND DISCUSSION

Changes of pH, nitrate nitrogen concentration and phosphoric ion concentration with time in this model test are shown in Figures 15, 16 and 17. Within about one to three hours after the start of deposition of the solution, the pH of the acid rain changes to a high alkalinity and the nitrate nitrogen concentration and the phosphoric acid concentration decrease, irrespective of variations in voids ratio. It can be said that the smaller voids ratio results in large variation of the pH and greater effect in adsorbing nitrate nitrogen and phosphoric acid.

Figure 18 shows the variation of the pH of the acid rain with the applied voltage, with varying sand layer thicknesses and voids ratios. With the sand layer thickness of 6 cm and voids ratio of 0.6, the pH tends to increase with an increase in applied voltage. At 120 V the variation of the pH increases with increasing thickness of the sand layer. The same result was obtained in the column test. It is considered that the phenomena result from the effect of the area and time of contact between the test solution and silica sand.

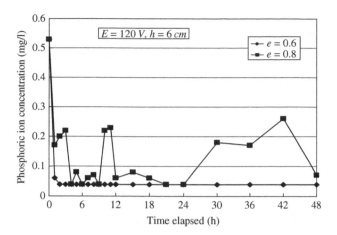

Figure 17. Variation of phosphoric ion concentration with time.

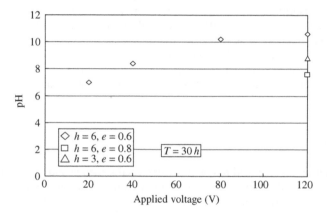

Figure 18. Variation of pH with applied voltage.

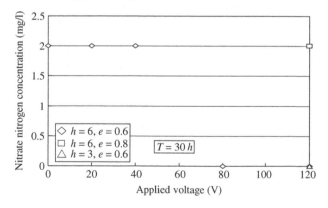

Figure 19. Variation of nitrate nitrogen concentration with applied voltage.

Figure 19 shows how the nitrate nitrogen concentration changes with the applied voltage. At applied voltages in excess of 20 V, the nitrate nitrogen adsorbing effect increases, irrespective of variations in sand layer thickness and voids ratio.

Figure 20 shows how the phosphoric ion concentration varies with the applied voltage, with varying sand layer thicknesses and voids ratios. With the sand layer thickness of 6 cm and voids

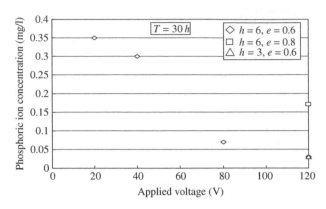

Figure 20.   Variation of phosphoric ion concentration with applied voltage.

ratio of 0.6, the phosphoric ion concentration tends to decrease with an increase in applied voltage. It is clear that the adsorption of phosphoric ions is not affected significantly by the voids ratio of silica sand at the applied voltage of 120 V. However, the phosphoric ion adsorbing effect is greater with the sand layer thickness of 6 cm than with 3 cm.

The model test results also reveal that silica sand used for the sand cushion layer has the effect of adsorbing ions in the presence of an electric field. Observations during the test show that the flow of electric current and the ion adsorbing effect are more stable with the voids ratio of 0.6 than with 0.8.

7   CONCLUSIONS

The overall conclusion is that acid rain and eutrophic salts such as nitrogen and phosphoric acid can be prevented from infiltrating into the ground by using silica sand for the sand cushion layer in permeable block pavement and applying a DC electric field to the sand layer.

From the study reported herein the following specific findings were obtained:

- The column test results showed clearly that silica sand under DC electric field changes the pH of acid rain and greatly affects the adsorption of anions, whether the sand is saturated or unsaturated.
- Pseudo-cid rain becomes alkaline immediately following the application of a DC voltage.
- Nitrate nitrogen and phosphoric acid, which are anions, are rapidly adsorbed by the silica sand layer under DC electric field.
- The column and the model test results demonstrated that the variation of the pH and the anion adsorbing effect tend to vary within the higher DC voltage.
- The column test results showed clearly that the variation of pH and the anion adsorbing effect are affected by the time of contact between silica sand and acid rain or a solution containing anions. This tendency was also be confirmed by the model test results.

REFERENCES

Anon (2002) *White Paper the Environment in Japan*, Ministry of the Environment, Tokyo.
Fukuda H and Hirata K (1999) *Basic Study on Movement Control of Calcium Ions Utilizing Electrokinetic Phenomenon*, Graduation thesis of Nihon University (in Japanese), Funabashi.
Iwai S, Miura Y, Todoroki Y, Morita Y and Fujii T (2001) *Prevention of diffusion of alkaline Components from Chemically Stabilized Soil*, Proc 3rd International Symposium on Geotechnics related to the environment, GREEN3, Thomas Telford, London, pp556–560.
Kunimatsu T and Muraoka K (1997) *Model Analysis of River Pollution* (in Japanese), Gihodo Syuppan, Tokyo.

Geotechnical and Environmental Aspects of Waste Disposal Sites – Sarsby & Felton (eds)
© 2007 Taylor & Francis Group, London, ISBN 978-0-415-42595-7

A multi-criterion approach for system and process design relating to waste management

A.A. Voronov

Scientific research centre for ecological safety, Saint-Petersburg, Russia

ABSTRACT: This paper describes an original technique for decision-making regarding solid waste management systems design including the process of choice with regard to complexity, location and economic aspects as well as a multi-criterion model for energy recovery of waste. The mathematical model is intended for the process of decision making about the system and process design and it is to a certain extent a universal solution. Study of the economic and environmental aspects of energy recovery from waste is also included in the paper. It includes analysis of investment effectiveness based on certain indices. Novel formulae are used for calculation and provision of the data for decision making according to the expenditure incurred.

1 INTRODUCTION

This paper relates to an examination of the approach for sustainable development of regions, in respect of waste management, based on multi-criterion analysis.

There are a lot of established programmes for deriving the most appropriate method of waste management (White et al, 1995). Energy recovery from urban and rural waste is among the most promising international developments in this area. At the same time there are some problems that prevent realisation of the whole advantages of the afore-mentioned programmes in Russia. Some plans and projects already exist for work in this area. However, a detailed analysis of these shows that most proposed solutions are fragmented and are based on incomplete technological measures. In this regard the following may be mentioned:

- Construction of incineration plants which are under consideration without taking into account the development of secondary materials industries.
- Lack of an appropriate legislation base which permits introduction of measures for economic stimulation of waste treatment together with investment attraction to this sector.
- Absence of an integrated approach to the problem which is shown by the low quality of educational and training programmes in this sphere.
- The absence of experience in preparing and making investment projects, especially for international ones, is a serious problem.

The foregoing list of problems for urgent consideration in the area of energy recovery from waste is not exhaustive. There are also several instances where programme planning has failed because it was impossible to satisfy requirements for professional management. All the foregoing points emphasize the low efficiency of current practice with regard to implementation of similar programmes because of incorrect account being taken of the multi-component specifics of the problem.

2 BACKGROUND

Rehabilitation of the economy of Northwest Russia will increase international trade in Nordic Region and will lead to inevitable growth of waste. This problem has already been noted by the

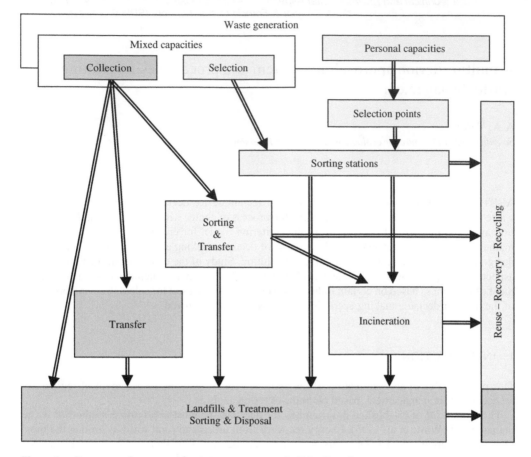

Figure 1.   Systems and processes for waste management in Saint Petersburg.

World Trade Organization (by its Committee for Environment and Trade), together with the needs for development of international collaboration, starting from regional and cross-border actions in particular markets to meet the approaching extensive challenge from the interface between environment and trade. In particular, there is a problem in the area of strategic design for solid municipal waste management (SMW) in NW Russia – to go from the current situation (97% of SMW is landfilled) to achieve enhancement of this situation, e.g. to have about 40% utilization within ten years.

A multi-criterion approach for waste management optimization is introduced in the context of Saint Petersburg. This city generates about 9 Mm³ of solid municipal waste per year (Voronov, 2003). Only 5% of this volume is available for utilization, i.e. the rest appears in dumps. The total budget for the waste management in 2003 was about 1,970 million Roubles.

It is known that development of the secondary materials industry has the most potential to minimise landfill. The scheme shown in Figure 1, for appropriate waste management, is currently under investigation. The proposed scheme of solid waste management, which includes energy recovery, is regarded as having all of the treatment components necessary in the context of environment and trade.

The design approach is based on a system of indicators for sustainable development as the criteria (Anon, 1999), with environmental, economic, social and institutional components. This provides an innovative and productive management technique to assess alternative schemes at a decision-making stage as illustrated in Table 1 (for N alternatives with four criterion groups). The approach is also suitable for creating a structure with the requisite capability to assess and supervise the projects in the field under discussion and to meet the key priorities of sustainable development of a region.

370

Table 1. Basics for multi-criterion analysis.

Alternatives 1 ÷ N	Criterion groups			
	Environmental	Economic	Social	Institution
1	$R_{1,1}$	$R_{1,2}$	$R_{1,3}$	$R_{1,4}$
...	...	...	...	...
k	$R_{k,1}$	$R_{k,2}$	$R_{k,3}$	$R_{k,4}$
...	...	...	...	...
N	$R_{N,1}$	$R_{N,2}$	$R_{N,3}$	$R_{N,4}$

The multi-criterion analysis technique is applied as follows: The alternative "I" is ranked above alternative "J" under criterion "K" if $R_{I,K} > R_{J,K}$, where $R_{L,M}$ is the quantitative assessment for the L-th alternative ranked under criterion "M". This leads to the ordering of the set of alternatives and provides the base for structure development as well as for project assessment and management.

About 130 indicators have been introduced by the United Nations Committee for Sustainable Development. Selection of the definitive criterion set will provide numerous tasks aimed at addressing specific circumstances.

3 CASE STUDY

New decisions in waste management are based on the use of various technological methods for complex processing of waste. The choice of best options is determined by local conditions in a particular country and even within a particular region.

Currently the plant in Saint Petersburg require modernisation in line with the programme of further development. The basic product of the plant is compost. Besides compost the plant produce some amount of pyrocarbon and polymeric granules, and ferrous and non-ferrous metals, broken glass, rags and wastepaper are recovered. In order to ensure that the quality of compost in compliance with relevant sanitary norms it is necessary to organize the selective collection of spent batteries in the city. Currently the compost is partly used as filling material to cover waste on landfills. However, stockpiles of compostable material are accumulating in the grounds of the waste recycling facilities.

3.1 Waste separation

Facilities for complete mechanized separation of waste are necessary in any case, since the benefits of separation are;

- firstly, recovery of additional value through sale of recyclable material,
- secondly, safe and competitive compost,
- thirdly, reduction of the content of toxic substances in flue gases resulting from incineration of waste.

The organization of selective collection requires serious organizational efforts, investment and development of public awareness, and this will require time. In the beginning the minimum requirement is to solve the problem of collection of the hazardous fraction of waste.

The separation of waste at refuse transfer stations is technologically feasible, but it will be necessary to solve certain related problems. For example, the introduction of on-site separation would change the status of the site to become a waste recycling enterprise and this change would incur an extension of the sanitary control zone by almost 100 times. The separation of waste at the plant has a number of advantages, since it enables the complex mechanized processing of waste to be undertaken in one place thereby reducing the processing costs.

3.2 Waste treatment

A composting facility enables up to 30% of waste (in terms of mass) to be transformed into a useful product that can be utilized as fertiliser, biofuel or biologically safe material for reclamation of rehabilitated land, landscaping and planting. In the absence of any demand for compost it can be recovered by thermal methods. In any case the compost is considerably less hazardous to the environment than dumped, unsorted waste.

Incineration facilities are necessary for salvaging the organic waste that cannot be utilized as recyclable material or cannot be composted, e.g. scraps of polyethylene film, wood chips, waste of rubber, etc. If the waste directed to incineration does not contain compounds of mercury, cadmium, arsenic or chlororganic substances (for instance PVC), then a properly designed combustion process will provide satisfactory cleansing of flue gases, if used in conjunction with standard methods and devices.

Since the organic material accounts for, on average, up to 70% of the dry matter of solid municipal waste, it is possible to utilize waste as a fuel or fuel aggregate for production of thermal or electrical power. There are two basic options for the use of waste as fuel:

- direct incineration of the waste or its components in power boilers;
- recovery and subsequent incineration of biogas from the waste masses.

The production of biogas from landfilled waste, its cleansing and use as a fuel is widely spread in many countries. The advantage of biogas incineration as opposite to incineration of solid waste is the generation of clean flue gases and complete combustion of gaseous fuel.

3.3 Waste management model

In the case study five alternatives were considered, as a result of the preceding review (Figure 2). The approach which is of interest for the strategic design is as follows:

- There are several 'alternatives' for the prospective system of waste management;
- Every 'alternative' has the structure modelled by the indicator system;
- Every 'alternative' can be assessed in regard to each indicator area.

Thus, one can operate with the model for each alternative using the system of indicator categories. After multi-criterion analysis the models can be ranked according to complexity. This ranking means

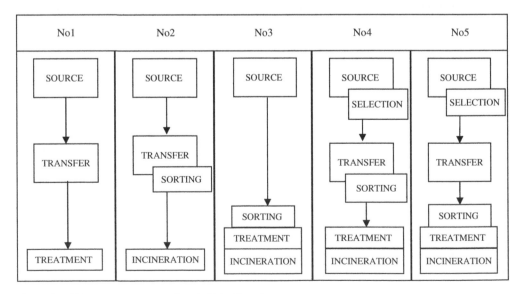

Figure 2.   Waste management options in Saint Petersburg.

that one proceeds from the lowest complexity up to the highest complexity. The indicator approach provides the universal basis for such modelling (not only at the technological level):

The model
Several (for instance, five) alternatives were used for the technological scheme for waste management in the city and every one has been modelled by the structure (Number S, S = 1 to 5), shown in Figure 3, containing:

- T (technology nodes)
- D (transportation interfaces)
- M (raw secondary material outputs)

The indicators
The following "Indicators" (the <complexity> of the system) were applied:

- IT = dimension { T }
- ID = dimension { D }
- IM = dimension { M }

Preferences
The matrix shown in Table 2 is derived. For every column there is a reason to provide the preference based on the principle exemplified by; NoI is better than NoJ according to the indicator "IT" if $R_{I1} > R_{J1}$ (where R_{I1} = dimension { T }, T – technology nodes in the alternative NoI, etc). The multi-criterion technique is used to aggregate the preference set (the matrix) into some order, e.g. partial order, semi-order, line order, between the alternatives.

Results
The outcome is the chain (Figure 4) of the alternatives starting from the 'simplest' up to the 'most complex' – this gives a way of looking at the relationship between a set of possible schemes for waste management. For example, the result can be presented at the form:

- No4 is the 'most complex'
- No3 is the 'simplest', as well as No2

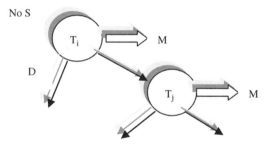

Figure 3.   Base model for waste management.

Table 2.   Basics for multi-criterion analysis of waste management complexity.

Alternatives	Criteria		
	I_T	I_D	I_M
No1	112	16	28
No2	80	48	12
No3	78	6	14
No4	900	72	90
No5	156	36	42

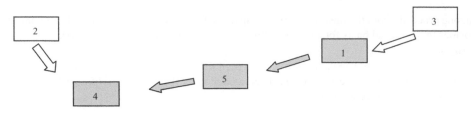

Figure 4. The multi-step approach to waste management system development.

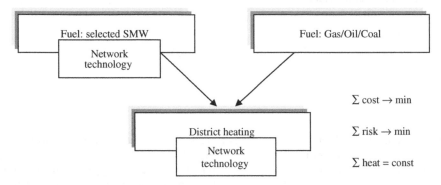

Figure 5. Base model for energy recovery.

• since No1 is the current situation, then the appropriate way forward could be a transition from No1 to No4 via No5

Energy recovery is the most promising direction for the city waste management, which in reality needs to be achieved through the consequential step-by-step approach.

4 ANALYSIS

To plan the energy production from landfills and waste recycling plants it is important to recognise that, in a market economy, the transfer of generated power (both thermal and electrical) to a potential consumer may encounter confront competition from the professional operators within the energy sector.

Currently the energy system in the city demonstrates unsustainable patterns of development. These unsustainable patterns are exacerbated in buildings and are intimately linked with the quality of life of citizens and with the many problems of communal systems. The challenge is to alleviate and reverse these adverse trends to achieve a truly sustainable energy system, while preserving the equilibrium of ecosystems and encouraging economic development. The strategic and policy objectives of further efforts for sustainable energy systems include reducing greenhouse gases and pollutant emissions, increasing the security of energy supplies, improving energy efficiency and increasing the use of renewable energy, as well as enhancing the competitiveness of communal industry and improving the quality of life.

The multi-criterion analysis may be used to investigate the market prospects of the network technology for energy recovery of solid municipal waste in the city (Figure 5). Two targets (the criteria) are tested:

(a) cost minimization,
(b) risk minimization.

The total energy produced in the network is the "heat" condition which has to be provided in the model. The following parameters are investigated;

• $P_{G,O,C,B}$ – operating-specific expenditures, modelled for example as the fuel specific cost per tonne (Gas, Oil-fuel, Coal, Bio-fuel),

Table 3. Data for energy recovery model.

Version	Parameters		
	P Rouble/tonne	E $10^{-6} \times$ Cub/tonne	T GJ/tonne
Bio-fuel, No1 Paper + Textile 0.8:0.2	900	217	10.6
Bio-fuel, No2 Leather + Plastic 0.08:0.92	900	266	24.4
Standard fuel, No1 Gas + Oil + Coal* 0.84:0.095:0.065	1156	167	35
Standard fuel, No2 Gas + Oil + Coal 0.64:0.07:0.29	1076	223	33.3
Bio-fuel, No3 Leather + Plastic 0.08:0.92	660	266	24.4
Standard fuel, No3 Gas + Oil + Coal 0.64:0.07:0.29	1184	223	33.3

* Gas – Serpuhov; Oil – mazut-100; Coal – Kuznetck

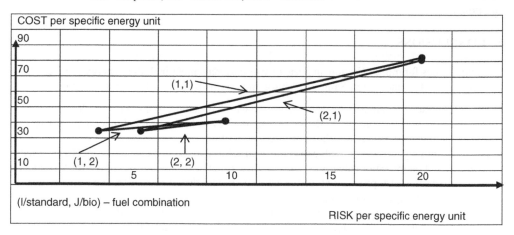

Figure 6. Assessment for energy recovery – possible sets (current situation).

- $E_{G,O,C,B}$ – environmental damage, modelled for example as the relative risk in cubic metres per tonne due to specific air pollution (averaged with inverse marginal density for dust, SO_2, CO, NO_2),
- $T_{G,O,C,B}$ – heat capacity, modelled for example as GJ per ton.

The variables in the optimization are the values of the fuels used in tonnes per year. The data used in the multi-criterion analysis are summarised in the Table 3. The data relate to bio-fuel mixtures (No1 and No2, in accordance with the morphology of solid municipal waste in the city) as well as standard fuel distributions (No1, currently in use; No2, considered for use). The bio-fuel mixture No3 is similar to No2 but with the nominal price reduced to the current cost in the household sector. The standard fuel No3 is similar to No2 but with a 10% price increase.

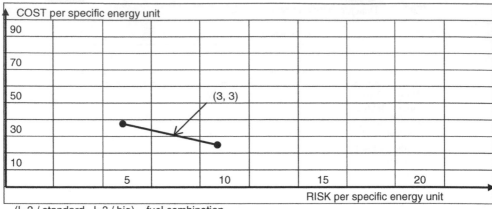

(I=3 / standard, J=3 / bio) – fuel combination

Figure 7.   Assessment for energy recovery: possible sets (potential situation).

The results of the multi-criterion analysis are summarised in the Figures 6 and 7. The first Figure shows that for compositions like 'Standard fuel NoI and Bio-fuel NoJ' (with $I = 1$ to 2, $J = 1$ to 2) application of energy recovery is not marketable. This is because of the position of the 'chosen set' which reduces the decision to the single point, i.e. the standard fuel NoI. However, from Figure7, the composition 'Standard fuel No3 and Bio-fuel No3' does provide the requisite stability for competitive energy recovery of waste. This results from the cost increase for standard fuel being compensated for by the cost decrease for bio-fuel together with the lower heat capacity of the first component.

5   CONCLUSIONS

The main issues for multi-criterion analysis of waste management including energy recovery have been introduced for the city of Saint Petersburg and are the subject of further developments:

- The solid waste problem is treated as a system in which the connections between elements define the key tasks of development. At the same time the key tasks are ordered in an appropriate way to provide the most effective path for problem resolution.
- The methodology introduced is able to create a structured model for a set of prospective thematic efforts based on indicator categories. At the same time the approach is effective for task identification which in turn provides objective assessment and the base for effective implementation of the waste management projects.
- Russia is in the process of joining international trading institutions and its economy is in a transitional state. These changes must be considered in the light of the modern approach to the solid waste problem and the interaction between the environment and trade. Under these conditions the multi-criterion development technique becomes of major interest.
- Models based on multi-criterion analysis of the waste management problem become important instruments to identify further marketable schemes in NW Russia. The projects assigned to its development will become the system core for appropriate development.

REFERENCES

Anon (1999) *Institutionalising Environmental Planning and Management*, Vol. 5, SCP Source Book Series, UN HABITAT/UNEP.
Voronov A (2003) *Sustainable development of Saint Petersburg and the tasks for municipal formations at the solid waste management*, Second Finnish-Russian landfill and waste management workshop, FEI, Finland.
White P, Franke M. and Hindle M (1995) *Integrated solid waste management: a lifecycle inventory*, Chapman & Hall.

Author index

Alshawabkeh, A.N. 133, 179

Behrens, M. 225
Berılgen, M.M. 217
Berılgen, S.A. 217
Bieberstein, A. 3
Bird, E. 311
Blunden, B. 349
Brierley, D.A. 305
Brink, A. 225

Cammaer, C. 187
Cohn, E.V.J. 101, 289
Czurda, K. 163

Davis, J. 133
De Ridder, L. 187
Debreczeni, A. 269
Dixon-Hardy, D. 233
Durn, G. 319

Engels, J. 233
Entenmann, W. 17

Fang, H.-Y. 29
Felton, A.J. 311
Frilander, R. 247
Fröschl, H. 319
Fullen, M.A. 43

Gaurina-Medimurec, N. 319
Gent, D. 133
Ghosh, S. 145
Glamore, W. 349
Golab, A. 349
Gruchot, A. 261

Hara, M. 331
Hara, Y. 331

Indraratna, B. 349
Ipekoglu, P. 217
Iwai, S. 359

Kaličanin, I.B. 345
Karri, R.S. 43
Kaya, A. 29
Kessler, A. 339
Kettunen, R.H. 119
Kim, T.-H. 29
Klein, A. 107, 149
Koliopoulos, T. 49
Koliopoulou, G. 49
Kollias, P. 49
Kollias, S. 49
Kollias, V. 49
Korzeniowska-Rejmer, E. 157, 339
Kozielska-Sroka, E. 285
Kreft-Burman, K. 247
Kruschwitz, S. 225

Leppänen, M.M. 119

Manning, D.A.C. 91
Martikkala, H.M. 119
Meggyes, T. 163, 269
Michalski, P. 261, 285
Millett, P. 101
Mott, A.P. 91
Mukherjee, S.N. 145
Muvrin, B. 345

Nguyen, M.V. 107
Niederleithinger, E. 225

Onitsuka, K. 331

Ray, R. 145
Reith, H. 3
Reyes, S.A. 59
Roberts, C. 207
Roehl, K.E. 163

Saarela, J. 247
Sarsby, R.W. 43
Saucke, U. 3
Serridge, C.J. 67
Shandyba, A.B. 173
Sheahan, T.C. 179
Simon, F.-G. 163
Svoboda, J. 81
Szczepaniak, Z. 339

Tang, G. 179
Tonks, D.M. 91, 107
Trueman, I.C. 101

Van Autenboer, T. 187
Vasicek, R. 199
Voronov, A.A. 369

Wesson, R. 207
Williams, C. 207
Woźniak, G. 289
Wu, X. 133

Xi, X.Z. 359

Zabielska-Adamska, K. 295
Zawisza, E. 261, 285

Author index

Printed and bound by CPI Group (UK) Ltd, Croydon, CR0 4YY

01/11/2024

01782636-0011